# Differential Equations for Engineers

## The Essentials

# Differential Equations for Engineers

## The Essentials

by
David V. Kalbaugh

CRC Press
Taylor & Francis Group
Boca Raton London New York

CRC Press is an imprint of the
Taylor & Francis Group, an **informa** business

CRC Press
Taylor & Francis Group
6000 Broken Sound Parkway NW, Suite 300
Boca Raton, FL 33487-2742

First issued in paperback 2021

© 2018 by Taylor & Francis Group, LLC
CRC Press is an imprint of Taylor & Francis Group, an Informa business

No claim to original U.S. Government works

ISBN 13: 978-1-03-224139-5 (pbk)
ISBN 13: 978-1-4987-9881-5 (hbk)

**Library of Congress Cataloging-in-Publication Data**

Names: Kalbaugh, David V., author.
Title: Differential equations for engineers : the essentials / David V. Kalbaugh.
Description: Boca Raton : CRC Press, [2017]
Identifiers: LCCN 2017016364 | ISBN 9781498798815 (hardback : acid-free paper) | ISBN 9781315154961 (ebook)
Subjects: LCSH: Differential equations. | Engineering mathematics.
Classification: LCC TA347.D45 K35 2017 | DDC 515/.35--dc23
LC record available at https://lccn.loc.gov/2017016364

**Visit the Taylor & Francis Web site at**
**http://www.taylorandfrancis.com**

**and the CRC Press Web site at**
**http://www.crcpress.com**

# Dedication

*To Angie, Laura and Trevor*

# Contents

# Preface

This book is intended as a tool for teaching differential equations to engineering students. Engineering students typically learn more readily when they understand how the material at hand is applied in the profession, so this book introduces most topics with an application. Also, the book recognizes that today's engineering students come from a spectrum of backgrounds. As modern society grows more technical, more engineers are needed, but many students come to the discipline inadequately prepared for its rigors. The book supports those students with reviews of necessary foundations, with deliberate, step-by-step explanations and additional examples of solved problems. However, through a number of *challenge problems*, the book encourages better prepared and intellectually curious students to delve deeper voluntarily into topics they may find interesting and rewarding. Finally, recognizing that most engineering students study differential equations in just a one semester course and not thereafter, the book surveys the broad landscape of differential equations, including elements of partial differential equations (PDEs), and presents concisely the topics of most use to engineers. In this regard, it is useful for self-study by practicing engineers who need a refresher course.

The book's focus is on applications, and its examples and problems cover a wide range of topics in electrical, mechanical, and aerospace engineering. The principal objective is to teach students how to *solve* differential equations, but an additional goal is to give the student a sense of the nature and behavior of solutions. The book endeavors to give students a physical feel for results and takes time to give concrete meaning to concepts students sometimes find abstract. Recognizing the overriding importance of computing in engineering, the book addresses numerical integration in concept and in practice, including several substantial projects. To take full advantage of the book, I strongly recommend instructors make a numerical integration program available to their students. The book incorporates use of the Internet, which has become a significant resource for engineers. The book also aims to strengthen the students' knowledge of, and ability to apply, sound engineering practices, for example, the art of approximation and methods to check one's own work.

Because the book is tailored for the engineer, it omits some traditional topics altogether, compresses others (e.g., it addresses power series solutions succinctly in the context of the PDEs in which they naturally arise), and adds some topics not normally taught, such as PDEs in problems without boundary conditions, which allow one to show (1) strong similarities with ordinary differential equations (ODEs) and (2) clearer illustrations of the nature of solutions (e.g., traveling waves). The aim, again, is to provide the essentials in a one-semester course.

There are four kinds of problems at the end of each chapter. The first kind is Solved Problems, which are example problems solved for the student, repeating to a degree what the book has said, but providing an additional help to students who need it. The second kind is Problems to Solve, typically purely mathematical, testing the students' basic understanding of the material. The third kind is Word Problems, requiring no more mathematical skill than Problems to Solve, but causing the students to read and comprehend a description of a situation and, usually, to deduce from this description the relevant equations that must then be solved. The last category is Challenge Problems, which are typically deeper and more difficult questions. Most of the Word Problems and Challenge Problems, together with the motivational introductory applications, are drawn in a balanced way from electrical, mechanical, and aerospace engineering.

Although the book's theme is utility, it strives to respect mathematical integrity. Engineers need to understand the conditions limiting the application of mathematics just they must know, when conducting experiments, under what conditions test equipment works. Furthermore, I believe that the discipline required to comprehend fundamental theorems and, to the extent possible in a one-semester course, their proofs is good training to develop critical thinking. An engineer's value to his or her organization is greatly enhanced when he or she is able to step in and find solutions to intractable problems, a situation where critical thinking is imperative.

Accompanying the book and available to instructors is an online Solutions Manual and package of visuals for use in the classroom. The Solutions Manual provides a fully explained answer to every problem in the book in a format that should permit instructors to post them easily for students to review online. The package of visuals consists of approximately 750 PowerPoint slides that parallel the basic topics of the book and another 140 or so that address supplementary material. Provided along with the visuals is an example course plan and a blank course plan template. The slides are organized in files that follow the example course plan, but the package is of course completely adaptable to the wishes of the instructor.

# Acknowledgments

I thank, first of all, Taylor & Francis Group/CRC Press, particularly Ms. Nora Konopka, for their confidence in this project.

I am particularly indebted to Frederick W. Riedel and John C. Sommerer, my friends and former colleagues at the Johns Hopkins University Applied Physics Laboratory, for reviewing a draft of this book. I very much appreciate their willingness to undertake the task. They caught many errors and gave me numerous constructive comments. Any errors that remain are mine.

Finally, I am very grateful for my wife Angie's strong encouragement and constant support throughout the effort.

# About the Author

**David V. Kalbaugh** earned a bachelor's degree in engineering science from the Johns Hopkins University, Baltimore, Maryland; a master's degree in aeronautics and astronautics from Stanford University, Stanford, California; and a PhD degree in electrical engineering from the University of Maryland, College Park, Maryland. He worked for most of his career at the Johns Hopkins University Applied Physics Laboratory (JHU/APL), where he was exposed to many practical problems across engineering and physics. Early in his career, he contributed to the development and testing of naval missile systems. He rose through the ranks of management and retired as assistant director, responsible for oversight and coordination of the laboratory's many and diverse programs.

In collateral professional experience, Dr. Kalbaugh served on numerous national boards and committees, including the National Research Council's Naval Studies Board and task forces of the Defense Science Board. He received the Secretary of Defense Medal for Outstanding Public Service for this work.

While working at JHU/APL, he also taught part-time for a decade in the Johns Hopkins University graduate engineering program in technical management. Dr. Kalbaugh's published research is on the subject of search theory.

After his JHU/APL career, Dr. Kalbaugh joined the adjunct faculty of the University of Maryland Eastern Shore, Princess Anne, Maryland, where he developed this book as he taught differential equations for engineers in the department of mathematics and computer science. At the same time, he taught courses for engineers in the department of engineering and aviation science there.

# 1 Introduction

In today's increasingly technical world, engineers are essential and differential equations are often among their most important tools. This book introduces elements of ordinary differential equations (ODEs) and partial differential equations (PDEs) through examples of their application in electrical, mechanical, and aerospace engineering.

A differential equation is an equation involving a function and its derivatives. We assume for the moment that the reader is familiar with these mathematical concepts.

Differential equations are crucial to the engineering profession because engineers must understand and control many phenomena that are described by differential equations. Differential equations are the language of much of physics. Examples of their applicability include the following:

- Motion of material bodies (e.g., dynamics of rigid bodies)
- Variation of voltage and current in electrical circuits
- Bending and stretching of solids
- Flow of fluids
- Propagation of electromagnetic, hydrodynamic, and acoustic waves
- Heat transfer

In addition, many large-scale systems designed and controlled by engineers are modeled by differential equations, for example:

- Industrial processes
- Traffic networks
- Electric power grids

Engineers must be able to solve elementary types of differential equations by hand and more complicated ones by computer. We must be able to examine an equation's form and, without solving it, determine qualitative characteristics of its solution—its nature and behavior. It is the purpose of this book to guide engineering students toward this knowledge. But before we begin the study of differential equations themselves we first review the mathematics on which they are based. Solving differential equations devolves into solving problems in arithmetic, algebra, trigonometry, differential calculus, and integral calculus. What follows is a quick review of some underlying math concepts that will be important in our study of differential equations.

## 1.1 FOUNDATIONS

Mathematics is a cumulative discipline. What we learn early in our lives is the basis for what we learn later. In mathematics we are always building on a foundation of earlier material.

### 1.1.1 ARITHMETIC

We first learned arithmetic: working with numbers themselves. We learned how to add, subtract, multiply, and divide numbers. At first those numbers were the integers; then we progressed to fractions and decimals. Today inexpensive hand-held electronic devices free us from tedious arithmetic calculations but hand-calculation can sometimes reveal insights that electronic calculation hides. For example, an engineer may be working with a sequence of numbers and with hand-calculation produce the sequence 1/7, 1/49, and 1/343. The engineer would likely recognize this as the sequence $1/7^n$. Meanwhile electronic calculation would yield the sequence: $1.429 \cdot 10^{-1}$, $2.041 \cdot 10^{-2}$, and $2.915 \cdot 10^{-3}$, offering little or no insight.

College students may still occasionally have difficulty calculating with fractions, stumbling over such operations as finding a common denominator or dividing by a fraction. For example:

$$\frac{\dfrac{2}{3}-\dfrac{3}{4}}{\dfrac{7}{2}-\dfrac{6}{5}} = \frac{\dfrac{2\cdot 4-3\cdot 3}{4\cdot 3}}{\dfrac{7\cdot 5-6\cdot 2}{2\cdot 5}} = \frac{\dfrac{8-9}{12}}{\dfrac{35-12}{10}} = \frac{\dfrac{-1}{12}}{\dfrac{23}{10}} = \frac{-1}{12}\cdot\frac{10}{23} = \frac{-1}{6}\cdot\frac{5}{23} = \frac{-5}{138}$$

Students who need to refresh their skills will find solved problems and exercises in these foundations at the end of this chapter.

### 1.1.2 ALGEBRA

From arithmetic, where we always worked with numbers, we advanced to algebra, where symbols sometimes took the place of numbers in equations and we learned methods to determine what the value of the symbol must be to satisfy the equation. Note that solutions to algebraic equations are numbers. (They may be complex, as defined and discussed momentarily).

Among the equations we learned how to solve were quadratic equations and systems of simultaneous linear equations.

#### 1.1.2.1 Quadratic Equations

Quadratic equations have the form

$$x^2 + bx + c = 0 \tag{1.1}$$

There is an explicit formula for the two solutions to this equation but it is not easy to remember. Of course the engineer can look it up in a textbook or on the web but there are quicker methods. If the coefficients $b$ and $c$ are integers then we can first see if we can factor the equation by finding two integers whose sum is $b$ and whose product is $c$. For example, consider

$$x^2 + 11x + 30 = 0 \tag{1.2}$$

If we do not see the solutions immediately we can quickly sort through the possibilities. The integers 10 and 1 sum to 11 but their product is 10. The integers 9 and 2 sum to 11 and their product is 18, closer than before. We are moving in the right direction, and the mind leaps readily to 5 and 6. That is, the quadratic in Equation 1.2 can be factored into

$$x^2 + 11x + 30 = (x+5)(x+6)$$

and we see that the roots (solutions) of the quadratic equation are

$$x_1 = -5 \quad \text{and} \quad x_2 = -6$$

The second method of solving quadratic equations is employed when the first fails or appears inconvenient because the coefficients are not integers. It is called *completing the square*, and it derives the hard-to-remember quadratic formula. Consider, for example,

$$x^2 + 6x + 7 = 0 \tag{1.3}$$

To begin solving Equation 1.3, we first note that

$$x^2 + 6x = (x+3)^2 - 9 \tag{1.4}$$

Substituting Equation 1.4 into Equation 1.3 we have

$$(x+3)^2 - 9 + 7 = 0$$
$$(x+3)^2 - 2 = 0$$
$$(x+3)^2 = 2$$
$$x + 3 = \pm\sqrt{2}$$
$$x = -3 \pm \sqrt{2}$$

That is, the solutions of Equation 1.3 are

$$x_1 = -3 + \sqrt{2} \quad \text{and} \quad x_2 = -3 - \sqrt{2}$$

We will work another example by the method of completing the square, this time with an ulterior motive. Consider the quadratic equation

$$x^2 + 14x + 65 = 0 \tag{1.5}$$

As before, we start by recognizing that

$$x^2 + 14x = (x+7)^2 - 49 \tag{1.6}$$

Substituting Equation 1.6 into Equation 1.5

$$(x+7)^2 - 49 + 65 = 0$$
$$(x+7)^2 + 16 = 0$$
$$(x+7)^2 = -16$$
$$x + 7 = \pm\sqrt{-16}$$
$$x = -7 \pm 4\sqrt{-1}$$

That is, the solution to Equation 1.5 involves the square root of a negative number. We call

$$\sqrt{-1} = i$$

and it is now obvious that the ulterior motive was to remind us that algebra introduced us to the realm of complex numbers, which, as in the case of solutions to Equation 1.5 have real and imaginary parts.

Complex numbers have an important place in applied mathematics. They simplify work in engineering, physics, and other mathematics-based disciplines. One aspect of their utility will become abundantly clear in Chapter 5.

### 1.1.2.2   Higher-Order Polynomial Equations

A polynomial equation has the form

$$x^n + a_1 x^{n-1} + a_2 x^{n-2} + \ldots + a_{n-1} x + a_n = 0 \tag{1.7}$$

A quadratic equation is a polynomial equation of order two.

A theorem is a mathematical statement that can be proven. The fundamental theorem of algebra states that polynomial equations of order $n$ have $n$ solutions. Another important theorem is that if the coefficients $a_i$ are all real then the solutions consist of real roots and roots that occur in complex conjugate pairs. That is, if $x = \alpha + i\beta$ is a root of Equation 1.7 ($\alpha$ and $\beta$ real) then so is $x = \alpha - i\beta$. Yet another important theorem is that if two polynomials of the same order are everywhere equal then their coefficients ($a_1, a_2, \ldots, a_{n-1}, a_n$ in Equation 1.7) are equal.

"Root calculators" that use iterative techniques can be found on the web for higher-order polynomial equations.

### 1.1.2.3   Systems of Simultaneous Linear Equations

Another type of equation we learned how to solve in algebra is $n$ simultaneous linear equations in $n$ unknowns. A simple two-by-two example is

$$4x_1 - 3x_2 = 26$$

$$5x_1 + 2x_2 = -2 \tag{1.8}$$

We find the solution by the process of *row reduction*.

$$x_2 = (-2 - 5x_1)/2 = -1 - (5/2)x_1$$

$$4x_1 - 3(-1 - (5/2)x_1) = 26$$

$$4x_1 + 3 + 3(5/2)x_1 = 26$$

$$(23/2)x_1 = 23$$

$$x_1 = 2$$

$$x_2 = -1 - (5/2)x_1 = -1 - (5/2)(2) = -6$$

For linear simultaneous equations of higher order there are handy "calculators" on the web.

We will be able to find unique solutions to simultaneous equations in the form of Equation 1.8 provided the *determinant* of the coefficients is not zero. The determinant of the two-by-two matrix

$$\underline{A} = \begin{pmatrix} a & b \\ c & d \end{pmatrix}$$

is given by

$$\det \underline{A} = ad - bc$$

For Equation 1.8 the determinant is

$$\det \begin{pmatrix} 4 & -3 \\ 5 & 2 \end{pmatrix} = 4 \cdot 2 - (-3) \cdot 5 = 23$$

The determinant of the three-by-three matrix

$$\underline{A} = \begin{pmatrix} A_{11} & A_{12} & A_{13} \\ A_{21} & A_{22} & A_{23} \\ A_{31} & A_{32} & A_{33} \end{pmatrix}$$

can be calculated by

$$\det \underline{A} = A_{11} \det \begin{pmatrix} A_{22} & A_{23} \\ A_{32} & A_{33} \end{pmatrix} - A_{12} \det \begin{pmatrix} A_{21} & A_{23} \\ A_{31} & A_{33} \end{pmatrix} + A_{13} \det \begin{pmatrix} A_{21} & A_{22} \\ A_{31} & A_{32} \end{pmatrix}$$

$$\det \underline{A} = A_{11}(A_{22}A_{33} - A_{23}A_{32}) - A_{12}(A_{21}A_{33} - A_{23}A_{31}) + A_{13}(A_{21}A_{32} - A_{22}A_{31})$$

We will have more to say about determinants in Chapter 8.

### 1.1.2.4  Logarithms and Exponentials

In the primitive notion of a logarithm, $x$ (an integer) is the base-10 logarithm of $z$ if

$$10^x = 10 \cdot 10 \cdot 10 \cdots 10 = z$$

That is, $z$ is 10 multiplied together $x$ times. From this notion come the rules

$$10^{\log_{10} z} = z \qquad \log_{10}(10^x) = x$$

$$10^x \cdot 10^y = 10^{x+y} \qquad \log_{10}(z \cdot y) = \log_{10} z + \log_{10} y$$

More rigorously and conveniently, using base $e$, mathematicians define

$$e^x = \lim_{N \to \infty} \left(1 + \frac{x}{N}\right)^N \tag{1.9}$$

$$\ln z = \lim_{h \to 0} \frac{z^h - 1}{h} \tag{1.10}$$

Rules similar to those for the primitive notions apply

$$e^{\ln z} = z \qquad \ln(e^x) = x$$

$$e^x \cdot e^y = e^{x+y} \qquad \ln(z \cdot y) = \ln z + \ln y$$

As an example to illustrate how these concepts will be used later, consider the equation

$$z = \frac{e^{-3t} e^{-2\tau} - e^{-2t} e^{-3\tau}}{e^{-5\tau}} \tag{1.11}$$

We simplify Equation 1.11 as follows:

$$z = \frac{e^{-3t} e^{-2\tau} - e^{-2t} e^{-3\tau}}{e^{-5\tau}}$$

$$z = (e^{-3t} e^{-2\tau} - e^{-2t} e^{-3\tau}) e^{5\tau}$$

$$z = e^{-3t} e^{-2\tau} e^{5\tau} - e^{-2t} e^{-3\tau} e^{5\tau}$$

$$z = e^{-3t} e^{(5-2)\tau} - e^{-2t} e^{(5-3)\tau}$$

$$z = e^{-3t} e^{3\tau} - e^{-2t} e^{2\tau}$$

$$z = e^{-3(t-\tau)} - e^{-2(t-\tau)}$$

The reader will understand the importance of performing these calculations correctly in Chapter 5.

An example that will prove significant in Chapter 2 is as follows:

$$z = \exp(2\ln(1+t)) = \exp(\ln(1+t)^2) = (1+t)^2$$

### 1.1.3   TRIGONOMETRY

Courses in geometry introduced us to trigonometric functions such as the sine and cosine. Given their simple definition and use in elementary figures such as the triangle we may be surprised to discover their prevalence in more advanced mathematics. We will find it important in our study of differential equations to recall basic trigonometric definitions and identities such as

$$\tan \theta = \sin \theta / \cos \theta$$

$$\sin(\theta + \varphi) = \sin \theta \cos \varphi + \cos \theta \sin \varphi$$

$$\cos(\theta + \varphi) = \cos \theta \cos \varphi - \sin \theta \sin \varphi$$

$$\sin^2 \theta = (1 - \cos 2\theta)/2$$

$$\cos^2 \theta = (1 + \cos 2\theta)/2$$

$$\sin \theta \cos \varphi = (\sin(\theta + \varphi) + \sin(\theta - \varphi))/2$$

### 1.1.4 DIFFERENTIAL CALCULUS

In algebra the solutions to our problems were always numbers (possibly complex). In differential calculus, we moved up a level of complexity to solutions that were *functions*.

We must first define *function*. Figure 1.1 presents a graph of an example function, in this case a variable $y$ whose value is a real number that depends on another variable $t$, which also has values among real numbers. In mathematical terms, a function is a one-to-one mapping, in this example, from the real line to the real line.

#### 1.1.4.1 Derivative

Differential calculus introduced us to the concept of *derivative*. The derivative is a rate of change. If a variable $y$ is a function of another variable $t$ then we write $dy/dt$ to denote the derivative of $y$ with respect to $t$, the rate of change of $y$ with respect to $t$. Graphically, the derivative represents the tangent or slope of the function $y$ at the point $t$, as illustrated in Figure 1.1. Mathematically, we define the derivative as

$$\frac{dy}{dt} = \lim_{\Delta t \to 0} \frac{y(t + \Delta t) - y(t)}{\Delta t} \tag{1.12}$$

The derivative $dy/dt$ is a new function of $t$. We can define yet another function, the derivative of $dy/dt$, which we denote as $d^2y/dt^2$, through the limiting process

$$\frac{d^2 y}{dt^2} = \lim_{\Delta t \to 0} \frac{\frac{dy}{dt}(t + \Delta t) - \frac{dy}{dt}(t)}{\Delta t}$$

We can continue in this way, depending on the smoothness of the function, defining higher-order derivatives of $y(t)$. The function $dy/dt$ is called the first derivative, $d^2y/dt^2$ the second derivative, and so on.

For example, let $y = t^2$. Then

$$\frac{dy}{dt} = \lim_{\Delta t \to 0} \frac{(t + \Delta t)^2 - t^2}{\Delta t} = \lim_{\Delta t \to 0} \frac{2t\Delta t + \Delta t^2}{\Delta t} = \lim_{\Delta t \to 0} (2t + \Delta t) = 2t$$

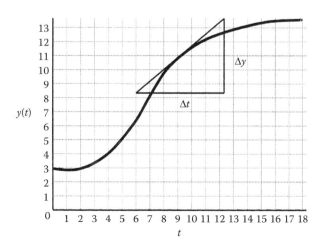

**FIGURE 1.1** A function and its slope at $t = 9$.

---

**TABLE 1.1**

**Common Functions
and Their Derivatives**

| $y(t)$ | $dy/dt$ |
|---|---|
| $t^n$ | $nt^{n-1}$ |
| $e^{kt}$ | $ke^{kt}$ |
| $\ln t$ | $1/t$ |
| $\sin t$ | $\cos t$ |
| $\cos t$ | $-\sin t$ |
| $\tan t$ | $\sec^2 t$ |
| $\arcsin t$ | $1/\sqrt{1-t^2}$ |
| $\arctan t$ | $1/(1+t^2)$ |

---

Although we could, in principle, use the aforementioned limiting process to determine a derivative of a given function, it is more practical to memorize the derivatives of common functions and to learn a few rules about how to differentiate combinations of functions. We should memorize at least the list in Table 1.1.

### 1.1.4.2 Rules of Differentiation

Let $y(t) = f(t) \cdot g(t)$. Then

$$\frac{dy}{dt} = \frac{df}{dt} g + f \frac{dg}{dt} \tag{1.13}$$

This is known as the *product rule*. As a concrete example, consider $y(t) = e^{3t} \cos 2t$. Then

$$\frac{dy}{dt} = \frac{d}{dt}(e^{3t} \cos 2t) = \frac{d}{dt}\left(e^{3t}\right)\cos 2t + e^{3t}\frac{d}{dt}(\cos 2t) = (3e^{3t})\cos 2t + e^{3t}(-2\sin 2t)$$

$$= e^{3t}(3\cos 2t - 2\sin 2t)$$

Now let $y(t) = f(g(t))$. That is, $y$ is written as a function of a variable $g$, which is in turn is a function of a variable $t$. Then

$$\frac{dy}{dt} = \frac{dy}{dg} \cdot \frac{dg}{dt} \tag{1.14}$$

This is called the *chain rule*. For example, let $y(t) = \ln(1+t^4)$. Then $f = \ln(g)$ and $g = 1+t^4$, and

$$\frac{dy}{dt} = \frac{d}{dt}(\ln(1+t^4)) = \frac{d}{dg}(\ln(g)) \cdot \frac{d}{dt}(1+t^4) = \frac{1}{g}(4t^3) = \frac{4t^3}{1+t^4}$$

From these two rules one can derive the *quotient rule.* Let $y(t) = f(t)/g(t)$. Then from the product rule we have

$$\frac{dy}{dt} = \frac{df}{dt}\frac{1}{g} + f\frac{d}{dt}\left(\frac{1}{g}\right)$$

From the chain rule,

$$\frac{d}{dt}\left(\frac{1}{g}\right) = \frac{d}{dg}\left(\frac{1}{g}\right)\cdot\frac{dg}{dt} = -\frac{1}{g^2}\cdot\frac{dg}{dt}$$

and then

$$\frac{dy}{dt} = \frac{df}{dt}\frac{1}{g} - \frac{f}{g^2}\frac{dg}{dt} = \frac{g\left(df/dt\right) - f\left(dg/dt\right)}{g^2} \tag{1.15}$$

For example, let $y(t) = 5t/(2 + 3t + 4t^2)$:

$$\frac{dy}{dt} = \frac{d}{dt}(5t/(2 + 3t + 4t^2)) = \frac{(2 + 3t + 4t^2)\dfrac{d}{dt}(5t) - 5t\dfrac{d}{dt}(2 + 3t + 4t^2)}{(2 + 3t + 4t^2)^2}$$

$$= \frac{(2 + 3t + 4t^2)(5) - 5t(3 + 8t)}{(2 + 3t + 4t^2)^2}$$

$$= \frac{5(2 - 4t^2)}{(2 + 3t + 4t^2)^2}$$

### 1.1.4.3   Infinite Series

We were typically introduced to the concept of infinite series in a course on differential calculus. We learned methods for determining whether a given series converged or diverged, and found series representations of common functions, for example:

$$e^t = 1 + t + \frac{t^2}{2} + \frac{t^3}{3\cdot 2} + \frac{t^4}{4\cdot 3\cdot 2} + \frac{t^5}{5\cdot 4\cdot 3\cdot 2}\ldots = \sum_{n=0}^{\infty}\frac{t^n}{n!} \tag{1.16}$$

$$\sin t = t - \frac{t^3}{3\cdot 2} + \frac{t^5}{5\cdot 4\cdot 3\cdot 2} - \ldots = \sum_{n=0}^{\infty}\frac{(-1)^n t^{2n+1}}{(2n+1)!} \tag{1.17}$$

$$\cos t = 1 - \frac{t^2}{2} + \frac{t^4}{4\cdot 3\cdot 2} - \ldots = \sum_{n=0}^{\infty}\frac{(-1)^n t^{2n}}{(2n)!} \tag{1.18}$$

You may see a relationship among these three series. Chapter 5 explores this.

Reviewing, the first problem in differential calculus was to determine a new function of $t$, the derivative, from a given function of $t$.

## 1.1.5  INTEGRAL CALCULUS

In integral calculus, we answered the question: If $dy/dt$ is a given function of $t$, what is its integral $y(t)$? That is, we solved problems of the form

$$\frac{dy}{dt} = f(t)$$

For example,

$$\frac{dy}{dt} = 2 + 4t$$

$$\frac{dy}{dt} = e^{3t} + e^{-3t}$$

$$\frac{dy}{dt} = (\sin t)(\sin 2t)$$

The integral is the inverse of the derivative. The fundamental theorem of calculus says that

$$\int_{t_0}^{t} \left(\frac{dy}{d\tau}\right) d\tau = y(t) - y(t_0) \tag{1.19}$$

The integral provides the area under the curve. Consider, for example, Figure 1.2.

The integral is derived by a limiting process: For example,

$$\int_{t_0}^{t} f(u) du = \lim_{N \to \infty} \sum_{i=1}^{N} f(t_i)\left(\frac{t - t_0}{N}\right)$$

where

$$t_i = t_0 + \frac{i}{N}(t - t_0)$$

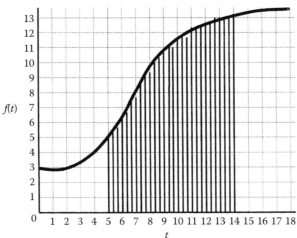

**FIGURE 1.2**   A function and an approximation to the area under its curve from $t = 5$ to $t = 14$.

For example, let $f(u) = 2u$

$$\int_{t_0}^{t} 2u\, du = \lim_{N \to \infty} \sum_{i=1}^{N} 2\left(t_0 + \frac{i}{N}(t - t_0)\right)\left(\frac{t - t_0}{N}\right)$$

$$= \lim_{N \to \infty} 2(t - t_0)t_0 + 2\frac{(t - t_0)^2}{N^2} \sum_{i=1}^{N} i$$

$$= \lim_{N \to \infty} 2(t - t_0)t_0 + (t - t_0)^2\left(1 + \frac{1}{N}\right) = t^2 - t_0^2$$

Table 1.2 lists integrals of common functions. If you memorize the previous list of derivatives of common functions then you will also know the integrals of most of them. You should be able to derive the others.

In integral calculus we also learned techniques for evaluating integrals involving combinations of functions. Two important techniques are (a) change of variable and (b) integration by parts.

### 1.1.5.1  Change of Variable

Consider the integral

$$I_1(t) = \int_0^t \frac{4\tau^2 d\tau}{(2 + 5\tau^3)^2}$$

The purpose of a change of variable is to transform an integral into a simple form that we can integrate using Table 1.2. The integral $I_1(t)$ has a $\tau^3$ in the denominator and a $\tau^2$ in the numerator. Recognizing that

$$\tau^2 = \frac{1}{3}\frac{d}{d\tau}(\tau^3)$$

suggests a substitution that is a function of $\tau^3$. We try

$$u = 2 + 5\tau^3$$

---

**TABLE 1.2**
**Common Functions and Their Integrals**

| $y(t)$ | $\int^t y(\tau)d\tau$ |
|---|---|
| $t^n$ $(n \neq -1)$ | $t^{n+1}/(n+1)$ |
| $1/t$ | $\ln t$ |
| $e^{kt}$ | $e^{kt}/k$ |
| $\ln t$ | $t \ln t - t$ |
| $\sin t$ | $-\cos t$ |
| $\cos t$ | $\sin t$ |
| $\tan t$ | $-\ln|\cos t|$ |

---

Then

$$du = 5 \cdot 3 \cdot \tau^2 d\tau$$

$$4\tau^2 d\tau = \frac{4}{5 \cdot 3} du$$

and

$$I_1(t) = \frac{4}{5 \cdot 3} \int_{u(0)}^{u(t)} \frac{du}{u^2}$$

Now we can use the first row of Table 1.2 and write

$$I_1(t) = \frac{4}{15}\left(\frac{1}{u(0)} - \frac{1}{u(t)}\right) = \frac{4}{15}\left(\frac{1}{2} - \frac{1}{2+5t^3}\right) = \frac{2}{3}\left(\frac{t^3}{2+5t^3}\right)$$

### 1.1.5.2   Integration by Parts

Consider the integral

$$I_2(t) = \int_0^t \tau \sin(4\tau) d\tau$$

The rule for integration by parts is

$$\int u\,dv = uv - \int v\,du$$

Here we let $u = \tau$ and then $du = d\tau$, $dv = \sin(4\tau)d\tau$, $v = -\cos(4\tau)/4$.
    Then

$$I_2(t) = \int_0^t \tau \sin(4\tau)d\tau = -\left(\frac{\tau\cos(4\tau)}{4}\right)\Big|_0^t + \frac{1}{4}\int_0^t \cos(4\tau)d\tau$$

$$I_2(t) = \frac{1}{16}(\sin 4t - 4t\cos 4t)$$

One can always check an integral by differentiating it and comparing with the integrand.

### 1.1.6   TRANSITIONING TO FIRST-ORDER ORDINARY DIFFERENTIAL EQUATIONS

As just discussed, in integral calculus we solved problems of the form

$$\frac{dy}{dt} = f(t)$$

In first-order ODEs we will solve problems of the form

$$\frac{dy}{dt} = f(t, y)$$

For example,

$$\frac{dy}{dt} = 2 + 4t - 5y$$

$$\frac{dy}{dt} = (e^{3t} + e^{-3t})y^2$$

$$\frac{dy}{dt} = \frac{(\sin t)(\sin 2t)}{\tan y}$$

Compare with the examples at the beginning of Section 1.1.5. Solutions will again be functions of $t$.

This completes our cursory review of the foundations on which we will build our study of differential equations. In Chapter 8 we will need vectors and matrices, but we postpone that review until then.

## 1.2   CLASSES OF DIFFERENTIAL EQUATIONS

A differential equation is an equation involving a function and its derivatives. An *ordinary* differential equation has one independent variable. A *partial* differential equation has two or more independent variables. For example, as we will see in Chapter 9, the PDE

$$\frac{\partial^4 y}{\partial x^4} + a \frac{\partial^2 y}{\partial t^2} = 0 \tag{1.20}$$

describes the deflection $y$ of a structural beam as a function of time $t$ and distance along the beam $x$. As another example, in Chapter 9 we will determine the temperature $T$ in a thin, straight rod as a function of time $t$ and distance along the rod $x$ with the PDE

$$\frac{\partial^2 T}{\partial x^2} - \frac{1}{\alpha} \frac{\partial T}{\partial t} = 0 \tag{1.21}$$

The *order* of a differential equation is the order of its highest derivative. For example,

$$\frac{dy}{dt} + y^2 = 0$$

is a first-order ODE;

$$\frac{d^2 y}{dt^2} + y = 0$$

is a second-order ODE.

A linear ODE has the form

$$a_0(t)\frac{d^n y}{dt^n} + a_1(t)\frac{d^{n-1} y}{dt^{n-1}} + ... + a_n(t)y = g(t) \tag{1.22}$$

Linear differential equations are especially important for engineers because theory and solution methods are highly developed for them. In Chapter 8 we will meet another way to represent nth-order linear ODEs, called the *state space representation*. With this exception, which we ignore for now, an ODE in any form other than Equation 1.22 is nonlinear.

The theory of nonlinear equations is less well developed and more complicated than that for linear equations. Nonlinear equations are often difficult to solve analytically.

A linear equation has the handy property that if $y_1(t)$ is the solution to

$$a_0(t)\frac{d^n y}{dt^n} + a_1(t)\frac{d^{n-1} y}{dt^{n-1}} + ... + a_n(t)y = g(t)$$

and $y_2(t)$ is the solution to

$$a_0(t)\frac{d^n y}{dt^n} + a_1(t)\frac{d^{n-1} y}{dt^{n-1}} + ... + a_n(t)y = h(t)$$

and $y(t)$ and its derivatives start from zero in both cases then the solution to

$$a_0(t)\frac{d^n y}{dt^n} + a_1(t)\frac{d^{n-1} y}{dt^{n-1}} + ... + a_n(t)y = bg(t) + ch(t)$$

where $b$ and $c$ are constants, is $by_1(t) + cy_2(t)$

Linear ODEs are called *time-invariant* if the coefficients $a_0(t), a_1(t), ..., a_n(t)$ in Equation 1.22 are constant:

$$a_0\frac{d^n y}{dt^n} + a_1\frac{d^{n-1} y}{dt^{n-1}} + ... + a_n y = g(t) \tag{1.23}$$

Note that an equation can be time-invariant and have a time-varying input $g(t)$. Linear time-invariant (LTI) ODEs have the strongest theoretical foundation and largest body of solution techniques. Many, but by no means all, ODEs important to engineers are LTI.

Engineers can sometimes, to a degree, circumvent difficulties with nonlinear equations by considering small perturbations about a position or trajectory of interest. This process, called linearization, derives an approximate linear equation from a nonlinear one and permits engineers to employ the powerful methods of linear equations and typically gain deeper insights than they otherwise would.

The professional standard in engineering for solving differential equations, especially nonlinear equations, is on a digital computer in a process known as numerical integration. Chapters 3 and 6 will define several substantive hands-on computing projects and Chapter 8 will describe the theoretical basis of several common numerical integration algorithms. In addition, a number of problems throughout the text invite students to compute a numerical solution. To take full advantage of this text, we strongly recommend that you find access to a numerical integration computer program.

A final classification of differential equations is in how additional conditions are imposed. A great many engineering applications of ODEs are posed as *initial value problems*, in which values of the dependent variable and its derivatives are specified at a specific value of the independent variable. In this book, for clarity, the symbol for the independent variable will usually be $t$, as it has been in this chapter. The symbol connotes time, as indeed many ODEs of engineering interest are written in derivatives with respect to time.

In Chapter 9, when we study PDEs, we will often find that in engineering applications one of the independent variables will be time and others will be spatial coordinates. In many of these applications the value of the dependent variable is specified at the boundaries (beginning and endpoints) of the spatial coordinates. These are called *boundary value problems*. We will see that solution of PDEs sometimes devolves into solution of an ODE with a spatial coordinate as independent variable and dependent variable subject to constraints at both beginning and endpoints. So, ODEs can sometimes have an independent variable other than time and be posed as a boundary value problem.

But we will not concern ourselves with boundary value problems until Chapter 9. Until then, all problems in this book will be posed as initial value problems.

## 1.3   A FEW OTHER THINGS

Digital computing is an omnipresent and beneficial factor in our lives today and this is certainly true in the profession of engineering. We have already mentioned numerical integration programs. In brief, these programs compute the value of a dependent variable as the independent variable marches in small incremental steps from its starting value.

You should be aware there are also software packages available to solve differential equations in an analytical sense—providing the user a solution in terms of standard functions such as the sine or exponential. This text does not rely on such packages. First, not all students have access to them. Second and more importantly, while they can be time-savers in your professional career, you should be able to do all that these programs can do in order to be fully responsible for your work. That is, you should be able to check that such programs are giving you correct results.

You will see that there are four kinds of problems at the end of each chapter. The first kind is Solved Problems. These are example problems solved for you, repeating to a degree what the text has said, but providing students an additional support. The second kind is Problems to Solve, typically purely mathematical, testing the student's basic understanding of the material. The third kind is Word Problems, requiring no more mathematical skill than Problems to Solve, but causing the students to read and comprehend a description of a situation, normally drawn from engineering, and to deduce from this description the relevant equations that must then be solved. The last category is Challenge Problems, which we offer to you to attack voluntarily if you are interested and if you are doing well enough in the course to warrant the time and energy they require. These are more difficult questions, often based on engineering issues. In most of these problems we provide fairly specific guidance. We recommend you to try to solve the Challenge Problems without referring to the guidance until you are stymied. You may find the effort rewarding.

A few words about units and the use of symbols instead of specific numerical values: Where, in examples or problems, we have sought numerical results, we have often opted for English units because we believe that most readers of this book will be more familiar and more comfortable with that system. We advise those readers, however, that they should also familiarize themselves with the metric system, and where it is not complicating we have stated parameter values in English and metric units in parallel. However, in most examples and problems we have derived and stated final results using symbols for parametric values. We recommend this as good practice; one can then (a) check the calculation by examining whether the units of the symbols combine to give the units of the desired result, (b) see how the final result depends on the various parameters of the problem, and (c) calculate the final result in either English or metric units. If one substitutes specific numbers at the beginning one loses all these capabilities.

One last word before we leave this introduction: We have thus far used the notation $dy/dt$, $d^2y/dt^2$, and $d^3y/dt^3$ for the first, second, and third derivative of $y$. A convenient and commonly-used shorthand, due to Isaac Newton, is $\dot{y}$, $\ddot{y}$, and $\dddot{y}$. We will use both notations interchangeably.

## 1.4   SUMMARY

In today's increasingly technical world engineers are essential and differential equations are often among their most important tools. A differential equation is an equation involving a function and its derivatives. Differential equations are crucial to the engineering profession because engineers must understand and control many phenomena that are described by differential equations.

Engineers must be able to solve elementary types of differential equations by hand and more complicated ones by computer. We must be able to examine an equation's form and, without solving it, determine qualitative characteristics of its solution—its nature and behavior.

Solving differential equations devolves into solving problems in arithmetic, algebra, trigonometry, differential calculus and integral calculus. The chapter provided a quick review of some underlying math concepts that are important in the study of differential equations.

There are different ways to classify differential equations. An ODE has one independent variable. A PDE has two or more independent variables. The order of a differential equation is the order of its highest derivative. A linear ODE has the form

$$a_0(t)\frac{d^n y}{dt^n} + a_1(t)\frac{d^{n-1}y}{dt^{n-1}} + ... + a_n(t)y = g(t)$$

The equation is called time-invariant if all the coefficients $a_i(t)$ are constant. The theory of nonlinear equations is less well developed and more complicated than that for linear equations. Nonlinear equations are often difficult to solve analytically. Engineers can sometimes derive an approximate linear equation from a nonlinear one by considering small perturbations about a position or trajectory of interest and thereby gaining deeper insights than one otherwise would. A final classification of differential equations is in how additional conditions are imposed. Many engineering applications of ODEs are posed as initial value problems, in which values of the dependent variable and its derivatives are specified at a given value of the independent variable. In many engineering applications of PDEs one of the independent variables will be time and others will be spatial coordinates. In many of these applications the value of the dependent variable is specified at the boundaries (endpoints) of the spatial coordinates. These are called boundary value problems.

**Solved Problems**

**SP1.1** Calculate $\dfrac{\dfrac{5}{8} - \dfrac{3}{5} + \dfrac{1}{2}}{\dfrac{2}{3} + \dfrac{5}{6} - \dfrac{1}{4}}$

*Solution*:

$$\frac{\dfrac{5}{8} - \dfrac{3}{5} + \dfrac{1}{2}}{\dfrac{2}{3} + \dfrac{5}{6} - \dfrac{1}{4}} = \frac{\dfrac{5(5 \cdot 2) - 3(8 \cdot 2) + 1(8 \cdot 5)}{8 \cdot 5 \cdot 2}}{\dfrac{2(6 \cdot 4) + 5(3 \cdot 4) - 1(3 \cdot 6)}{3 \cdot 6 \cdot 4}}$$

$$= \frac{\dfrac{50 - 48 + 40}{80}}{\dfrac{48 + 60 - 18}{72}} = \frac{\dfrac{42}{80}}{\dfrac{90}{72}}$$

$$= \frac{42}{80} \cdot \frac{72}{90} = \frac{21}{40} \cdot \frac{4}{5} = \frac{84}{200} = \frac{21}{50}$$

**SP1.2** Find the solutions to $x^2 + 3x - 18 = 0$.

*Solution*: Because the coefficients are integers we factor the quadratic, looking for two integers whose sum is +3 and whose product is −18. Because the product is negative, one of the two integers must be negative. (They cannot both be negative.) Because the sum is positive the larger in magnitude of the two integers must be positive. Integer pairs satisfying all but the first of these four conditions are (−1,18), (−2,9), and (−3,6). The sum of the first pair is +17, too large. The sum of the second pair is +7, still too large. The sum of the third pair is +3, as desired. The factors are therefore −3 and +6; that is,

$$x^2 + 3x - 18 = (x - 3)(x + 6)$$

Then the two solutions to $x^2 + 3x - 18 = 0$ are $x = 3$ and $x = -6$. Note the change in sign.

**SP1.3** Find the solutions to $x^2 + 10x + 29 = 0$.

*Solution*: Because the coefficients are integers, we try to factor the quadratic, as in Problem 1.2, looking for two integers whose sum is +10 and whose product is +29. But 29 is a prime number, so its only factors are 1 and 29. The sum of these is 30, too high, so we try the second approach: completing the square. Here,

$$x^2 + 10x = (x + 5)^2 - 25$$

Then

$$x^2 + 10x + 29 = (x + 5)^2 - 25 + 29 = (x + 5)^2 + 4 = 0$$

$$x + 5 = \pm\sqrt{-4}$$

$$x = -5 \pm 2i$$

**SP1.4** Find the solutions to

$$3x_1 + 8x_2 = -4$$

$$-2x_1 + 7x_2 = -22$$

*Solution:*
By row reduction:

$$x_2 = (-4 - 3x_1)/8 = -1/2 - (3/8)x_1$$

$$-2x_1 + 7(-1/2 - (3/8)x_1) = -22$$

$$(-2 - (21/8))x_1 - 7/2 = -22$$

$$(8 \cdot (-2) - 21)x_1 - 8 \cdot 7/2 = 8(-22)$$

$$(-16 - 21)x_1 = -176 + 28$$

$$x_1 = \frac{-148}{-37} = 4$$

$$x_2 = -1/2 - (3/8)x_1 = -1/2 - 3/2 = -2$$

**SP1.5** Simplify $\exp\big(2\ln(\cos 2t) + 3t + \ln 4\big)/\exp(6t)$.

*Solution*: To begin, $\exp(A)/\exp(B) = \exp(A - B)$ so,

$$\frac{\exp\big(2\ln(\cos 2t) + 3t + \ln 4\big)}{\exp(6t)} = \exp\big(2\ln(\cos 2t) + 3t + \ln 4 - 6t\big)$$

$$= \exp\big(2\ln(\cos 2t) - 3t + \ln 4\big)$$

Because $\exp(A + B + C) = \exp(A) \cdot \exp(B) \cdot \exp(C)$,

$$\exp(2\ln(\cos 2t) - 3t + \ln 4) = \exp(2\ln(\cos 2t)) \cdot \exp(-3t) \cdot \exp(\ln 4)$$

Now $2\ln(\cos 2t) = \ln(\cos^2 2t)$   and   $\exp(\ln(z)) = z$ so,

$$\exp(\ln(\cos^2 2t)) = \cos^2 2t$$

$$\exp(\ln 4) = 4$$

and the expression simplifies to $4e^{-3t}\cos^2 2t$

**SP1.6** Show that $\sin^4\theta = \dfrac{3}{8} - \dfrac{1}{2}\cos 2\theta + \dfrac{1}{8}\cos 4\theta$

*Solution*:

$$\sin^4\theta = \sin^2\theta\cdot\sin^2\theta = \frac{(1-\cos 2\theta)(1-\cos 2\theta)}{4}$$

$$= \frac{1}{4} - \frac{1}{2}\cos 2\theta + \frac{1}{4}\cos^2 2\theta$$

$$= \frac{1}{4} - \frac{1}{2}\cos 2\theta + \frac{1}{8}(1+\cos 4\theta)$$

$$= \frac{3}{8} - \frac{1}{2}\cos 2\theta + \frac{1}{8}\cos 4\theta$$

**SP1.7** Find the derivative of $\cos 3t / \sqrt{1-(3t)^2}$.

*Solution:*
Let $x = 1-(3t)^2$, $y = \sqrt{x}$, and $z = 1/y$. Then

$$\frac{d}{dt}\left(\frac{\cos 3t}{\sqrt{1-(3t)^2}}\right) = \frac{d}{dt}(z\cos 3t)$$

$$= \frac{dz}{dt}\cos 3t - 3z\sin 3t$$

$$= \frac{dz}{dy}\cdot\frac{dy}{dx}\cdot\frac{dx}{dt}\cos 3t - 3z\sin 3t$$

Now

$$\frac{dz}{dy} = \frac{d}{dy}\left(\frac{1}{y}\right) = -\frac{1}{y^2}, \qquad \frac{dy}{dx} = \frac{d}{dx}\left(\sqrt{x}\right) = \frac{1}{2\sqrt{x}}$$

$$\frac{dx}{dt} = \frac{d}{dt}(1-(3t)^2) = -2\cdot 3\cdot(3t) = -18t$$

Hence

$$\frac{dz}{dy}\cdot\frac{dy}{dx}\cdot\frac{dx}{dt}\cos 3t = \frac{9t}{y^2\sqrt{x}}\cos 3t = \frac{9t}{x^{3/2}}\cos 3t = \frac{9t\cdot\cos 3t}{(1-(3t)^2)^{3/2}}$$

and

$$\frac{d}{dt}\left(\frac{\cos 3t}{\sqrt{1-(3t)^2}}\right) = \frac{9t\cdot\cos 3t}{(1-(3t)^2)^{3/2}} - \frac{3\sin 3t}{(1-(3t)^2)^{1/2}}$$

**SP1.8** Determine $I(t) = \int_0^t \tau^2 e^{-6\tau} d\tau$.

*Solution*:
We integrate by parts:

$$I(t) = \int_0^t \tau^2 e^{-6\tau} d\tau = \int u\, dv = uv - \int v\, du$$

Let $u = \tau^2$, $dv = e^{-6\tau} d\tau$. Then $du = 2\tau d\tau$, $v = -e^{-6\tau}/6$, and

$$I(t) = \int_0^t \tau^2 e^{-6\tau} d\tau = \int u\, dv = -\frac{\tau^2 e^{-6\tau}}{6} \Big|_0^t + \frac{1}{3}\int_0^t \tau e^{-6\tau} d\tau$$

Integrating by parts again, letting $u = \tau$, $dv = e^{-6\tau} d\tau$, we have

$$I(t) = -\frac{t^2 e^{-6t}}{6} + \frac{1}{3}\int_0^t \tau e^{-6\tau} d\tau$$

$$= -\frac{t^2 e^{-6t}}{6} - \frac{\tau e^{-6\tau}}{18}\Big|_0^t + \frac{1}{18}\int_0^t e^{-6\tau} d\tau$$

$$= -\frac{t^2 e^{-6t}}{6} - \frac{te^{-6t}}{18} + \frac{1}{108}(1 - e^{-6t})$$

Checks: (a) $I(0) = 0$, as required.

(b) $\dfrac{dI}{dt} = t^2 e^{-6t} - \dfrac{te^{-6t}}{3} + \dfrac{te^{-6t}}{3} - \dfrac{e^{-6t}}{18} + \dfrac{e^{-6t}}{18} = t^2 e^{-6t}$, also as required.

**SP1.9** Examine the following equations *carefully* and determine the order of the equation and whether the equation is nonlinear, linear time-varying, or LTI.

1. $\dfrac{dy}{dt} + e^y y = 0$

*Solution to* (1): The equation is first order because the highest derivative is a first derivative, and nonlinear because the coefficient of $y$ is a function of $y$.

2. $\dfrac{dy}{dt} + e^t y = 0$

*Solution to* (2): The equation is first order because the highest derivative is a first derivative, and linear time-varying because the coefficient of $y$ is a function of $t$.

3. $\dfrac{d^2 y}{dt^2} + 4y\dfrac{dy}{dt} + 2y = 0$

*Solution to* (3): The equation is second order because the highest derivative is a second derivative and nonlinear because the coefficient of $dy/dt$ is a function of $y$.

4. $\left(\dfrac{dy}{dt}\right)^2 + 4\dfrac{dy}{dt} + 2y = 0$

*Solution to* (4): The first term is the square of the first derivative, not the second derivative. Hence the equation is first order and nonlinear.

5. $\dfrac{d^3y}{dt^3} + \dfrac{d^2y}{dt^2} + 4\dfrac{dy}{dt} + 2y = e^{-t^2}$

*Solution to* (5): The equation is third order because the highest derivative is a third derivative and LTI. The term $e^{-t^2}$ is an input. All the coefficients are constant.

**Problems to Solve**

**P1.1** Calculate $\dfrac{\dfrac{2}{5} + \dfrac{1}{6} - \dfrac{4}{15}}{\dfrac{2}{3} - \dfrac{3}{4} + \dfrac{5}{6}}$

**P1.2** Calculate $\dfrac{\dfrac{5}{4} + \dfrac{1}{9} - \dfrac{1}{3}}{\dfrac{2}{7} + \dfrac{3}{4} - \dfrac{3}{8}}$

**P1.3** Calculate $\dfrac{\dfrac{7}{12} - \dfrac{3}{5} - \dfrac{1}{4}}{\dfrac{2}{5} + \dfrac{5}{6} - \dfrac{7}{10}}$

**P1.4** Find the solutions to $x^2 + 10x + 21 = 0$

**P1.5** Find the solutions to $x^2 + 22x + 96 = 0$

**P1.6** Find the solutions to $x^2 + 5x - 36 = 0$

**P1.7** Find the solutions to $x^2 + 8x + 52 = 0$

**P1.8** Find the solutions to $x^2 + 18x + 76 = 0$

**P1.9** Find the solutions to $x^2 + 12x + 100 = 0$

**P1.10** Find the solutions to

$$5x_1 - 2x_2 = -20$$
$$3x_1 + 6x_2 = 24$$

**P1.11** Find the solutions to

$$3x_1 + 2x_2 = -31$$
$$-5x_1 + 4x_2 = -7$$

**P1.12** Find the solutions to

$$7x_1 + 9x_2 = 26$$
$$8x_1 + 3x_2 = -14$$

**P1.13** Simplify

$$z = \frac{(\exp(-3t)\exp(-5\tau) - \exp(-5t)\exp(-3\tau))}{\exp(-8\tau - 2\ln(t - \tau))}$$

**P1.14** Simplify

$$z = \ln\left(\frac{\exp(a)\exp(b)\exp(c)}{\exp(-a)\exp(-b)\exp(-c)}\right) + \exp\left(\ln\left(\frac{a}{b}\right) + \ln\left(\frac{b}{c}\right) + \ln\left(\frac{c}{a}\right)\right)$$

**P1.15** Simplify

$$z = \exp(2\ln(\sin\theta)) + \exp(2\ln(\cos\theta)) + \exp(\ln(\sin\theta)) - \ln(\cos\theta)$$

**P1.16** Show that $\sin\theta\sin\varphi = \dfrac{1}{2}(\cos(\theta - \varphi) - \cos(\theta + \varphi))$

**P1.17** Show that $\sin\theta\cos\varphi = \dfrac{1}{2}(\sin(\theta + \varphi) + \sin(\theta - \varphi))$

**P1.18** Show that $\sin^6\theta = (10 - 15\cos 2\theta + 6\cos 4\theta - \cos 6\theta)/32$

**P1.19** Find the derivative of $te^{-4t}\sin 8t$

**P1.20** Find the derivative of $\exp(-5\tan 2t)/(1 + t^2)^3$

**P1.21** Find the derivative of $\arcsin(\arctan\omega t)$

**P1.22** Determine $I(t) = \displaystyle\int_0^t \tau e^{-2\tau} d\tau$

**P1.23** Determine $I(t) = \displaystyle\int_1^t \tau\ln\tau d\tau$

**P1.24** Determine $I(t) = \displaystyle\int_0^t \frac{e^{k\tau} - e^{-k\tau}}{(e^{k\tau} + e^{-k\tau})^2} d\tau$

**P1.25** Examine the following equations *carefully* and determine the order of the equation and whether the equation is nonlinear, linear time-varying, or LTI.

1. $\dfrac{dy}{dt} + \sqrt{\ln 2}\, y = e^t$

2. $5\dfrac{d^4 y}{dt^4} + 4\dfrac{d^3 y}{dt^3} + 3\dfrac{d^2 y}{dt^2} + 2t\dfrac{dy}{dt} + y = 0$

3.  $\dfrac{d^5 y}{dt^5} + e^\pi \dfrac{dy}{dt} + \left(\sin\left(\dfrac{\pi}{8}\right)\right) y = t$

4.  $\left(\dfrac{dy}{dt}\right)^3 + 2\dfrac{d^2 y}{dt^2} + 6\dfrac{dy}{dt} + 9 y = 7\cos 3t$

5.  $\dfrac{d^2 y}{dt^2} + ty\dfrac{dy}{dt} + ty = 0$

6.  $\dfrac{d^3 y}{dt^3} + (\sin t)\dfrac{d^2 y}{dt^2} + (\cos y)\dfrac{dy}{dt} + (\tan t) y = 1$

**Word Problems**

**WP1.1** An aircraft flies at an altitude $h$. Its line of sight to an object on the surface of the Earth is limited to a distance $d$ because of the Earth's curvature. See Figure 1.3. Given that the radius of the earth is 3,960 statute miles and that there are 5,280 ft in a statute mile, show that the line of sight can be approximated by $d = \sqrt{3h/2}$ where $h$ is in feet and $d$ is in statute miles.

**WP1.2** A factory cuts the corners out of a thin sheet of metal and folds it into a box. See Figure 1.4. The sheet is initially square with side $l$ and the cutouts are square

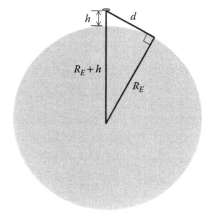

**FIGURE 1.3**   Line of sight $d$ at an altitude $h$ above Earth's surface.

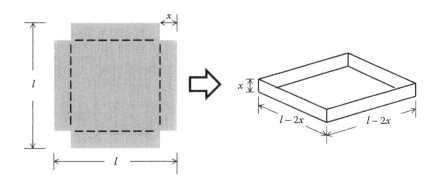

**FIGURE 1.4**   Folding a square sheet into a box.

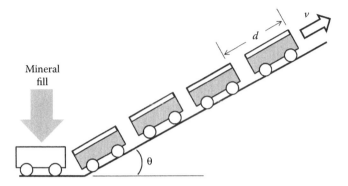

**FIGURE 1.5**   Railcars in mining operation.

with side $x$. Find the length of the cutout (the value of $x$) maximizing the volume
of the box. What is the maximum volume of the box?

**WP1.3** A mining operation transports its mineral output via a narrow-gauge railway
up a slope of constant angle $\theta$ to the Earth's surface. See Figure 1.5. The operation
will be most profitable if it moves as much mineral per unit time as possible. The
number of railcars completing the climb per unit time is given by their speed $v$
divided by their separation $d$. The speed of the railcars is limited by the maximum
power $P$ that the operation can provide to each. (The railcars are independently
powered; that is, not interconnected.) Recall that power equals force times speed.
A force must be exerted to counter gravity up the climb. A heavier load moves at a
slower speed. If each railcar weighs $W_0$ lbs empty and has the capacity for a weight
$W_M$ of mineral, what is the optimum weight of mineral the operation should load
onto each railcar? What is the maximum amount of mineral the operation can
move per unit time?

**WP1.4** A circuit designer must choose the value of a resistor $R_p$ in order to maximize
the electrical power $P$ expended over it. See Figure 1.6. The power expended
over the resistor will be $P = I^2 R_p$ where, by Ohm's law, the current $I$ flowing through
the circuit in Figure 1.6 will satisfy $V_0 = I(R_0 + R_p)$ since the total resistance of
resistors in series is the sum of their individual resistances. If $V_0 = 10$ volts and
$R_0 = 10^3$ ohms, find the value of $R_p$ maximizing $P$.

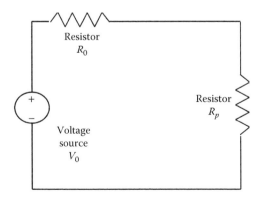

**FIGURE 1.6**   Electrical circuit with known voltage source $V_0$, known resistor $R_0$, and resistor $R_p$
to be determined.

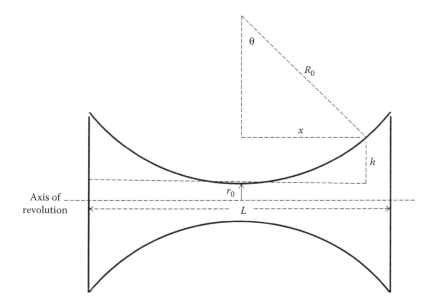

**FIGURE 1.7**    Chamber in the form of a body of revolution.

**WP1.5** Figure 1.7 depicts in profile a chamber in the form of a body of revolution, meaning that it is radially symmetric; that is, its cross section is circular everywhere. The figure of the body is defined by an arc of constant radius of curvature $R_0$. The length of the body is $L$ and its minimum radius is $r_0$. Determine the volume of the chamber. Hints: $x = R_0 \sin\theta$; $h = R_0(1 - \cos\theta)$.

**WP1.6** The contour of the bottom of a certain pond fed only by rainwater is radially symmetric; that is, its cross section is circular everywhere. Water evaporates from the pond at a daily rate proportional to its surface area. Show that, in a dry period, the depth of the pond at its center falls at a constant daily rate.

**Challenge Problems**

**CP1.1** A water clock is a tank shaped in such a way that when water is released from the bottom of the tank the level of the water in the tank decreases at a constant rate. See Figure 1.8. Archaeologists believe mankind has used the water clock for timekeeping about as long as it has the sundial, which is many thousands of years.

If the exit portal is at the very bottom of the tank, Torricelli's law states that the stream of water leaving the tank has speed $v = \sqrt{2gh}$, where $h$ is the height of water in the tank. Assume that the tank is to run continuously without being filled for a time $T$, the height of the tank is $H$, the tank is radially symmetric, the area of the exit portal is $a_E$, and the exit stream fills the exit portal. Under these conditions:

1. Show that in order for the level of the water to decrease at a constant rate, the functional relationship that must exist between the radius of the tank $r$ and its height at that point is

$$r = \left( \frac{a_E T \sqrt{2g}}{\pi H} \right)^{1/2} h^{1/4}$$

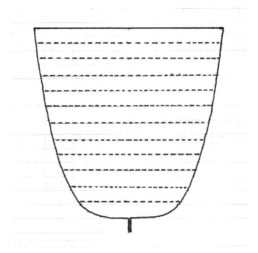

**FIGURE 1.8**  A water clock.

and that the radius $R$ at the top of the tank is given by

$$R = \left( \frac{a_E T \sqrt{2g}}{\pi} \right)^{1/2} \frac{1}{H^{1/4}} \tag{1.24}$$

Note that the radius at the top increases as the tank height decreases.

2.  Show that the volume of water to fill the tank is

$$V = \left( \frac{2}{3} \right) T a_E \sqrt{2g} H^{1/2} \tag{1.25}$$

Note that the volume of water required is minimized when the area of the exit portal is as small as possible. Also note that volume increases with height. From Equations 1.24 and 1.25 we see that a tank minimizing volume would be short and wide.

3.  Show that the volume of water to fill a water clock of height $H$ and radius $R$ at its top is two-thirds the volume of a tank of height $H$ and *constant* radius $R$.

4.  Suppose experiments show that, for proper stream flow, the smallest practical diameter of the exit portal is $d$. Also suppose that for aesthetic purposes (to create a nicely shaped tank to sit in the ancient town square) we want the diameter of the top of the tank to equal its height. Show that the height of the tank is then determined:

$$H = d^{4/5} (T \sqrt{2g})^{2/5}$$

5.  Assuming we want the clock to run for 12 h without refilling and that $d = 3/32$ of an inch, what is the height of the tank? Recalling that one cubic foot equals 7.48 gallons, how many gallons of water does it take to fill the tank?

**FIGURE 1.9**   Trajectory for a soft rocket landing.

**CP1.2** At this writing, several aerospace companies are demonstrating a reusable rocket for spacecraft launches. The rocket's motor is throttled and steered to enable it to achieve a soft vertical landing. See Figure 1.9. We envision here that the rocket comes to a point in its trajectory where it is in a vertical orientation at an altitude $h_0$ with vertical velocity $-V_0$ directly above its intended landing location. The challenge: devise a thrust profile to bring the rocket in a vertical descent safely to the Earth. Determine practical limits on $h_0$ and $V_0$ to be compatible with the rocket's maximum thrust-to-weight ratio and the available fuel.

*Guidance:* The rocket's trajectory is governed by Newton's laws of motion. Begin the analysis at the moment when the rocket is at an altitude $h_0$ above the Earth's surface and prepared for its vertical final descent. Assume that (a) vertical orientation is maintained perfectly; (b) the weight of fuel used in final descent is small compared to the vehicle weight; (c) air resistance is negligible; and (d) the rocket motor is initially off for a time $t_S$, and then full-on until touchdown with a thrust-to-weight ratio $k$. Newton's law says

$$m\ddot{h} = T - mg \tag{1.26}$$

where:
   $m$ is the rocket mass
   $T$ is its thrust
   $g$ is the acceleration of gravity

Use Equation 1.26 and apply initial conditions to obtain expressions for $\dot{h}(t)$ and $h(t)$. Setting $\dot{h}(t)$ and $h(t)$ to zero, determine the time at touchdown $t_D$ and the rocket start time $t_S$. Show that

$$t_D = \left( \frac{k}{g(k-1)} \left( 2h_0 + \frac{V_0^2}{g} \right) \right)^{1/2} - \left( \frac{V_0}{g} \right) \tag{1.27}$$

Show that for $t_S$ to be greater than zero (i.e., for the thrust profile to be realizable) the thrust-to-weight ratio must satisfy

$$k \geq 1 + \frac{V_0^2}{2gh_0} \tag{1.28}$$

We will see in Chapter 3, Section 3.1.1, that the mass of fuel used by the rocket is given by

$$m_F = \frac{1}{c} \int_{t_S}^{t_D} T(t)dt \tag{1.29}$$

where $c$ is the speed of the rocket's exit gases. Assume that the ratio of $m_F$ to vehicle mass is limited to $r$. Use Equations 1.26, 1.27, and 1.29 to show that we must have

$$r \geq \frac{1}{c} \left( \frac{k}{k-1} \right)^{1/2} \left( V_0^2 + 2gh_0 \right)^{1/2} \tag{1.30}$$

Equations 1.28 and 1.30 together provide constraints on rocket motor design and on initial conditions $h_0$ and $V_0$ for final descent.

# 2 First-Order Linear Ordinary Differential Equations

As discussed in Chapter 1, linear equations may be time-invariant or time-varying. They may also be homogenous or nonhomogeneous. A nonhomogeneous first-order linear ordinary differential equation (ODE) has the form

$$\frac{dy}{dt} + p(t)y = g(t)$$

where $g(t)$ is nonzero. A homogeneous equation has $g(t) = 0$. The homogeneous equation has an interesting solution only if its initial condition is nonzero.

## 2.1 FIRST-ORDER LINEAR HOMOGENEOUS ORDINARY DIFFERENTIAL EQUATIONS

### 2.1.1 EXAMPLE: RC CIRCUIT

Consider the electrical circuit with resistor and capacitor diagrammed in Figure 2.1. The values $R$ and $C$ are known constants. By Kirchhoff's voltage law, the sum of the voltage drops around a closed circuit is zero. In this example, a voltage drop $V_0$ exists over the capacitor's terminals before the circuit is closed. The charge on the capacitor is $q = CV$. By Ohm's law, the voltage drop over the resistor is $IR$. Finally, by the conservation of electrical charge (Kirchhoff's current law) the current through the resistor must equal the rate of change of the charge on the capacitor. That is, $I = dq/dt = CdV/dt$.

Then the sum of the voltage drops around the circuit after it is closed at $t = 0$ can be expressed as

$$IR + V = 0$$

$$RC\frac{dV}{dt} + V = 0$$

$$\frac{dV}{dt} + \left(\frac{1}{RC}\right)V = 0, \qquad V(0) = V_0 \tag{2.1}$$

A first-order ODE posed as an initial value problem must have one stated initial condition to yield a unique solution. In this case we are given that $V(0) = V_0$.

We will solve Equation 2.1 with a useful, easy-to-remember procedure that takes some license with mathematical rigor. The next Section 2.1.2 will justify this simplified, short-cut procedure with a more mathematically rigorous approach.

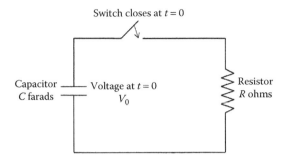

**FIGURE 2.1**   First-order electrical circuit; an RC (resistor–capacitor) circuit.

We solve Equation 2.1 by *separating variables*; that is, moving all terms involving $V$ to the left-hand side and all terms involving $t$ to the other:

$$\frac{dV}{V} = -\frac{dt}{RC}$$

Then we perform the definite integrals from $V(0)$ to $V(t)$ and 0 to $t$:

$$\int_{V(0)}^{V(t)} \frac{dV}{V} = -\int_{0}^{t} \frac{dt}{RC}$$

$$\ln\left(\frac{V(t)}{V(0)}\right) = -t/RC$$

Taking the exponential of both sides:

$$\exp\left(\ln\left(\frac{V(t)}{V(0)}\right)\right) = \exp\left(-t/RC\right)$$

$$\frac{V(t)}{V(0)} = e^{-t/RC}$$

$$V(t) = V(0)e^{-t/RC}$$

$$V(t) = V_0 e^{-t/RC} \tag{2.2}$$

Equation 2.2 is the solution to Equation 2.1.

Figure 2.2 graphs Equation 2.2. Note that the *time constant* of the circuit, $RC$, can be determined graphically as the intercept on the time axis of the voltage's initial slope.

Equation 2.1 is a time-invariant ODE. From Equation 2.2 we can see that the solution to *any* first-order linear time-invariant (LTI) ODE of the form

$$\frac{dy}{dt} + \lambda y = 0, \qquad y(0) = y_0 \quad \text{is}$$

$$y(t) = y_0 e^{-\lambda t}$$

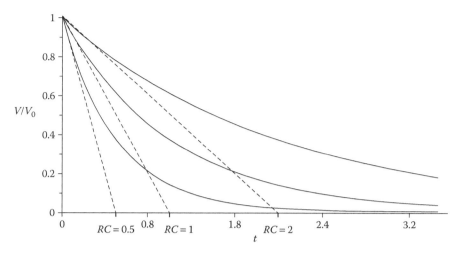

**FIGURE 2.2** Graph of $V(t) = V_0 e^{-t/RC}$ for $RC = 0.5$, 1, and 2.

### 2.1.2 TIME-VARYING EQUATIONS AND JUSTIFICATION OF SEPARATION OF VARIABLES PROCEDURE

We solve *time-varying* first-order homogeneous ODEs in the same way we did the time-invariant one in the previous Section 2.1.1; that is, via separation of variables. Consider the general form:

$$\frac{dy}{dt} + p(t)y = 0, \qquad y(t_0) = y_0 \tag{2.3}$$

We separate variables

$$\frac{dy}{y} = -p(t)dt$$

and perform definite integrals on the left from $y(t_0) = y_0$ to $y(t)$ and on the right from $t_0$ to $t$:

$$\int_{y_0}^{y(t)} \frac{dy}{y} = -\int_{t_0}^{t} p(\tau)d\tau$$

$$\ln\left(\frac{y(t)}{y_0}\right) = -\int_{t_0}^{t} p(\tau)d\tau$$

Then we take the exponential of both sides:

$$\exp\left(\ln\left(\frac{y(t)}{y_0}\right)\right) = \exp\left(-\int_{t_0}^{t} p(\tau)d\tau\right)$$

$$\frac{y(t)}{y_0} = \exp\left(-\int_{t_0}^{t} p(\tau)d\tau\right)$$

$$y(t) = y_0 \exp\left(-\int_{t_0}^{t} p(\tau)d\tau\right) \tag{2.4}$$

Equation 2.4 is the solution to Equation 2.3. Success depends entirely on being able to perform the integration

$$\int_{t_0}^{t} p(\tau)d\tau$$

We will now justify the shortcut separation of variables procedure by solving Equation 2.3 with more attention to mathematical rigor. Dividing Equation 2.3 by $y$:

$$\frac{1}{y}\frac{dy}{dt} + p(t) = 0 \tag{2.5}$$

Integrating Equation 2.5 over time from $t_0$ to $t$:

$$\int_{t_0}^{t}\left(\frac{1}{y}\frac{dy}{d\tau} + p(\tau)\right)d\tau = \int_{t_0}^{t} (0)d\tau = 0.$$

$$\int_{t_0}^{t}\left(\frac{1}{y}\frac{dy}{d\tau}\right)d\tau = -\int_{t_0}^{t} p(\tau)\,d\tau \tag{2.6}$$

Now

$$\int_{t_0}^{t}\left(\frac{1}{y}\frac{dy}{d\tau}\right)d\tau = \int_{t_0}^{t}\left(\frac{d}{d\tau}(\ln y(\tau))\right)d\tau = \ln(y(t)) - \ln(y(t_0)) = \ln\left(\frac{y(t)}{y_0}\right) \tag{2.7}$$

From Equations 2.6 and 2.7, taking the exponential of both sides, we arrive at

$$y(t) = y_0 \exp\left(-\int_{t_0}^{t} p(\tau)d\tau\right) \tag{2.8}$$

Equation 2.8 matches Equation 2.4, so the separation of variables procedure gives the correct answer.

## 2.2   FIRST-ORDER LINEAR NONHOMOGENEOUS ORDINARY DIFFERENTIAL EQUATIONS

A *nonhomogeneous* first-order linear ODE has the form

$$\frac{dy}{dt} + p(t)y = g(t) \tag{2.9}$$

where $g(t)$ is nonzero.

**Example:**

Consider an uninsulated thin-walled pipe in a ventilated and unheated area, such as a crawl space under a house (Figure 2.3). The pipe is full of water and the water does not run for days. Over this time the pipe experiences diurnal variation in the ambient air temperature (warmer in the day, cooler at night).

We model the air temperature variation as sinusoidal:

$$T_\infty = \bar{T} + T_0 \sin \omega t \tag{2.10}$$

where:

$T_\infty$ is the ambient air temperature (the air temperature at locations distant from the pipe)
$\bar{T}$ is the daily mean temperature
$T_0$ is the maximum variation from the mean

We assume that ventilation causes a slight draft of air over the pipe and that the pipe's wall thickness $d$ is much less than its outer radius $r$. Under these circumstances, by Isaac Newton's law of cooling, the rate of heat going into the water from the airstream is

$$q = hA(T_\infty - T_W) \tag{2.11}$$

where:

$A = 2\pi r L$ is the exposed surface area of the pipe
$T_W$ is the temperature of the water
$h$ is a constant known as the *convection coefficient*

Also, assuming that the water does not freeze, the thermal energy stored in the water is given by

$$E = mcT_W \tag{2.12}$$

where:

$m$ is the mass
$c$ is the specific heat of water

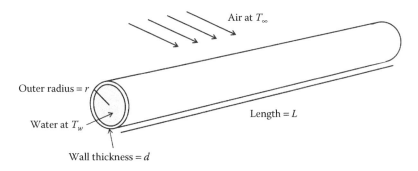

**FIGURE 2.3**  Uninsulated water pipe in air draft.

Now let

$$T_W = \bar{T} + T \tag{2.13}$$

Then, using Equations 2.10 and 2.13, Equation 2.11 becomes

$$q = hA(T_0 \sin \omega t - T) \tag{2.14}$$

Now we apply the key physical principle: the rate of change of the thermal energy in the water equals the rate of heat exchanged between airstream and water:

$$\frac{dE}{dt} = q \tag{2.15}$$

From Equations 2.12 through 2.15,

$$mc\frac{dT}{dt} = hA(T_0 \sin \omega t - T) \tag{2.16}$$

Rearranging Equation 2.16,

$$\frac{dT}{dt} + \lambda T = \lambda T_0 \sin \omega t \tag{2.17}$$

where

$$\lambda = \frac{hA}{mc} \tag{2.18}$$

## 2.2.1 INTEGRATING FACTOR

How do we solve Equation 2.17? Let us be more inclusive and ask how we solve a linear nonhomogeneous first-order ODE in its general form:

$$\frac{dy}{dt} + p(t)y = g(t), \quad y(t_0) = y_0 \tag{2.19}$$

For reasons that will be clear momentarily, we seek an *integrating factor* $\mu(t)$ that, when multiplied into the equation turns the left-hand side into a derivative. That is, we seek a function $\mu(t)$ such that

$$\mu(t)\left(\frac{dy}{dt} + p(t)y\right) = \frac{d}{dt}(\mu y) \tag{2.20}$$

Multiplying through on the left-hand side of Equation 2.20:

$$\mu(t)\left(\frac{dy}{dt} + p(t)y\right) = \mu(t)\frac{dy}{dt} + \mu(t)p(t)y \tag{2.21}$$

Now, by the product rule of differentiation, the right-hand side of Equation 2.20 becomes

$$\frac{d}{dt}\big(\mu(t)y\big) = \mu(t)\frac{dy}{dt} + \frac{d\mu}{dt}\,y \tag{2.22}$$

We want to find a function $\mu$ such that the right-hand sides of Equations 2.21 and 2.22 are equal. The first term on the right-hand side is common to both equations. Therefore, the integrating factor we seek must satisfy

$$\mu(t)p(t)y = \frac{d\mu}{dt}\,y$$

or

$$\frac{d\mu}{dt} = p(t)\mu(t) \tag{2.23}$$

We can apply the results of Section 2.1 (Equations 2.3 and 2.4) to Equation 2.23 and write

$$\mu(t) = \mu(t_0)\exp\left(\int_{t_0}^{t} p(\tau)d\tau\right) \tag{2.24}$$

We do not know what value to assign to $\mu(t_0)$, but it turns out not to matter for an integrating factor (as we will see, the value cancels out), so we set

$$\mu(t_0) = 1 \tag{2.25}$$

Furthermore, it suffices, and is most convenient, to use the indefinite integral form for the integrating factor:

$$\mu(t) = \exp\left(\int^{t} p(\tau)d\tau\right) \tag{2.26}$$

From Equations 2.19 and 2.20 we arrive at

$$\frac{d}{dt}\big(\mu(t)y(t)\big) = \mu(t)g(t) \tag{2.27}$$

The value of the integrating factor is now clear: with $\mu(t)$ given by Equation 2.26 we can solve Equation 2.27 with an integration.

For the waterpipe temperature example, using Equations 2.17, 2.19, and 2.26,

$$\mu(t) = e^{\lambda t} \tag{2.28}$$

and we have

$$\frac{d}{dt}(e^{\lambda t}T) = e^{\lambda t}\lambda T_0 \sin\omega t \tag{2.29}$$

Integrating both sides,

$$\int_0^t \frac{d}{d\tau}(e^{\lambda\tau}T)d\tau = \int_0^t e^{\lambda\tau}\lambda T_0 \sin(\omega\tau)d\tau \tag{2.30}$$

From the fundamental theorem of calculus and assuming zero initial condition, the left-hand side becomes

$$\int_0^t \frac{d}{d\tau}(e^{\lambda\tau}T)d\tau = e^{\lambda t}T(t) \tag{2.31}$$

From Equations 2.30 and 2.31,

$$T(t) = e^{-\lambda t}\int_0^t e^{\lambda\tau}\lambda T_0 \sin(\omega\tau)d\tau \tag{2.32}$$

Performing the integral in Equation 2.32 is a little involved and it is convenient to postpone dealing with it explicitly until Chapter 7. Meanwhile we offer it to the student as challenge problem CP2.2 and state without proof here that

$$T(t) = e^{-\lambda t}\lambda T_0\int_0^t e^{\lambda\tau}\sin(\omega\tau)d\tau = \left(\frac{\lambda}{\lambda^2 + \omega^2}\right)T_0(\lambda\sin\omega t - \omega\cos\omega t + \omega e^{-\lambda t}) \tag{2.33}$$

Solved Problems SP2.3 and SP2.4 at the end of this chapter are additional examples of how to solve ODEs with an integrating factor.

If we carry through the integration of

$$\frac{d}{dt}\big(\mu(t)y(t)\big) = \mu(t)g(t)$$

in the general case we find

$$\mu(t)y(t) - \mu(t_0)y(t_0) = \int_{t_0}^t \mu(\tau)g(\tau)d\tau$$

$$y(t) = \frac{\mu(t_0)}{\mu(t)}y(t_0) + \int_{t_0}^t \left(\frac{\mu(\tau)}{\mu(t)}\right)g(\tau)d\tau$$

Recalling Equation 2.24, we have the general solution

$$y(t) = y_0\exp\left(-\int_{t_0}^t p(\tau)d\tau\right) + \int_{t_0}^t \exp\left(-\int_\tau^t p(\tau)d\tau\right)g(\tau)d\tau.$$

Note that the first term is just the homogeneous solution, Equation 2.8.

It is important for engineers to find ways to check their own work. We can always substitute a supposed solution into the ODE to verify that the ODE is satisfied, including the initial conditions. Doing so with the above equation verifies that it satisfies Equations 2.19. We will return often to the issue of checking our work.

## 2.2.2 KERNEL AND CONVOLUTION

Before leaving this topic, we have several important observations. First, note that Equation 2.32 can be written as

$$T(t) = \lambda T_0 \int_0^t e^{-\lambda(t-\tau)} \sin(\omega\tau) d\tau \tag{2.34}$$

This is of the form

$$y(t) = \int_0^t K(t-\tau) g(\tau) d\tau \tag{2.35}$$

where:
  $y(t)$ is the solution of the ODE
  $g(\tau)$ is the input
  $K(t-\tau)$ is called the *kernel* for the ODE
In this example, $g(\tau) = \lambda T_0 \sin(\omega\tau)$ (recall Equation 2.17) and

$$K(t-\tau) = e^{-\lambda(t-\tau)}$$

Equation 2.35 is a particularly important equation. It applies to much more than our water-pipe temperature example; in fact, the solution to *any* LTI nonhomogeneous ODE (recall Equation 1.23) with zero initial conditions, regardless of its order, can be written in the form of Equation 2.35. The form is called the *convolution* of $K(t)$ and $g(t)$.

For LTI nonhomogeneous ODEs, the convolution and kernel determine the degree to which the present value of the solution depends on values of past inputs. The value of the solution at any time is a weighted sum of values of the input at times past. In the waterpipe temperature example (examining Equation 2.34), the weight given to an input decreases exponentially with the distance in time of that input from the current time.

## 2.2.3 SYSTEM VIEWPOINT

For engineers, whatever is being modeled by the ODE is the *system*. A system is defined by its inputs, outputs, and the rules by which the inputs determine the outputs. In the waterpipe temperature example, the system has one input ($T_0 \sin \omega t$, the temperature variation of the ambient air about the daily mean), one output ($T(t)$, the temperature variation of the water about the daily mean), and Equation 2.17 prescribes the input–output rule.

Systems governed by linear ODEs lend themselves to *block diagrams* as a means of depicting clearly the inputs, outputs, and interaction rules. Figure 2.4 is a block diagram for the waterpipe temperature example.

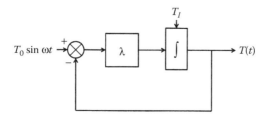

**FIGURE 2.4** Block diagram for waterpipe temperature example.

Had there been an initial variation $T_I$ in the water temperature from the daily mean, the output of the system would have been

$$T(t) = \left(\frac{\lambda}{\lambda^2 + \omega^2}\right) T_0(\lambda \sin(\omega t) - \omega \cos(\omega t) + \omega e^{-\lambda t}) + T_I e^{-\lambda t} \qquad (2.36)$$

Equation 2.36 can be parsed in two ways, each of them important, depending on the engineer's interest at the time.

The first way parses the solution into *forced response* and *natural response*. The forced response arises as a result of the input and the natural response results from initial conditions. In Equation 2.36, the forced response is

$$\left(\frac{\lambda}{\lambda^2 + \omega^2}\right) T_0(\lambda \sin(\omega t) - \omega \cos(\omega t)) + \left(\frac{\lambda \omega}{\lambda^2 + \omega^2}\right) T_0 e^{-\lambda t}$$

and the natural response is $T_I e^{-\lambda t}$.

A second, equally valid, way parses the solution into a *steady-state response* and *transient response*. The transient response dies away in time and the steady-state response remains. In Equation 2.36, the steady-state response is

$$\left(\frac{\lambda}{\lambda^2 + \omega^2}\right) T_0(\lambda \sin(\omega t) - \omega \cos(\omega t)) \qquad (2.37)$$

and the transient response is

$$T_I e^{-\lambda t} + \left(\frac{\lambda \omega}{\lambda^2 + \omega^2}\right) T_0 e^{-\lambda t}$$

For engineers, Equation 2.37 is an inadequate expression for the steady-state response. For example, an engineer may wonder if the amplitude of the output is greater or less than the amplitude of the input, and how this ratio depends on $\omega$, the frequency of the input. To improve on Equation 2.37 we can define

$$\cos \theta = \frac{\lambda}{\sqrt{\lambda^2 + \omega^2}}$$

$$\qquad (2.38)$$

$$\sin \theta = \frac{\omega}{\sqrt{\lambda^2 + \omega^2}}$$

Substituting Equations 2.38 into Equation 2.37 results in the expression

$$T_{SS}(t) = \frac{\lambda}{\sqrt{\lambda^2 + \omega^2}} T_0 \sin(\omega t - \theta) \qquad (2.39)$$

for the steady-state response of the waterpipe temperature. This can be written as

$$T_{SS}(t) = GT_0 \sin(\omega t - \theta) \qquad (2.40)$$

where:

$$G = \frac{\lambda}{\sqrt{\lambda^2 + \omega^2}}$$ (2.41)

$$\theta = \arctan\left(\frac{\omega}{\lambda}\right)$$ (2.42)

The coefficient $G$ is called the *gain* of the system and $\theta$ is called the *phase lag*. These are the defining elements in the *frequency response* of the system.

From Equations 2.40 and 2.41 engineers can immediately see that the gain is always less than one and hence the steady-state amplitude of the output is always less than amplitude of the input. Moreover, Equation 2.41 tells us that the gain diminishes monotonically from one to zero as $\omega$, the frequency of the input, goes from zero to infinity. At the same time, from Equation 2.42, the phase lag increases monotonically from zero to $\pi/2$.

We will find in Chapter 5 that second-order LTI ODEs can have a more complex and more interesting frequency response.

## 2.3  SYSTEM STABILITY

A linear system is said to be *stable* if its response to bounded initial conditions is bounded. An LTI system is said to be *damped* if its response to bounded initial conditions goes to zero as time goes to infinity. A first-order time-invariant system described by the differential equation

$$\frac{dy}{dt} + \lambda y = 0$$

is stable and damped if and only if $\lambda > 0$.

The definition of stability just given suffices for our purposes for linear systems, but not for nonlinear ones. We will refine our definition of stability in Chapter 4. Also, engineers reserve the term "damped" for LTI systems. For other systems the term used is *asymptotically stable*, as will also be discussed in Chapter 4.

## 2.4  SUMMARY

We solve the equation

$$\frac{dy}{dt} + p(t)y = 0, \quad y(t_0) = y_0$$

with a useful, easy-to-remember procedure that we needed to justify because it takes some license with mathematical rigor. We separate variables

$$\frac{dy}{y} = -p(t)dt$$

and perform definite integrals on the left from $y(t_0)$ to $y(t)$ and on the right from $t_0$ to $t$:

$$\int_{y(t_0)}^{y(t)} \frac{dy}{y} = -\int_{t_0}^{t} p(\tau)d\tau$$

$$\ln\left(\frac{y(t)}{y_0}\right) = -\int_{t_0}^{t} p(\tau)d\tau$$

leading to a solution that can be written as

$$y(t) = y_0 \exp\left(-\int_{t_0}^{t} p(\tau)d\tau\right)$$

To solve

$$\frac{dy}{dt} + p(t)y = g(t), \qquad y(t_0) = y_0$$

we look for an integrating factor $\mu(t)$ such that

$$\mu(t)\left(\frac{dy}{dt} + p(t)y\right) = \frac{d}{dt}(\mu y)$$

We find that

$$\mu(t) = \exp\left(\int^{t} p(\tau)d\tau\right)$$

in its indefinite integral form suffices and we integrate

$$\frac{d}{dt}(\mu y) = \mu g$$

to a solution that can be written as

$$y(t) = y_0 \exp\left(-\int_{t_0}^{t} p(\tau)d\tau\right) + \int_{t_0}^{t} \exp\left(-\int_{\tau}^{t} p(\tau)d\tau\right) g(\tau)d\tau.$$

The term $\exp\left(-\int_{\tau}^{t} p(\tau)d\tau\right) = K(t,\tau)$ is called the *kernel* of the ODE. For an LTI ODE the kernel has the form $K(t-\tau)$.

In an engineering-oriented viewpoint, one considers whatever is being modeled by the ODE as a system defined by its inputs, outputs, and the rules (the ODE) by which the inputs determine the outputs. The output of a system can be parsed into a forced response and natural response, or alternatively into a steady-state response and transient response. When a system's steady-state response to a unit–amplitude sine-wave input is

$$y_{ss}(t) = A\sin \omega t - B\cos \omega t$$

it can be written as

$$y_{SS}(t) = G\sin(\omega t - \theta)$$

where $G = \sqrt{A^2 + B^2}$ and $\theta = \arctan(B, A)$ are the system's gain and phase lag, which are defining elements in the system's frequency response. The gain of a first-order time-invariant system with sine-wave input in the form

$$\frac{dy}{dt} + \lambda y = C\sin\omega t$$

decreases monotonically to zero as $\omega$ goes to infinity, and its phase lag will increase monotonically from zero to $\pi/2$.

A linear system is said to be *stable* if its response to bounded initial conditions is bounded. An LTI system is said to be *damped* if its response to bounded initial conditions goes to zero as time goes to infinity. A first-order time-invariant system described by

$$\frac{dy}{dt} + \lambda y = 0$$

is stable and damped if and only if $\lambda > 0$.

**Solved Problems**

**SP2.1** Solve

$$\frac{dy}{dt} + \left(\frac{4t}{1+t^2}\right)y = 0, \qquad y(0) = 3$$

*Solution:*
Since the equation is homogeneous we separate variables:

$$\frac{dy}{y} = \frac{-4t\,dt}{1+t^2}$$

Integrating both sides and using the dummy variable $\tau$ under the integral on the right-hand side to avoid confusion with the integration end point $t$:

$$\int_{y(0)}^{y(t)} \frac{dy}{y} = -\int_0^t \frac{4\tau\,d\tau}{1+\tau^2}$$

$$\ln y(t) - \ln y(0) = -2\ln(1+t^2) + 2\ln(1)$$

$$\ln\left(\frac{y(t)}{3}\right) = -\ln(1+t^2)^2 = \ln\left(\frac{1}{(1+t^2)^2}\right)$$

Taking the exponent of both sides:

$$\exp\left(\ln\left(\frac{y(t)}{3}\right)\right) = \exp\left(\ln\left(\frac{1}{(1+t^2)^2}\right)\right)$$

$$\frac{y(t)}{3} = \frac{1}{(1+t^2)^2}$$

$$y(t) = \frac{3}{(1+t^2)^2}$$

**SP2.2** Solve

$$\frac{dy}{dt} + (2 - \tan t)y = 0, \qquad y(0) = 4$$

*Solution:*
Since the equation is homogeneous we separate variables:

$$\frac{dy}{y} = -(2 - \tan t)dt$$

Integrating both sides:

$$\int_{y(0)}^{y(t)} \frac{dy}{y} = -\int_0^t (2 - \tan \tau)d\tau$$

$$\ln\left(\frac{y(t)}{y(0)}\right) = -2t - \ln\left(\frac{\cos t}{1}\right) = -2t - \ln(\cos t) = -2t + \ln\left(\frac{1}{\cos t}\right)$$

Taking the exponent of both sides:

$$\exp\left(\ln\left(\frac{y(t)}{y(0)}\right)\right) = \exp\left(-2t + \ln\left(\frac{1}{\cos t}\right)\right)$$

$$\frac{y(t)}{4} = \exp(-2t) \cdot \exp\left(\ln\left(\frac{1}{\cos t}\right)\right) = \frac{e^{-2t}}{\cos t}$$

$$y(t) = \frac{4e^{-2t}}{\cos t}$$

**SP2.3** Solve

$$\frac{dy}{dt} + \left(\frac{3}{t}\right)y = t^2, \qquad y(1) = 2 \tag{2.43}$$

*Solution:*

Since Equation 2.43 is nonhomogeneous, we search for an integrating factor $\mu(t)$, such that

$$\mu(t)\left(\frac{dy}{dt} + \left(\frac{3}{t}\right)y\right) = \frac{d}{dt}(\mu(t)y)$$

We can determine $\mu(t)$ either by working through the requirements on it, as we did in Equations 2.20 through 2.23 earlier, or by recognizing that $p(t) = 3/t$ in Equation 2.43 and applying the formula given in Equation 2.26:

$$\mu(t) = \exp\left(\int^t p(\tau)d\tau\right) = \exp\left(\int^t \frac{3d\tau}{\tau}\right) = \exp(3\ln t) = \exp\left(\ln(t^3)\right) = t^3$$

Note that we are using the indefinite integral here.

Multiplying both sides of Equation 2.43 by $\mu(t) = t^3$,

$$t^3\left(\frac{dy}{dt} + \frac{3}{t}y\right) = t^3 \cdot t^2$$

$$\frac{d}{dt}(t^3 y) = t^{3+2} = t^5$$

Integrating both sides:

$$\int_1^t \frac{d}{d\tau}(\tau^3 y)d\tau = \int_1^t \tau^5 d\tau$$

$$\tau^3 y\Big|_1^t = \frac{\tau^6}{6}\Big|_1^t$$

$$t^3 y(t) - 1 \cdot y(1) = \frac{t^6}{6} - \frac{1}{6}$$

$$t^3 y(t) - 2 = \frac{t^6}{6} - \frac{1}{6}$$

$$y(t) = \frac{t^3}{6} + \frac{11}{6t^3}$$

**SP2.4** Solve

$$\frac{dy}{dt} + e^{\lambda t} y = ke^{\lambda t}, \qquad y(0) = y_0 \tag{2.44}$$

*Solution*:
Since Equation 2.44 is nonhomogeneous we search for an integrating factor $\mu(t)$ such that

$$\mu(t)\left(\frac{dy}{dt} + e^{\lambda t} y\right) = \frac{d}{dt}\left(\mu(t) y\right)$$

Recognizing that $p(t) = e^{\lambda t}$ in Equation 2.44 and applying Equation 2.26,

$$\mu(t) = \exp\left(\int^t e^{\lambda \tau} d\tau\right) = \exp\left(\frac{e^{\lambda t}}{\lambda}\right)$$

Multiplying both sides of Equation 2.44 by $\mu(t) = \exp(e^{\lambda t}/\lambda)$,

$$\exp\left(\frac{e^{\lambda t}}{\lambda}\right)\left(\frac{dy}{dt} + e^{\lambda t} y\right) = \exp\left(\frac{e^{\lambda t}}{\lambda}\right)\left(ke^{\lambda t}\right)$$

$$\frac{d}{dt}\left(\exp\left(\frac{e^{\lambda t}}{\lambda}\right) y\right) = ke^{\lambda t} \exp\left(\frac{e^{\lambda t}}{\lambda}\right)$$

Integrating both sides:

$$\int_0^t \frac{d}{d\tau}\left(y \exp\left(\frac{e^{\lambda \tau}}{\lambda}\right)\right) d\tau = \int_0^t ke^{\lambda \tau} \exp\left(\frac{e^{\lambda \tau}}{\lambda}\right) d\tau$$

$$y \exp\left(\frac{e^{\lambda \tau}}{\lambda}\right)\bigg|_0^t = k \exp\left(\frac{e^{\lambda \tau}}{\lambda}\right)\bigg|_0^t$$

$$y(t) \exp\left(\frac{e^{\lambda t}}{\lambda}\right) - y(0) \exp\left(\frac{1}{\lambda}\right) = k\left(\exp\left(\frac{e^{\lambda t}}{\lambda}\right) - \exp\left(\frac{1}{\lambda}\right)\right)$$

$$y(t) = k + (y_0 - k)\exp(-(e^{\lambda t} - 1)/\lambda)$$

## Problems to Solve

In Problems 2.1 through 2.14, find $y(t)$.

**P2.1** $\dfrac{dy}{dt} + 5y = 0, \qquad y(0) = 3$

**P2.2** $\dfrac{dy}{dt} - \left(\dfrac{1}{\sqrt{1-t}}\right)y = 0,$ $\qquad y(0) = 5$

**P2.3** $\dfrac{dy}{dt} + 2\left(t + \dfrac{1}{t}\right)y = 0,$ $\qquad y(1) = 2$

**P2.4** $\dfrac{dy}{dt} + (4 + \sin 2t)y = 0,$ $\qquad y(0) = -2$

**P2.5** $\dfrac{dy}{dt} + \left(\dfrac{t - \sin 2t}{t^2 + \cos 2t}\right)y = 0,$ $\qquad y(0) = 4$

**P2.6** $\dfrac{dy}{dt} + 2e^{-4t}y = 0,$ $\qquad y(0) = -3$

**P2.7** $\dfrac{dy}{dt} + (t \ln t)y = 0,$ $\qquad y(1) = 7$

**P2.8** $\dfrac{dy}{dt} + 4y = 7e^{-2t},$ $\qquad y(0) = 3$

**P2.9** $\dfrac{dy}{dt} - 3y = 2t,$ $\qquad y(0) = -4$

**P2.10** $\dfrac{dy}{dt} + 3\left(\dfrac{e^{3t} - e^{-3t}}{e^{3t} + e^{-3t}}\right)y = 4e^{2t},$ $\qquad y(0) = 6$

**P2.11** $\dfrac{dy}{dt} + \left(\dfrac{1}{t}\right)y = 6\sin 2\pi t,$ $\qquad y(1) = 1$

**P2.12** $\dfrac{dy}{dt} + \left(\dfrac{2}{2+t}\right)y = 3,$ $\qquad y(0) = 2$

**P2.13** $\dfrac{dy}{dt} + \left(\dfrac{t}{1-t^2}\right)y = 9,$ $\qquad y(0) = 4$

**P2.14** $\dfrac{dy}{dt} + \dfrac{3}{(1+t^2/4)^{3/2}}y = \dfrac{2t}{(1+t^2/4)^2},$ $\qquad y(0) = 8$

## Word Problems

**WP2.1** A mass $m$ of a radioactive element disintegrates into lighter elements at an average rate proportional to $m$. The *half-life* of a radioactive element is the average time required for half of the original atoms in mass $m$ to decay. If the half-life of an element is $T_{1/2}$, what is its rate of disintegration per unit mass? How long does it take for the mass of the element to reach 10% of its original value, in terms of $T_{1/2}$?

**WP2.2** A pond of water initially contains a volume of water $V_0$ with a mass of salt $m_0$. Salt enters the pond in a water flow with constant volume rate $\dot{V}_{IN}$ and salinity (mass of salt per unit volume of water) $\rho_{IN}$. Salt leaves the pond in a water flow with constant volume rate $\dot{V}_{OUT}$ and salinity that equals the average salinity in the pond $\rho = m/V(t)$. (We are assuming that the pond water is well mixed). Derive and solve the differential equation for the mass of salt in the pond for two cases: (a) the water flow volume rate in equals the water flow volume rate out, in which case the volume of water in the pond is constant; and (b) the flow rates in and out differ, in which case the volume of water in the pond is not constant. Show that, for case (a):

$$m(t) = (m_0 - \rho_{IN}V_0)\exp\left(\frac{-\dot{V}t}{V_0}\right) + \rho_{IN}V_0 \tag{2.45}$$

and for case (b):

$$m(t) = (m_0 - \rho_{IN}V_0)\left(1 + \frac{\alpha\dot{V}_{OUT}t}{V_0}\right)^{-1/\alpha} + \rho_{IN}V_0\left(1 + \frac{\alpha\dot{V}_{OUT}t}{V_0}\right) \tag{2.46}$$

where $\alpha = \dfrac{\dot{V}_{IN} - \dot{V}_{OUT}}{\dot{V}_{OUT}}$.

Show that Equation 2.46 reduces to Equation 2.45 when $\alpha = 0$.

**WP2.3** Consider the electrical circuit shown in Figure 2.5. It contains a resistor, inductor, and sine-wave voltage source. The values $R$ and $L$ are known constants.

As in the example in Section 2.1.1, by Ohm's law the voltage drop over the resistor in the direction of current is $V = IR$. The voltage drop over an inductor is $LdI/dt = (L/R)(dV/dt)$ and the voltage *drop* over the sine-wave source is $-V_0\sin\omega t$. (Notice the sense of the terminals.) Here $V_0$ is the amplitude of the sine wave, in volts, and $\omega$ is the frequency of the oscillation in radians per second. Assume that the sine-wave source turns on at $t = 0$ and that no current is flowing in the circuit at that time. Kirchhoff's voltage law is that the sum of voltage drops around a circuit is zero. The input is the voltage of the sine-wave source. The output is the voltage over the resistor.

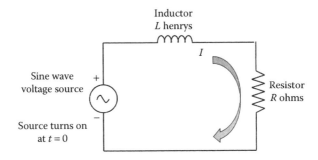

**FIGURE 2.5** First-order electrical circuit; an RL (resistor–inductor) circuit with a sine-wave voltage source.

With these facts in hand, derive the first-order LTI differential equation for this circuit. Assuming that $R = 2 \cdot 10^3$ and $L = 5 \cdot 10^{-4}$ in consistent units and $\omega = 8 \cdot 10^6$ radians per second, determine the gain and phase of the circuit.

**WP2.4** An object moving slowly in a highly viscous (sticky) fluid experiences resistance in proportion to its velocity. Consider an object falling vertically from rest in such a fluid under the influence of gravity. The force on the object due to gravity is $F = mg$ where $m$ is the mass of the object and $g$ the acceleration of gravity. The resistance due to the viscous fluid is $F = -\beta v$ where $\beta$ is a constant and $v$ is the velocity. Using Newton's second law, force equals mass times acceleration, derive and solve the differential equation for the object's motion. Determine the object's steady-state velocity.

**WP2.5** Consider a metal object at initial temperature $T_0$ brought into the center of an evacuated chamber and hung by a slender cord. According to the Stefan–Boltzmann law, the heat lost from the object due to *radiation* is

$$q = -\varepsilon\sigma A(T^4 - T_S^4)$$

where:
$\sigma$ is the Stefan–Boltzmann constant (a constant of nature)
$A$ is the surface area of the object
$\varepsilon$ is its emissivity ($0 < \varepsilon < 1$)
$T$ is the temperature of the object's surface
$T_S$ is the temperature of the surroundings, relative to absolute zero

Given that the metal has good heat conduction properties, the object's temperature is nearly uniform throughout, and the thermal energy in the object is given by $E = mcT$ where $m$ is the mass of the object and $c$ is the specific heat of the metal. The key physical principle is that the rate of change of the thermal energy in the object equals the heat lost due to radiation. (Because the chamber is evacuated, convection is minimal. We also assume that conduction through the suspension cord can be ignored.) Assuming that $T_0 - T_S \ll T_S$, we can approximate

$$q = -\varepsilon\sigma A(T^4 - T_S^4) = -\varepsilon\sigma A(T^2 + T_S^2)(T + T_S)(T - T_S) \cong -4\varepsilon\sigma A T_S^3 (T - T_S)$$

Given these facts, derive and solve a first-order time-invariant differential equation for the temperature $T$. Given that $4\varepsilon\sigma A T_S^3/mc = 1.2$ per hour, how long does it take for the difference between the object's temperature and the temperature of the surroundings to fall to one-fifth of its original value?

**WP2.6** The object here is to derive equations describing the variation of air density with altitude from sea level to 82,000 ft. Reference [1] has tables of atmospheric variables as a function of altitude for a "standard atmosphere," an average over time and geography. In the standard atmosphere air temperature $T$ varies linearly from 519° Rankine at sea level to 390° at 36,000 ft and then is constant to 82,000 ft. Air density $\rho$ is tabulated, too, varying from $2.3769 \cdot 10^{-3}$ slugs per cubic foot at sea level to $7.1028 \cdot 10^{-4}$ at 36,000 ft to $7.8931 \cdot 10^{-5}$ at 82,000 ft, but the variation is never linear or constant. From physical principles, can you derive differential equations for air density as a function of altitude for the two altitude regimes (0 to 36 Kft and then 36 to 82 Kft) and solve them to match the tabulated results?

We can model air density as satisfying the hydrostatic equation

$$\frac{dp}{dh} = -\rho g$$

and the perfect gas law

$$p = \rho RT$$

where:
  $h$ is the altitude above sea level
  $p$ is the atmospheric pressure
  $g$ is the acceleration of gravity (32.18 ft/sec$^2$)
  $R$ is the gas constant for air (1,716 in consistent units)

From these facts and temperature's dependence on altitude, derive and solve differential equations for air density $\rho$ as a function of altitude for the two altitude regimes. Given the standard atmosphere's value of air density at sea level as an initial condition, how do your results compare with the standard atmosphere's value of air density at 36,000 ft and 82,000 ft altitude?

## Challenge Problems

**CP2.1** Figure 2.6 depicts an industrial process in which newly cast metal plates at an initial temperature $T_0$ are conveyed from a heat-treatment process through the side of a large, tall tube of rectangular cross section where the plates are cooled by a stream of air at an initial temperature $T_\infty$.

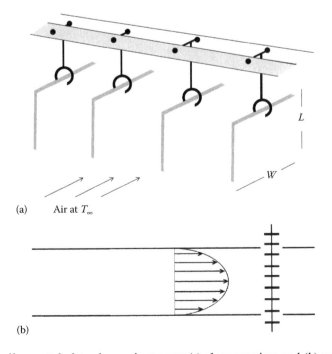

(a)      Air at $T_\infty$

(b)

**FIGURE 2.6**  Cooling metal plates in an air stream: (a) close-up view and (b) overhead view.

Under certain conditions, heat is conducted quickly enough through each plate that its internal temperature is essentially uniform. The heat lost by the plate causes a drop in its temperature:

$$mc\frac{dT}{dt} = -q \tag{2.47}$$

where:
$m$ is the mass of the plate
$c$ is its specific heat

Isaac Newton's law of cooling by convection is

$$q = h(u)A(T - T_\infty) \tag{2.48}$$

where $h(u)$ is the convection coefficient, which depends on the velocity of the air stream, and $A = 2LW$. Substituting Equation 2.48 into Equation 2.47 and rearranging we have

$$\frac{dT}{dt} + \left(\frac{h(u)A}{mc}\right)(T - T_\infty) = 0 \tag{2.49}$$

The problem is simplified if we define $T_* = T - T_\infty$ and use $T_*$ as our independent variable. Then Equation 2.49 becomes

$$\frac{dT_*}{dt} + \left(\frac{h(u)A}{mc}\right)T_* = 0 \tag{2.50}$$

with initial condition $T_*(0) = T_0 - T_\infty$.

If each plate were held stationary in the center of the tube then the convection coefficient would be a constant value, call it $h_C$. Then Equation 2.50 would be time invariant, the methods we used and applied in Section 2.1.1 would apply here and its solution would be $T_*(t) = (T_0 - T_\infty)e^{-\lambda t}$ where

$$\lambda = \frac{h_C A}{mc} \tag{2.51}$$

As it is, however, the conveyor carries each plate through the tube and the velocity of the air stream varies across the tube. As shown in Figure 2.7, the air velocity is zero at the sides of the tube and a maximum in the center. The velocity profile across the tube is given by

$$u(x) = u_C(1 - (x/x_0)^2)$$

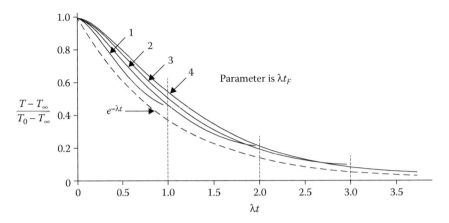

**FIGURE 2.7**  Temperature of a hot metal plate carried on a conveyor belt at varying speeds through a cooling air stream, showing solutions to Equation 2.52. The function $e^{-\lambda t}$ is shown for reference.

where:

    $x = 0$ is the center of the tube

    $2x_0$ is the width of the tube

    $u_C$ is the airspeed in the center of the tube

Under certain conditions, if the airspeed is not too high and the conveyor speed is slow enough, the convection coefficient is approximately proportional to the square root of the airspeed. Assume that the conveyor speed is constant.

With these facts in hand, show that

1. The convection coefficient over a given plate varies as a function of the time elapsed since the plate entered the tube by

$$h(t) = 2h_C \sqrt{t(t_F - t)}/t_F$$

    where $t_F$ is the time required for the conveyor to carry each plate through the tube.

2. The differential equation describing the plate's temperature is

$$\frac{dT_*}{dt} + \left( \frac{2\lambda}{t_F} \sqrt{t(t_F - t)} \right) T_* = 0, \qquad T_*(0) = T_0 - T_\infty \qquad (2.52)$$

    where $\lambda$ is given by Equation 2.51.

3. The final temperature of each plate is

$$T(t_F) = T_\infty + (T_0 - T_\infty)\exp\left( -\frac{\pi}{4}\lambda t_F \right)$$

4. The graph of the (nondimensionalized) temperature as a function of the (nondimensionalized) time is as shown in Figure 2.7. The (nondimensionalized) time required for the conveyor to carry each plate through the tube is a parameter of the curves. Figure 2.7 shows for reference the function $e^{-\lambda t}$, which represents the temperature profile for a stationary plate at the center of the tube. To create your graph use either the closed form solution or the numerical integration.

## CP2.2

Recalling Equation 2.33, show that

$$T(t) = e^{-\lambda t}\lambda T_0 \int_0^t e^{\lambda \tau} \sin(\omega \tau)d\tau = \left(\frac{\lambda}{\lambda^2 + \omega^2}\right)T_0(\lambda \sin \omega t - \omega \cos \omega t + \omega e^{-\lambda t})$$

Hint: Integrate by parts twice and combine terms.

# 3 First-Order Nonlinear Ordinary Differential Equations

Nonlinear ordinary differential equations (ODEs) are typically difficult to solve analytically. This chapter presents three techniques for doing so that have a modest range of applicability: (1) separating variables; (2) transforming the nonlinear ODE into a linear one; and (3) for almost linear systems, successive approximations. The great majority of the times, engineers solve nonlinear differential equations by numerical integration on digital computers. At the end of this chapter are the first phases of several extensive, continuing computing projects intended to give students first-hand experience with their use. The exercises provide you an opportunity to program a computer to achieve a practical result. Chapter 8 will examine algorithms used in these computer routines.

## 3.1 NONLINEAR SEPARABLE ORDINARY DIFFERENTIAL EQUATIONS

The first method applies only to first-order separable equations; that is, to ODEs of the form

$$\frac{dy}{dt} = f(t)g(y), \qquad y(t_0) = y_0 \tag{3.1}$$

To solve Equations 3.1 we separate variables by dividing through by $g(y)$:

$$\frac{1}{g(y)}\frac{dy}{dt} = f(t)$$

Then we employ the shortcut procedure introduced and justified in Section 2.1:

$$\int_{y(t_0)}^{y} \frac{dy}{g(y)} = \int_{t_0}^{t} f(\tau)d\tau \tag{3.2}$$

Sometimes we can perform the integration on both sides of Equation 3.2 and arrive at an implicit equation

$$G(y) = F(t) \tag{3.3}$$

In rare circumstances we are then able to solve Equation 3.3 for $y$ as an explicit function of $t$. In the typical case when we cannot, we can still use Equation 3.3 to determine the values of the function through graphical techniques. The nonlinear problems in this text have been selected so that they can be solved explicitly.

### 3.1.1   EXAMPLE: SOUNDING ROCKET TRAJECTORY

As an example of the application of nonlinear equations in engineering, we consider a *sounding rocket* that flies vertically to carry instruments into the upper atmosphere or edge of space for scientific measurements. The National Aeronautics and Space Agency (NASA) operates a sounding rocket program as part of its activities.

We ask three fundamental questions about the sounding rocket's trajectory:

- What is its maximum speed going up?
- What altitude does it reach?
- What is its maximum speed coming down?

To simplify analysis we initially make a number of assumptions. We will later relax a number of these:

- Flight is perfectly vertical.
- The rocket's turn at the top (apogee) of its trajectory is instantaneous.
- Rocket thrust is so large and operates over such a short period of time that gravity and drag can be ignored during the high-thrust segment.
- Time elapsed and height achieved during the high-thrust segment can be ignored.
- Air density is constant.
- Drag coefficient (defined shortly) is constant.

The last two assumptions are too stringent for NASA's sounding rockets; air density is roughly constant only for trajectories with maximum altitudes less than about 5000 ft (about 1500 m) and the rocket's drag coefficient is constant only for trajectories with maximum speeds less than about 500 ft (about 150 m) per second. The assumptions are valid for a hobbyist's rocket. As we will see, however, consideration of this simple case can be useful for checking more realistic and more complex computations.

### 3.1.1.1   Maximum Speed Going Up

The equation of motion for the thrusting phase (i.e., when the rocket motor is on) can be derived from the conservation of momentum and mass. The total rate of change of the momentum of the vehicle and the gases exiting from its engine (considered as a unit) is zero. The vehicle accelerates in reaction to the hot gases exiting it at high speed. The mass of the vehicle is reduced as the gases leave. The equation of motion is

$$m\frac{dV}{dt} + \frac{dm}{dt}c = 0$$

or

$$\frac{dV}{dt} = -\frac{1}{m}\frac{dm}{dt}c \qquad (3.4)$$

where:
    $V$ is the speed of the vehicle
    $m$ is the mass
    $c$ is the speed of the exit gases relative to the rocket

There are three variables in this equation. We can eliminate time by writing

$$\frac{dV}{dt} = \frac{dV}{dm}\frac{dm}{dt}$$

and then, cancelling $dm/dt$ from both sides, Equation 3.4 becomes the separable equation

$$\frac{dV}{dm} = -\frac{c}{m}$$

Following the procedure outlined in Equations 3.1 through 3.3,

$$dV = -c\frac{dm}{m}$$

$$\int_0^{V(m)} dV = -c\int_{m_0}^m \frac{dm}{m}$$

$$V(m) = c\ln\left(\frac{m_0}{m}\right)$$

and in particular, at rocket burnout, that is, the moment when the rocket fuel is exhausted,

$$V_b = c\ln\left(\frac{m_0}{m_b}\right) \tag{3.5}$$

We will see shortly that $V_b$, the velocity at burnout, is the maximum speed the sounding rocket attains on its way up. Equation 3.5 is the so-called *rocket equation*.

### 3.1.1.2  Maximum Altitude

Newton's second law of motion is $F = ma$, force equals mass times acceleration. After burnout and on the way up, the sounding rocket's trajectory is governed by the nonlinear differential equation

$$m_b\frac{dV}{dt} = -m_b g - \left(\frac{1}{2}\rho S C_D\right)V^2 \tag{3.6}$$

Acceleration is $dV/dt$, the derivative of velocity with respect to time, and the forces are weight and air resistance (drag), both acting in the downward direction. Drag is proportional to the air density $\rho$, the rocket's cross-sectional area $S$, its drag coefficient $C_D$ (at low speeds a constant typically around 0.02 or 0.03), and the square of the velocity. We can rewrite Equation 3.6 as

$$\frac{dV}{dt} = -g - kV^2 \tag{3.7}$$

where

$$k = \frac{\rho S C_D}{2m_b} \tag{3.8}$$

and then separate variables:

$$\frac{dV}{g + kV^2} = -dt \tag{3.9}$$

We want to make a substitution to change the left-hand side of Equation 3.9 into the form $dv/(1 + v^2)$, which we know how to integrate. We accomplish this by defining $v = V\sqrt{k/g}$. After a little algebra, Equation 3.9 becomes

$$\frac{dv}{1 + v^2} = -\sqrt{gk}\,dt \tag{3.10}$$

Integrating the right-hand side of Equation 3.10 from 0 to $t$, and the left-hand side correspondingly from $v_b = V_b\sqrt{k/g}$ to $v(t) = V(t)\sqrt{k/g}$:

$$\arctan(v(t)) - \arctan(v_b) = -\sqrt{gk}\,t \tag{3.11}$$

Solving Equation 3.11

$$v(t) = \tan\left(\arctan(v_b) - \sqrt{gk}\,t\right)$$

$$V(t) = \sqrt{g/k}\,\tan\left(\arctan\left(\sqrt{k/g}V_b\right) - \sqrt{gk}\,t\right) \tag{3.12}$$

Equation 3.12 tells us that the maximum speed on the way up is $V_b$. This is as we would expect from physical intuition; drag and gravity can only reduce the speed gained with the rocket engine.

Now the altitude $h(t)$ is given by

$$h(t) = \int_0^t V(\tau)\,d\tau = \int_0^t \sqrt{g/k}\,\tan\left(\arctan\left(\sqrt{k/g}V_b\right) - \sqrt{gk}\,\tau\right)d\tau \tag{3.13}$$

Equation 3.13 may look formidable but it is not difficult to integrate. Let $t' = \sqrt{gk}\,\tau$ and

$$\arctan\left(\sqrt{k/g}V_b\right) = t_a' \tag{3.14}$$

Then Equation 3.13 becomes

$$h(t) = \frac{1}{k}\int_0^{\sqrt{gk}t}\tan(t_a' - t')dt' = \frac{1}{k}\int_0^{\sqrt{gk}t}\frac{\sin(t_a' - t')}{\cos(t_a' - t')}dt' = \frac{1}{k}\ln\left(\frac{\cos\left(t_a' - \sqrt{gk}\,t\right)}{\cos(t_a')}\right) \tag{3.15}$$

The maximum altitude (apogee) is reached when $V(t) = 0$. From Equations 3.12 and 3.14 this occurs when $\sqrt{gk}\,t = t_a'$. Then, from Equation 3.15, the maximum altitude $h_a$ is given by

$$h_a = \frac{1}{k}\ln\left(\frac{1}{\cos(t_a')}\right) \tag{3.16}$$

From Equation 3.14, $\tan(t'_a) = \sqrt{kV_b^2/g}$, so

$$h_a = \frac{1}{k}\ln\left(\frac{1}{\cos(t'_a)}\right) = \frac{1}{k}\ln(\sec(t'_a)) = \frac{1}{k}\ln\left(\sqrt{1+kV_b^2/g}\right) = \frac{1}{2k}\ln\left(1+kV_b^2/g\right) \quad (3.17)$$

Summarizing, we have found that the sounding rocket's maximum altitude is given by

$$h_a = \frac{1}{2k}\ln\left(1+kV_b^2/g\right) \quad\quad (3.18)$$

where $k = \rho SC_D/2m_b$.

As a check on this result, consider the case where the effect of drag is very small; that is, when $k$ is small. Recalling that $\ln(1+x) \cong x$ when $x$ is small, Equation 3.18 becomes

$$h_a = \frac{1}{2k}\ln\left(1+kV_b^2/g\right) \cong \frac{V_b^2}{2g} \quad\quad (3.19)$$

which is the classic drag-free result.

### 3.1.1.3   Maximum Speed Coming Down

When the sounding rocket is on the way down, its equation of motion is

$$m_b\frac{dV}{dt} = -m_b g + \left(\frac{1}{2}\rho SC_D\right)V^2 \quad\quad (3.20)$$

With the same definitions and substitutions as in the previous section, Equation 3.20 can be separated into the form

$$\frac{dv}{1-v^2} = -\sqrt{gk}\,dt \quad\quad (3.21)$$

The left-hand side of Equation 3.21, in contrast to that in Equation 3.10, is not a familiar integrand. We proceed via a partial fraction expansion:

$$\frac{dv}{1-v^2} = \left(\frac{1}{1-v^2}\right)dv = \left(\frac{c_1}{1-v} + \frac{c_2}{1+v}\right)dv = -\sqrt{gk}\,dt \quad\quad (3.22)$$

Several methods exist to find the coefficients $c_1$ and $c_2$. One way is to recombine over the common denominator:

$$\frac{c_1}{1-v} + \frac{c_2}{1+v} = \frac{1}{1-v^2}$$

$$\frac{c_1(1+v)+c_2(1-v)}{1-v^2} = \frac{(c_1+c_2)+(c_1-c_2)v}{1-v^2} = \frac{1}{1-v^2}$$

and then match coefficients of powers of $v$ in the numerators. As reviewed in Section 1.1.2.2, if two polynomials of the same order are everywhere equal then their coefficients are equal.

One numerator is $(c_1 + c_2) + (c_1 - c_2)v$, the other is 1. Both numerators can be considered polynomials of the form $a + bv$. Matching coefficients in powers of $v$ yields two linear equations in two unknowns:

$$c_1 + c_2 = 1$$
$$c_1 - c_2 = 0$$

The solutions are $c_1 = c_2 = 1/2$.

Returning to Equation 3.22, we have

$$\left( \frac{1/2}{1-v} + \frac{1/2}{1+v} \right) dv = -\sqrt{gk}\, dt = -dt' \tag{3.23}$$

Integrating the right-hand side of Equation 3.23 from $t' = t'_a$ and the left-hand side correspondingly from $v = 0$,

$$-(1/2)\ln(1-v) + (1/2)\ln(1+v) = -(t' - t'_a)$$

$$\ln\left( \frac{1+v}{1-v} \right) = -2(t' - t'_a)$$

$$\frac{1+v}{1-v} = e^{-2(t'-t'_a)}$$

$$1 + v = (1-v)e^{-2(t'-t'_a)}$$

$$-v(1 + e^{-2(t'-t'_a)}) = 1 - e^{-2(t'-t'_a)}$$

$$v = -\frac{1 - e^{-2(t'-t'_a)}}{1 + e^{-2(t'-t'_a)}}$$

Multiplying numerator and denominator by $e^{t'-t'_a}$ yields a more favored form

$$v = -\frac{e^{\theta} - e^{-\theta}}{e^{\theta} + e^{-\theta}} \tag{3.24}$$

where $\theta = t' - t'_a = \sqrt{gk}\,(t - t_a)$.

The right-hand side of Equation 3.24 is the negative of a function called the *hyperbolic tangent* of $\theta$, written as $\tanh(\theta)$.

In dimensional form, from Equation 3.24, the sounding rocket's velocity on the way down is given by

$$V = -\sqrt{g/k}\left( \frac{e^{\theta} - e^{-\theta}}{e^{\theta} + e^{-\theta}} \right) = -\sqrt{g/k}\, \tanh(\theta) \tag{3.25}$$

where $\theta = \sqrt{gk}\,(t - t_a)$ and $t_a$ is the time of apogee, the instant before the rocket begins to fall.

To determine the maximum speed, we first integrate Equation 3.25 to obtain $h(t)$, find the time $t_I$ when the rocket impacts the ground, and then use Equation 3.25 to calculate the speed at that point. The student is invited to carry out this approach in *Challenge Problem* 3.2.

The result is that, on its way down, the sounding rocket reaches its maximum speed just before it strikes the ground and its value is

$$V_m = \frac{-V_b}{\sqrt{1 + kV_b^2/g}} \tag{3.26}$$

As a check, note that for vanishing $k$, that is, for very low drag, Equation 3.26 becomes

$$V_m = -V_b \tag{3.27}$$

as we expect from conservation of energy in the case where there is no atmosphere. On the other hand, when $kV_b^2/g \gg 1$ Equation 3.26 becomes

$$V_m = -\sqrt{g/k} \tag{3.28}$$

In this case $V_m$ is independent of $V_b$. The value $\sqrt{g/k}$ is called the *terminal velocity*.


## 3.2 POSSIBILITY OF TRANSFORMING NONLINEAR EQUATIONS INTO LINEAR ONES

As it happens, the nonlinear equations for both upward and downward rocket flight (Equations 3.6 and 3.20) can be transformed into linear ones by changing the independent variable from $t$ to $h$ and the dependent variable from $V$ to $V^2/2$. This shortcut method cannot determine the way velocity varies with time, but it can yield both the maximum altitude and the maximum speed on the way down. Word Problem WP3.1 outlines this approach. Word Problems WP3.2 and 3.3 are variations on this approach for two different and important applications. Problem 3.13 presents another transformation with some generality. Finally, Challenge Problem CP3.1 employs a qualitatively different transformation to derive a first-order linear equation that we can solve to gain valuable insight into a system's dynamics. It is good to know that the possibility of transforming nonlinear ODEs into linear ones exists but, unfortunately, such shortcuts are unavailable for most nonlinear problems and little can be said in general about when it is reasonable to look for one.

Chapter 1 mentioned the process of linearization, in which we derive a linear equation from a nonlinear one by considering small perturbations about a position or trajectory of interest. This is different from the transformations discussed in the preceding paragraph. The solution of a linearized equation is an approximation good only for small perturbations, whereas the solution of a transformed equation is (in principle) exact and valid for any magnitude of the dependent variable. We will devote more time to linearization in later chapters, beginning in Chapter 5.

## 3.3   SUCCESSIVE APPROXIMATIONS FOR ALMOST LINEAR SYSTEMS

Consider the electrical circuit shown in Figure 3.1. It contains a resistor, inductor, and sine-wave voltage source.

As in Chapter 2, Word Problem 2.3, by Ohm's law the voltage drop over the resistor in the direction of current is $V = IR$, the voltage drop over an inductor is

$$LdI/dt = (L/R)(dV/dt)$$

and the voltage *drop* over the sine-wave source is $-V_0 \sin \omega t$. (Notice the sense of the terminals.) Here $V_0$ is the amplitude of the sine wave, in volts, and $\omega$ is the frequency of the oscillation in radians per second. Assume that the sine wave source turns on at $t = 0$ and that no current is flowing in the circuit at that time. Applying Kirchhoff's voltage law to this circuit results in

$$L\frac{dI}{dt} + IR = V_0 \sin \omega t \qquad (3.29)$$

Dividing through by $L$ and defining $\lambda = R/L$ and $I_0 = V_0/R$, Equation 3.29 can be written as

$$\frac{dI}{dt} + \lambda I = \lambda I_0 \sin \omega t \qquad (3.30)$$

Suppose now that resistor in the circuit is slightly nonlinear, as illustrated in Figure 3.2. The resistance increase can occur when higher voltages raise the resistor's temperature. Then the circuit equation is modified to

$$\frac{dI}{dt} + \lambda I + \lambda \varepsilon I^3 = \lambda I_0 \sin \omega t \qquad (3.31)$$

Nonlinear Equation 3.31 cannot be solved in closed form in the same way that the linear Equation 3.30 can. However, if the nonlinear term $\varepsilon I^3$ is not too large compared to $I_0$, that is, if the system modeled by Equation 3.31 is *almost linear*, then we can employ a technique called *successive approximations*. We solve the nonlinear problem as a sequence of linear problems.

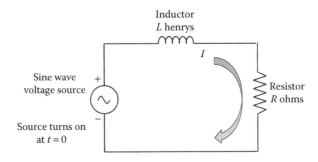

**FIGURE 3.1**   An RL (resistor–inductor) electrical circuit with a sine-wave voltage source.

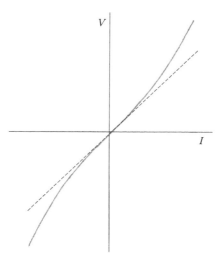

**FIGURE 3.2**   Nonlinear resistor characteristic.

The term *successive approximations* is applied to many processes. The process we will use here is limited to "almost linear" systems. We will meet a more general and theoretically more powerful form of successive approximations in Chapter 4.

The first approximation ignores the nonlinearity altogether and is found by solving Equation 3.30. Call that solution $I_1$. The next solution is obtained by approximating the nonlinear term as the *known function of time* $\lambda \varepsilon I_1^3$. Then we obtain $I_2$ through

$$\frac{dI_2}{dt} + \lambda I_2 = \lambda I_0 \sin \omega t - \lambda \varepsilon I_1^3 \tag{3.32}$$

$$I_2(t) = \int_0^t e^{-\lambda(t-\tau)} (\lambda I_0 \sin \omega \tau - \lambda \varepsilon I_1^3(\tau)) d\tau \tag{3.33}$$

(Recall Equations 2.17 and 2.34.)

Now

$$\int_0^t e^{-\lambda(t-\tau)} \lambda I_0 \sin(\omega \tau) d\tau = I_1(t) \tag{3.34}$$

so, Equation 3.33 becomes

$$I_2(t) = I_1(t) - \lambda \varepsilon \int_0^t e^{-\lambda(t-\tau)} I_1^3(\tau) d\tau \tag{3.35}$$

Proceeding in this way, we calculate $I_{n+1}(t)$ via

$$I_{n+1}(t) = I_1(t) - \lambda \varepsilon \int_0^t e^{-\lambda(t-\tau)} I_n^3(\tau) d\tau \tag{3.36}$$

Focusing on the steady-state solution only, *Challenge Problem 3.3* invites the student to show that performing the integral in Equation 3.35 leads to

$$I_2(t) = G_1 I_0 (\sin \omega t - \eta_2(t)) \tag{3.37}$$

where the effect of the nonlinearity is captured in $\eta_2(t)$:

$$\eta_2(t) = \frac{\alpha}{4}\left(3G_1 \sin(\omega t - 2\theta_1) - G_3 \sin(3\omega t - 3\theta_1 - \theta_3)\right) \tag{3.38}$$

and

$$G_n = \frac{\lambda}{\sqrt{\lambda^2 + (n\omega)^2}} \tag{3.39}$$

$$\theta_n = \arctan\left(\frac{n\omega}{\lambda}\right) \tag{3.40}$$

$$\alpha = \varepsilon\left(G_1 I_0\right)^2 \tag{3.41}$$

If we were to continue the sequence we would find

$$I_n(t) = I_1(t) - G_1 I_0 \eta_n(t) \tag{3.42}$$

where $\eta_n(t)$ is of the form

$$\eta_n(t) = \sum_{m=1}^{N_n} k_{mn} \sin(m\omega t - \phi_{mn}) \tag{3.43}$$

Equations 3.42 and 3.43 are specific examples of an important general fact. When driven by a sine-wave source of frequency $\omega$, stable and damped time-invariant linear systems respond in steady state with oscillations of that frequency only. Nonlinear systems, on the other hand, as we have just seen, introduce multiples of that frequency, called *harmonics*. The phenomenon is called *harmonic distortion* and engineers usually strive to avoid it by operating a system within the linear range of all of its analog components.

It is easy to generalize the process described in this section to systems of higher order. As mentioned in Section 2.2.2 and discussed further in Chapter 6, a system described by the differential equation

$$a_0 \frac{d^n y}{dt^n} + a_1 \frac{d^{n-1} y}{dt^{n-1}} + \ldots + a_n y = g(t) \tag{3.44}$$

with zero initial conditions has a solution that can be written as

$$y(t) = \int_0^t K(t - \tau) g(\tau) d\tau$$

where $K(t - \tau)$ is the kernel for the system. Suppose now a system is described by the ODE

$$a_0 \frac{d^n y}{dt^n} + a_1 \frac{d^{n-1} y}{dt^{n-1}} + \ldots + a_n y + f(t, y) = g(t) \tag{3.45}$$

where $f(t, y)$ is nonlinear in the variable $y$. If $y_1(t)$ is a solution of Equation 3.44 for $t_0 < t < T$ and $f(t, y_1(t))$ is sufficiently small for that same time interval then we can sometimes calculate a solution to Equation 3.45 in the successive approximations

$$y_{n+1}(t) = \int_{t_0}^{t} K(t - \tau)(g(\tau) - f(\tau, y_n(\tau)) d\tau \tag{3.46}$$

$$y_{n+1}(t) = y_1(t) - \int_{t_0}^{t} K(t - \tau) f(\tau, y_n(\tau)) d\tau \tag{3.47}$$

Note that Equation 3.47 can apply to homogeneous almost linear problems as well.

The professional standard in engineering for solving nonlinear differential equations is numerical integration, as we will meet in practice in the Computer Projects that begin at the end of this chapter and we address in theory in Chapter 8. Approximate analytical solutions such as those obtained by Equation 3.47 can be valuable for providing insight and for checking for coding errors in differential equations to be solved by numerical integration.

## 3.4 SUMMARY

Nonlinear ODEs are typically difficult to solve analytically. This chapter presented three techniques for doing so that have a modest range of applicability: (1) separating variables; (2) transforming the nonlinear ODE into a linear one; and (3) for almost linear systems, successive approximations.

Separable equations have the form

$$\frac{dy}{dt} = f(t)g(y), \qquad y(t_0) = y_0$$

We separate variables by dividing through by $g(y)$:

$$\frac{1}{g(y)} \frac{dy}{dt} = f(t)$$

Then we employ the shortcut procedure introduced and justified in Section 2.1:

$$\int_{y(t_0)}^{y(t)} \frac{dy}{g(y)} = \int_{t_0}^{t} f(\tau) d\tau$$

Sometimes we can perform the integration on both sides and arrive at an implicit equation

$$G(y) = F(t)$$

In rare circumstances we are then able to solve for $y$ as an explicit function of $t$.

This chapter presented several nonlinear ODEs that could be transformed into linear ones and pointed to more examples in the problem sets at the end of the chapter.

The transformations can involve changing either the independent or dependent variable or both. It is good to know that the possibility exists but, unfortunately, such shortcuts are unavailable for most nonlinear problems and little can be said, in general, about when it is reasonable to look for one.

Successive approximations on almost linear systems turns a nonlinear problem into a sequence of linear ones.

If

$$\frac{dy}{dt} + p(t)y + f(t,y) = g(t), \qquad y(t_0) = y_0$$

and the nonlinear term $f(t,y)$ is not too large then solving the sequence

$$\frac{dy_1}{dt} + p(t)y_1 = g(t) - f(t,y_0), \qquad y_1(t_0) = y_0$$

$$\frac{dy_2}{dt} + p(t)y_2 = g(t) - f(t,y_1), \qquad y_2(t_0) = y_0$$

$$\cdots$$

$$\frac{dy_n}{dt} + p(t)y_n = g(t) - f(t,y_{n-1}), \qquad y_n(t_0) = y_0$$

can sometimes yield a solution approximating the nonlinear answer. The idea generalizes to nth order ODEs.

The great majority of the times, engineers solve nonlinear differential equations by numerical integration on digital computers. At the end of this chapter are the first phases of several extensive, continuing computing projects.

**Solved Problems**

**SP3.1** Solve

$$\frac{dy}{dt} = 3(2+t)\sqrt{4-y^2}, \qquad y(1) = 0$$

*Solution*:

We separate variables, putting everything dealing with $y$ on the left-hand side of the equation and everything dealing with $t$ on the right:

$$\frac{dy}{\sqrt{4-y^2}} = 3(2+t)dt$$

Then we integrate both sides from beginning to endpoint. These are definite integrals. On the right-hand side the beginning is 1 and its endpoint is the independent variable $t$. On the left-hand side the beginning is $y(1) = 0$ and the endpoint the dependent variable $y$. To avoid confusion we use the dummy integration variables $u$ and $\tau$ underneath the integral signs to distinguish them from the endpoint variables:

$$\int_0^y \frac{du}{\sqrt{4-u^2}} = 3\int_1^t (2+\tau)d\tau \tag{3.48}$$

Now, on the left, let $u = 2v$. Then

$$\int_0^y \frac{du}{\sqrt{4-u^2}} = \int_0^{y/2} \frac{2dv}{\sqrt{4-4v^2}} = \int_0^{y/2} \frac{dv}{\sqrt{1-v^2}} = \arcsin\left(\frac{y}{2}\right) \tag{3.49}$$

And on the right

$$3\int_1^t (2+\tau)d\tau = 3\left(2t + \frac{t^2}{2} - 2(1) - \frac{(1)^2}{2}\right) = 6t + \frac{3t^2}{2} - \frac{15}{2} \tag{3.50}$$

Using Equations 3.48, 3.49, and 3.50:

$$\arcsin\left(\frac{y}{2}\right) = 6t + \frac{3t^2}{2} - \frac{15}{2}$$

Hence

$$y(t) = 2\sin\left(6t + \frac{3t^2}{2} - \frac{15}{2}\right)$$

**SP3.2** Solve

$$\frac{dy}{dt} = 5e^{2t+4y}, \qquad y(t_0) = y_0$$

*Solution*:
Separating variables:

$$e^{-4y}dy = 5e^{2t}dt$$

Integrating both sides and changing to dummy variables under the integral signs:

$$\int_{y_0}^y e^{-4u}du = 5\int_{t_0}^t e^{2\tau}d\tau$$

$$\left(\frac{1}{4}\right)\left(e^{-4y_0} - e^{-4y}\right) = 5\left(\frac{1}{4}\right)\left(e^{2t} - e^{2t_0}\right)$$

$$e^{-4y} = e^{-4y_0} - 10\left(e^{2t} - e^{2t_0}\right)$$

$$-4y = \ln\left(e^{-4y_0} - 10\left(e^{2t} - e^{2t_0}\right)\right)$$

$$y = -\left(\frac{1}{4}\right)\ln\left(e^{-4y_0} - 10\left(e^{2t} - e^{2t_0}\right)\right)$$

**SP3.3** Solve

$$\frac{dy}{dt} = a\cos^2 y\cos^2 t, \qquad y(t_0) = y_0$$

*Solution*:
Separating variables:

$$\frac{dy}{\cos^2 y} = a\cos^2 t\,dt$$

Integrating both sides and changing to dummy variables under the integral sign:

$$\int_{y_0}^{y} \frac{du}{\cos^2 u} = \int_{t_0}^{t} a\cos^2 \tau\,d\tau \tag{3.51}$$

On the left-hand side, recalling our trigonometric identities:

$$\int_{y_0}^{y} \frac{du}{\cos^2 u} = \int_{y_0}^{y} \sec^2 u\,du \tag{3.52}$$

Now remembering the derivatives of common functions in Table 1.1:

$$\frac{d}{du}\left(\tan u\right) = \sec^2 u$$

so Equation 3.52 becomes, by the fundamental theorem of calculus

$$\int_{y_0}^{y} \frac{du}{\cos^2 u} = \int_{y_0}^{y} \sec^2 u\,du = \int_{y_0}^{y} \frac{d}{du}\left(\tan u\right)du = \tan y - \tan y_0 \tag{3.53}$$

On the right-hand side of Equation 3.51, we recall the trigonometric identity

$$\cos^2 \tau = (1 + \cos 2\tau)/2 \tag{3.54}$$

and then

$$\int_{t_0}^{t} a\cos^2 \tau\,d\tau = \frac{a}{2}\int_{t_0}^{t} (1 + \cos 2\tau)d\tau = \frac{a}{2}(t - t_0) + \frac{a}{4}\left(\sin 2t - \sin 2t_0\right) \tag{3.55}$$

From Equations 3.51, 3.53, and 3.55:

$$\tan y - \tan y_0 = \frac{a}{2}(t - t_0) + \frac{a}{4}\left(\sin 2t - \sin 2t_0\right)$$

$$y = \arctan\left(\tan y_0 + \frac{a}{2}(t - t_0) + \frac{a}{4}\left(\sin 2t - \sin 2t_0\right)\right)$$

**SP3.4** Solve

$$\frac{dy}{dt} + 8y + y^2 = -15, \qquad y(0) = 0$$

*Solution*

Rewriting:

$$\frac{dy}{dt} = -(y^2 + 8y + 15)$$

Separating variables:

$$\frac{dy}{y^2 + 8y + 15} = -dt$$

Integrating both sides, changing to dummy variables under the integral sign,

$$\int_0^y \frac{du}{u^2 + 8u + 15} = -\int_0^t d\tau = -t \tag{3.56}$$

Using partial fraction expansion:

$$\frac{1}{u^2 + 8u + 15} = \frac{1}{(u+3)(u+5)} = \frac{c_1}{u+3} + \frac{c_2}{u+5} \tag{3.57}$$

Recombining:

$$\frac{c_1}{u+3} + \frac{c_2}{u+5} = \frac{c_1(u+5) + c_2(u+3)}{(u+3)(u+5)} \tag{3.58}$$

Collecting coefficients in powers of $u$ in the numerator:

$$\frac{c_1(u+5) + c_2(u+3)}{(u+3)(u+5)} = \frac{(c_1 + c_2)u + 5c_1 + 3c_2}{(u+3)(u+5)} \tag{3.59}$$

From Equations 3.57, 3.58, and 3.59:

$$\frac{1}{(u+3)(u+5)} = \frac{(c_1 + c_2)u + 5c_1 + 3c_2}{(u+3)(u+5)}$$

which implies that

$$1 = (c_1 + c_2)u + 5c_1 + 3c_2 \tag{3.60}$$

We can consider both sides of Equation 3.60 as a polynomial in $u$. As discussed in Section 1.1.2.2, if two polynomials are everywhere equal then their coefficients are equal. Then $5c_1 + 3c_2$, the coefficient of $u^0 = 1$ on the right-hand side, equals the coefficient of $u^0$ on the left, which is 1. Similarly, $c_1 + c_2$, the coefficient of $u^1 = u$ on the right-hand side, equals the coefficient of $u^1$ on the left, which is zero. This gives us two equations in two unknowns:

$$5c_1 + 3c_2 = 1$$

$$c_1 + c_2 = 0$$

Solving:

$$c_2 = -c_1$$
$$5c_1 + 3(-c_1) = 1$$
$$c_1 = \frac{1}{2}$$
$$c_2 = \frac{-1}{2}$$

Then Equation 3.56 can be written as

$$\int_0^y \frac{(1/2)du}{u+3} - \int_0^y \frac{(1/2)du}{u+5} = -t$$

$$\int_0^y \frac{du}{u+3} - \int_0^y \frac{du}{u+5} = -2t$$

$$\ln(y+3) - \ln 3 - \ln(y+5) + \ln(5) = -2t$$

$$\ln\left(\frac{y+3}{3} \cdot \frac{5}{y+5}\right) = -2t$$

$$\exp\left(\ln\left(\frac{y+3}{3} \cdot \frac{5}{y+5}\right)\right) = \exp(-2t)$$

$$\frac{y+3}{3} \cdot \frac{5}{y+5} = \exp(-2t)$$

$$5(y+3) = 3(y+5)e^{-2t}$$

$$(5 - 3e^{-2t})y = -15(1 - e^{-2t})$$

$$y = -\frac{3(1 - e^{-2t})}{1 - (3/5)e^{-2t}}$$

**SP3.5** Find the second in the sequence of successive approximations to the solution of

$$\frac{dy}{dt} + 3y + \varepsilon y^2 = 3, \qquad y(0) = 0$$

given that the first is the solution when $\varepsilon = 0$.

*Solution*:

We first find the solution $y_1(t)$ when $\varepsilon = 0$. The equations are

$$\frac{dy_1}{dt} + 3y_1 = 3, \qquad y_1(0) = 0$$

Using the integrating factor,

$$e^{3t}\left(\frac{dy_1}{dt} + 3y_1\right) = 3e^{3t}$$

$$\frac{d}{dt}\left(e^{3t}y_1\right) = 3e^{3t}$$

$$\int_0^t \frac{d}{d\tau}(e^{3\tau}y_1)d\tau = \int_0^t 3e^{3\tau}d\tau$$

$$e^{3t}y_1(t) = e^{3t} - 1$$

$$y_1(t) = 1 - e^{-3t}$$

We find the second approximation $y_2(t)$ by solving the equations

$$\frac{dy_2}{dt} + 3y_2 = 3 - \varepsilon y_1^2(t), \qquad y_2(0) = 0$$

$$\frac{dy_2}{dt} + 3y_2 = 3 - \varepsilon(1 - e^{-3t})^2$$

$$\frac{dy_2}{dt} + 3y_2 = 3 - \varepsilon(1 - 2e^{-3t} + e^{-6t})$$

Again using the integration factor,

$$e^{3t}\left(\frac{dy_2}{dt} + 3y_2\right) = 3e^{3t} - \varepsilon e^{3t}(1 - 2e^{-3t} + e^{-6t})$$

$$\frac{d}{dt}\left(e^{3t}y_2\right) = 3e^{3t} - \varepsilon\left(e^{3t} - 2 + e^{-3t}\right)$$

$$\int_0^t \frac{d}{d\tau}\left(e^{3\tau}y_2\right) = \int_0^t \left[3e^{3\tau} - \varepsilon\left(e^{3\tau} - 2 + e^{-3\tau}\right)\right]d\tau$$

$$e^{3t}y_2(t) = e^{3t} - 1 - \varepsilon\left[(e^{3t} - 1)/3 - 2t + (1 - e^{-3t})/3\right]$$

$$y_2(t) = 1 - e^{-3t} - \varepsilon\left(1 - 6te^{-3t} - e^{-6t}\right)/3$$

**Problems to Solve**

In Problems P3.1 through P3.6 find $y(t)$

**P3.1** $\dfrac{dy}{dt} + (\cos 2t) y^3 = 0, \qquad y(0) = 1$

**P3.2** $\dfrac{dy}{dt} + \dfrac{te^{-4t}}{1+y} = 0, \qquad y(0) = 0$

**P3.3** $\dfrac{dy}{dt} - \dfrac{2y \ln y}{t} = 0, \qquad y(1) = 2$

**P3.4** $\dfrac{dy}{dt} + \sqrt{3+y} \tan 2t = 0, \qquad y(0) = 0$

**P3.5** $\dfrac{dy}{dt} - 10y - y^2 = 21, \qquad y(0) = 0$

**P3.6** $\dfrac{dy}{dt} + 18y + 3y^2 = -24, \qquad y(0) = 0$

**P3.7** Find the second in the sequence of successive approximations to the solution of

$$\frac{dy}{dt} + \left(\frac{2}{t}\right) y + \varepsilon y^2 = t, \qquad y(1) = \frac{1}{4},$$

given that the first is the solution when $\varepsilon = 0$.

**P3.8** Find the second in the sequence of successive approximations to the solution of

$$\frac{dy}{dt} - (\tan t) y + \varepsilon y^2 = 1, \qquad 0 \le t < \pi/2, \qquad y(0) = 0$$

given that the first is the solution when $\varepsilon = 0$. Hint:

$$\int_0^t \sec \tau \, d\tau = \int_0^t \frac{d\tau}{\cos \tau} = \int_0^t \frac{\cos \tau \, d\tau}{\cos^2 \tau} = \int_0^t \frac{\cos \tau \, d\tau}{1 - \sin^2 \tau} = \int_0^{\sin t} \frac{dx}{1 - x^2}$$

Use partial fraction expansion on the last integral.

**P3.9** Find the second in the sequence of successive approximations to the solution of

$$\frac{dy}{dt} + 2y + \varepsilon \ln y = 0, \qquad y(0) = 1$$

given that the first is the solution when $\varepsilon = 0$.

**P3.10** Find the third in the sequence of successive approximations to the solution of

$$\frac{dy}{dt} - \left(\frac{t}{1-t^2}\right) y + \varepsilon y^2 = 0, \qquad y(0) = 1$$

given that the first is the solution when $\varepsilon = 0$. Ignore terms involving $\varepsilon^3$. From the sequence of approximations can you guess the exact solution? See Problem 3.12.

**P3.11** Find the third in the sequence of successive approximations to the solution of

$$\frac{dy}{dt} + \left(\frac{2t}{1+t^2}\right)y + \varepsilon y^2 = 0, \qquad y(0) = 1$$

given that the first is the solution when $\varepsilon = 0$. Ignore terms involving $\varepsilon^3$. From the sequence of approximations can you guess the exact solution? See Problem 3.12.

**P3.12** Show by direct substitution that if $y_1(t)$ is the solution to

$$\frac{dy}{dt} + p(t)y = 0, \qquad y(0) = y_0$$

then the solution to

$$\frac{dy}{dt} + p(t)y + \varepsilon y^2 = 0, \qquad y(0) = y_0$$

is

$$y_\infty(t) = \frac{y_1(t)}{1 + \varepsilon q(t)}$$

where

$$q(t) = \int_0^t y_1(\tau)d\tau$$

Note that $\varepsilon$ need not be small. See also Problem 3.13.

**P3.13** Consider again the circuit with nonlinear resistor analyzed in Section 3.2. We examine here the dynamics of the circuit with an initial condition but no input voltage. The differential equation for the current through the resistor is

$$\frac{dI}{dt} + \lambda I + \varepsilon \lambda I^3 = 0, \qquad I(0) = I_0$$

This is an example of a Bernoulli equation, which has the general form

$$\frac{dy}{dt} + p(t)y = q(t)y^n$$

Gottfried Leibniz showed in 1696 that the substitution $z = y^{1-n}$ transforms the Bernoulli equation into a linear one. Use this technique to find the current through the resistor as a function of time.

**Word Problems**

**WP3.1** In this problem we will use a different approach to obtain the same basic results we found in Section 3.1. With appropriate transformations we will convert the nonlinear equations into linear ones. We begin with the sounding rocket's equation of motion on the way up, after burnout:

$$\frac{dV}{dt} = -g - kV^2, \qquad V(0) = V_b \tag{3.61}$$

First we transform from the independent variable $t$ to the independent variable $h$ (altitude). Then

$$\frac{dV}{dt} = \frac{dV}{dh}\frac{dh}{dt} = \frac{dV}{dh}V = \frac{d}{dh}\left(\frac{V^2}{2}\right)$$

Next we transform from the dependent variable $V$ to the dependent variable $V^2/2 = E$. Then Equations 3.61 becomes

$$\frac{dE}{dh} + 2kE = -g, \qquad E(0) = \frac{V_b^2}{2} \tag{3.62}$$

Part (a): Solve Equations 3.62 to obtain $E$ as a function of $h$. Set $E = 0$ and determine the rocket's maximum altitude.

Part (b): For the rocket's trajectory down, use the same transformation as in part (a), then substitute $z = h_a - h$ and obtain

$$\frac{dE}{dz} + 2kE = g, \qquad E(0) = 0 \tag{3.63}$$

Solve Equations 3.63, set $z = h_a$, (i.e., $h = 0$) and find the value of $E$, and hence the maximum speed.

**WP3.2** Consider a projectile launched vertically from the Earth's surface with initial velocity $v_0$. Ignoring air resistance, the only force it experiences is that of gravity, which varies with altitude, according to Newton's law of gravitation, by

$$F = -m\frac{GM}{R^2}$$

where:
    $m$ is the mass of the projectile
    $G$ is the gravitational constant
    $M$ is the mass of the Earth
    $R$ is the distance of the projectile from the Earth's center

For ease of calculation this can be written as

$$F = -m\frac{gR_E^2}{R^2}$$

where:

$g = 32.2 \, \text{ft}/\text{s}^2$ is the acceleration of gravity at the Earth's surface
$R_E = 20.91 \cdot 10^6 \, \text{ft}$ is the radius of the Earth

Newton's second law of motion is $F = ma$, where $a = dv/dt$ is the projectile's acceleration.

We can write

$$\frac{dv}{dt} = \frac{dv}{dR} \cdot \frac{dR}{dt} = \frac{dv}{dR} v$$

let $R$ be the independent variable and define a new dependent variable $E = v^2/2$.

a.  Show that projectile's equation of motion is then

$$\frac{dE}{dR} = -\frac{gR_E^2}{R^2}$$

b.  Solve the equation of motion and show that the projectile's velocity is given by

$$v = \sqrt{v_0^2 + (2gR_E^2)\left(\frac{1}{R} - \frac{1}{R_E}\right)}$$

c.  Show that, if $v_0$ is not too large, the apogee of the projectile's trajectory (its maximum distance from the center of the Earth) is

$$R_{\text{MAX}} = \frac{R_E}{1 - (v_0^2/2gR_E)}$$

d.  Show that the *escape velocity*, the initial velocity $v_0$ large enough that the projectile does not fall back to Earth, is 36.7 thousand feet per second.

**WP3.3** A spacecraft reenters Earth's atmosphere at a high speed $V_E$ and shallow flight path angle $\gamma$ with respect to the horizon. During much of the reentry, aerodynamic resistance (drag) is so large that gravity may be ignored. Under these conditions the equation of motion for the spacecraft is

$$m\frac{dV}{dt} = -\frac{1}{2}\rho S C_D V^2 \tag{3.64}$$

Here $\rho$ is the air density, which varies with altitude according to

$$\rho = \rho_0 \exp(-Zh) \tag{3.65}$$

where we will model the coefficient $Z$ as constant, the parameter $S$ is the spacecraft's (constant) cross-sectional area, and $C_D$ is its drag coefficient, which we will also assume here to be a constant.

Equation 3.64 is hard to solve in this form because of the air density's variation with altitude. However, we can make several transformations to enable a solution. First, we change the dependent variable from $V$ to $E = V^2/2$.

From Equation 3.64

$$m\frac{dV}{dt} = -\rho SC_D\left(\frac{V^2}{2}\right) = -\rho SC_D E \tag{3.66}$$

Then we change the independent variable, several times, in steps. Note first that if we denote the distance along the flight path as $s$ and recognize that $ds/dt = V$ we can write

$$m\frac{dV}{dt} = m\frac{dV}{ds}\cdot\frac{ds}{dt} = mV\frac{dV}{ds} = m\frac{d}{ds}\left(\frac{1}{2}V^2\right) = m\frac{dE}{ds} \tag{3.67}$$

Next we observe that the spacecraft's altitude $h$ changes as it progresses along the flight path according to

$$\frac{dh}{ds} = -\sin\gamma$$

and so, from Equation 3.67,

$$m\frac{dV}{dt} = m\frac{dE}{ds} = m\frac{dE}{dh}\cdot\frac{dh}{ds} = -m\sin\gamma\frac{dE}{dh} \tag{3.68}$$

Finally, we note that

$$\frac{dE}{dh} = \frac{dE}{d\rho}\frac{d\rho}{dh} \tag{3.69}$$

and from Equation 3.65,

$$\frac{d\rho}{dh} = \frac{d}{dh}\left(\rho_0\exp(-Zh)\right) = -Z\rho_0\exp(-Zh) = -Z\rho \tag{3.70}$$

Use Equations 3.66 through 3.70 to derive a first-order LTI ODE for $E$ as a function of $\rho$. Solve the equation. The parameter $m/C_D S$ is called the spacecraft's ballistic coefficient. If the ballistic coefficient is 2 slugs/ft$^2$, $Z = 3\cdot10^{-5}$/ft, $\rho_0 = 2.377\cdot10^{-3}$ slugs/ft$^3$, and $\gamma = 5°$, find the altitude at which the spacecraft's speed is $V_E/100$.

**WP3.4** Consider a metal slab brought immediately from a furnace at temperature $T_0$ and hung by a slender cord in the center of an evacuated refrigerator. According to the Stefan–Boltzmann law, the rate of heat lost from the slab due to *radiation* is

$$q = -\varepsilon\sigma A\left(T^4 - T_S^4\right)$$

where:
   $\sigma$ is the Stefan–Boltzmann constant (a constant of nature)
   $A$ is the surface area of the slab
   $\varepsilon$ is the emissivity ($0 < \varepsilon < 1$)

$T$ is the temperature of the slab
$T_S$ is the temperature of the surroundings

The thermal energy in the slab is given by

$$E = mcT$$

where:
$m$ is the mass of the slab
$c$ is the specific heat of the slab material

The key physical principle is that the rate of change of the thermal energy in the slab equals the heat lost due to radiation. (Because the refrigerator is evacuated, convection is minimal. We also assume that conduction through the suspension cord can be ignored.) Assuming that the temperature of the surroundings can be ignored in the Stefan–Boltzmann equation, write the first-order nonlinear ODE governing the slab's temperature as a function of time. Defining for simplicity the parameter $\gamma = \varepsilon\sigma A/mc$, solve the ODE, and determine the slab temperature as a function of $t$, $\gamma$, and $T_0$. If $T_0 = 1500$ K and $\gamma = 3.0 \cdot 10^{-12}$ $K^{-3}$/s, find the time required for the slab to cool to 500K.

**WP3.5** Consider a radially symmetric tank of water of height $H$ releasing its contents through a small portal of area $a_E$. (See Figure 3.3) If the exit portal is at the very bottom of the tank, Torricelli's law states that the stream of water leaving the tank has speed $v = \sqrt{2gh}$, where $h$ is the height of water in the tank. Assuming that the stream fills the portal completely, the volume of water leaving the tank per unit time is given by the product of $a_E$ and $v$. The rate of change of the volume of water can be described as the surface area at the water level times the rate of change of the water level. Derive and solve the differential equation describing the height of the water in the tank as a function of time for the following two cases: (a) the tank is a cylinder of constant radius $r_0$ and (b) the radius of the tank increases with height according to $r = \sqrt{r_0^2 + bh^{1/2}}$. Assume that the tank is initially completely filled in both cases. Show that the time required for a completely filled tank to empty is, for case (a):

$$T_0 = \left(\frac{\pi r_0^2}{a_E}\right)\left(\frac{2H}{g}\right)^{1/2}$$

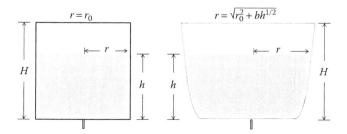

**FIGURE 3.3** Radially symmetric tanks releasing water through portals at their bottoms. (a) constant radius and (b) radius increasing with height.

and for case (b):

$$T = T_0 + \frac{b\pi H}{(2g)^{1/2} a_E}$$

See also Challenge Problem CP1.1.

**WP3.6** Consider again the RC (resistor–capacitor) circuit examined in Chapter 2, Section 2.1.1, but this time suppose the capacitor is nonlinear in the way depicted in Figure 3.4. The opposing electrical charges cause the capacitor plates to attract one another. Suppose the plates are not held rigid, but rather have a compliance that can be modeled as the stretching of a strong spring: $F = kx$, where $x$ is the displacement of each plate toward the other. It can be shown that, for small displacements, the capacitance increases with voltage according to $C = C_0(1 + C_0 V^2/kd^2)$, where $C_0$ is the capacitance and $d$ the separation between the plates at zero voltage. If there is initially a voltage over the capacitor, the rate of change of that voltage $V$ with respect to time after the switch is thrown is approximated by

$$-\left(\frac{1}{RC_0}\right)V + \left(\frac{1}{RC_0}\right)\varepsilon'V^3$$

where:
  $R$ is the resistor value
  $\varepsilon' = \dfrac{3C_0}{kd^2}$

Use the Leibniz substitution described in Problem to Solve P3.13 to determine the value of the voltage $V$ as a function of time.

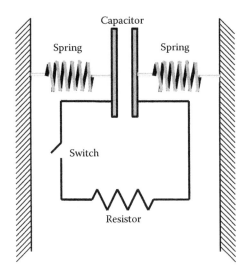

Capacitor

Spring                    Spring

Switch

Resistor

**FIGURE 3.4**   A nonlinear RC (resistor–capacitor) circuit.

## Challenge Problems

**CP3.1** (From [2]) Consider a particle projected along the horizontal axis with velocity $v_0$ on a rough inclined plane (Figure 3.5). The roughness of the surface causes friction with magnitude proportional to the particle's normal force on the plane and direction opposite the particle's velocity vector. The equations of motion are

$$m\frac{du}{dt} = mg\sin\theta - \mu mg\cos\theta\frac{u}{\sqrt{u^2+v^2}} \tag{3.71}$$

$$m\frac{dv}{dt} = -\mu mg\cos\theta\frac{v}{\sqrt{u^2+v^2}} \tag{3.72}$$

where:
$u$ and $v$ are the components of velocity
$m$ is the mass of the particle
$\theta$ is the angle of inclination of the plane
$\mu$ is the coefficient of friction of the plane

We will not challenge you to solve these equations directly but rather to derive a first-order linear equation from them that you then solve to gain valuable insight into the system's dynamics.

First, let us examine the problem qualitatively. Depending on the values of $\mu$ and $\theta$, we can see intuitively that there are three possible final conditions for the motion. If $\mu$ is sufficiently large and $\theta$ sufficiently small the particle will stop. If $\mu$ is sufficiently small and $\theta$ sufficiently large the particle will travel an infinite distance in the $v$ direction. (To see this consider the motion if $\mu = 0$.) Finally, if the values of $\mu$ and $\theta$ are "just right" (we might call them the "Goldilocks values"); the particle travels a finite distance in the $v$ direction and approaches asymptotically a line in the $u$ direction.

Part (a): Using Equation 3.71, show that the particle will not stop if $\mu < \tan\theta$.

Part (b): Using Equations 3.71 and 3.72, derive a first-order differential equation for $v$ as it depends on the ratio $r = u/v$ rather than $t$. Hints:

$$\frac{dv}{dr} = \frac{dv}{dt}\cdot\frac{dt}{dr}, \qquad \frac{dt}{dr} = \frac{1}{(dr/dt)}, \qquad \frac{dr}{dt} = \frac{d}{dt}\left(\frac{u}{v}\right) = \frac{1}{v}\frac{du}{dt} - \frac{u}{v^2}\frac{dv}{dt}$$

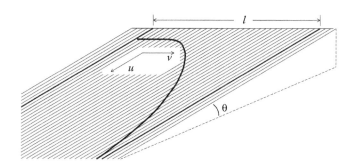

**FIGURE 3.5** A particle sliding on a rough inclined plane.

Part (c): Solve the equation you derived in Part (b). Hints: Let $r = \tan\phi$ in your integral:

$$\int_0^r \frac{dr}{\sqrt{1+r^2}} = \int_0^{\tan^{-1}\phi} \sec\phi \, d\phi = \int_0^{\tan^{-1}\phi} \frac{d\phi}{\cos\phi} = \int_0^{\tan^{-1}\phi} \frac{\cos\phi \, d\phi}{\cos^2\phi} = \int_0^{\tan^{-1}\phi} \frac{\cos\phi \, d\phi}{1-\sin^2\phi} = \int_0^z \frac{dx}{1-x^2}$$

where $z = \sin\phi = r/\sqrt{1+r^2}$. Then use partial fraction expansion on $\int_0^z dx/(1-x^2)$. Simplify your result by recognizing that $(1/(\sqrt{1+r^2} - r)) = \sqrt{1+r^2} + r$.

Part (d): Using the solution $v = v(r)$ you found in Part (c), show that if $\mu > (1/2)\tan\theta$ the particle travels a finite distance $l$ in the $v$ direction and that

$$l = \frac{v_0^2}{g}\left( \frac{2\mu\cos\theta}{4\mu^2\cos^2\theta - \sin^2\theta} \right)$$

Hints:

$$l = \lim_{t'\to\infty} \int_0^{t'} v(t)dt = \lim_{r'\to\infty} \int_0^{r'} v(r) \cdot \frac{dt}{dr} \cdot dr$$

Also, in your integral of the form $\int f(r + \sqrt{1+r^2})dr$ let $w = r + \sqrt{1+r^2}$. Then $r = w/2 - 1/(2w)$ and $dr = (1/2)(1 + 1/w^2)dw$ and the integral becomes

$$\left(\frac{1}{2}\right)\int f(w)dw + \left(\frac{1}{2}\right)\int \frac{f(w)}{w^2} dw$$

**CP3.2** In this problem we will continue the approach used in Section 3.1.1.3 and determine the maximum speed the sounding rocket achieves on its way down. We start with the dimensional version of the velocity calculated in that section:

$$V = -\sqrt{g/k}\left( \frac{e^\theta - e^{-\theta}}{e^\theta + e^{-\theta}} \right) \tag{3.73}$$

where:
$$\theta = \sqrt{gk}(t - t_a)$$

$t_a$ is the time of apogee (the time the rocket reaches its highest altitude)

Part (a): Integrate Equation 3.73 to obtain the altitude as a function of time:

$$h(t) = h_a - \frac{1}{k}\ln\left( \frac{e^\theta + e^{-\theta}}{2} \right) \tag{3.74}$$

where $h_a$ is the maximum altitude (altitude at apogee).

Part (b): Set $h(t) = 0$ in Equation 3.74 to obtain an implicit equation for the time $t_I$ when the sounding rocket impacts the Earth. Solve the implicit equation in two steps. First, let $x = e^{\theta_I}$, where $\theta_I = \sqrt{gk}(t_I - t_a)$, recall the equation for $h_a$, and arrive at

$$2\sqrt{1 + v_b^2} = x + 1/x \tag{3.75}$$

where $v_b = V_b \sqrt{k/g}$. Then solve Equation 3.75 for $x$ and obtain

$$x = \sqrt{1 + v_b^2} + v_b \tag{3.76}$$

Part (c): Express $V$ in terms of $x$ in Equation 3.73, then use Equation 3.76 to arrive at the desired result.

**CP3.3** Prove that the steady-state solution to Equation 3.35 is Equations 3.37 through 3.41.

*Guidance*: Use the list of geometric identities in Chapter 1, Section 1.1.3, to derive an equation of the form

$$\sin^3 \phi = c_1 \sin \phi + c_2 \sin 2\phi + c_3 \sin 3\phi$$

Solved Problem SP1.6 in Chapter 1 solves a similar problem. Take advantage of Equations 2.39 through 2.42 in Chapter 2, replacing $\omega$ with $n\omega$ as appropriate.

## COMPUTING PROJECTS: PHASE 1

The great majority of the times, engineers solve nonlinear differential equations by numerical integration on digital computers. Section 8.3 in Chapter 8 examines algorithms used in these computer routines, but we define here the first phases of several extensive, continuing computing projects intended to give the student first-hand experience with their use. The exercises provide you an opportunity to program a computer to achieve a practical result.

Each project is divided into two phases. The objective of the first phase is to demonstrate an effective method for checking basic elements of your program. It is imperative that professional engineers find ways to check their own work. We cannot expect our supervisors or coworkers to do this for us. The need to check results is particularly acute when we rely on a computer program. We must not fall into the trap of blindly accepting a computer program's output. In our professional lives, when we are programming a computer to determine important quantitative information, we will of course carefully review the lines of code we have written, but this is insufficient. It is often difficult to recognize our own errors. We must find independent means to check results.

Phase 1 of the computing projects checks basic elements of our programs by comparing numerically computed results with hand-calculations. The full equations, which we will solve in Phase 2, set at the end of Chapter 6, are too complicated for hand-calculation. In Phase 1 we consider our equations in a modified setting, simple enough to permit hand-calculation yet realistic enough to permit testing of a significant portion of our program.

We leave it to you, or to your instructor, to decide which of the projects to undertake and which of the many available software packages for numerical integration you will employ.

## PROJECT A: PHASE 1

Project A uses numerical integration to explore further the sounding rocket trajectory analyzed in Section 3.1.1. In Phase 1 we will program the basic equations of motion and the basic relations for thrust and drag. We will program constant values for drag coefficient and air density, matching the simple, approximate model we used in Section 3.1.1.

We will analyze the trajectory of a small model rocket for which the simple model should work well. We will compare the numerically integrated results with the closed form solutions we derived. This comparison will test that we have programmed the basic equations of motion and the basic equations for thrust and drag correctly.

In Phase 2, which Chapter 6 defines, we will program equations for drag coefficient and air density that more closely resemble reality. At that time we will run the program with the parameters of the small model rocket and two much larger rockets more like those used in NASA's science program.

### Desired Output

Run your numerical integration program with the equations given below. Print out and hand in a copy of a graph of the computed rocket trajectory (altitude versus time) and your program, including all functions that you define. *Save your program for use in Phase 2.* Hand in copies of the computer's instant-by-instant tabulated results at the appropriate times to record the maximum speed during ascent, the maximum height, and the maximum speed during descent. Compare the maximum speed during ascent, the maximum altitude, and the maximum speed during descent as computed by numerical integration and as given by the closed form equations in Section 3.1. You should expect numerically computed solutions to be within 2% or better of closed-form-calculated values.

### Definition of Terms

Table 3.1 defines terms used in Phase 1 of Project A.

### Equations

Basic equations of motion:

$$\dot{h} = V \qquad\qquad\qquad h(0) = 0$$

$$\dot{V} = (T - D)/m - g \qquad V(0) = 0$$

$$\dot{m} = -T/c \qquad\qquad m(0) = m_0$$

Basic equations for thrust and drag:

$$T = \rho_{rg} A c^2 \quad m > m_b$$

$$T = 0 \qquad\quad m \le m_b$$

$$D = \frac{1}{2}\rho_a S C_D V^2 \qquad V > 0$$

$$D = -\frac{1}{2}\rho_a S C_D V^2 \quad V \le 0$$

**TABLE 3.1**

**Definition of Terms for Project A; Phase 1**

| Term | Definition |
|------|-----------|
| $h$ | Altitude (ft) |
| $V$ | Vertical velocity (ft/s) |
| $m$ | Mass (slugs) |
| $m_0$ | Mass at launch (slugs) |
| $m_b$ | Mass at burnout (slugs) |
| $g$ | Acceleration of gravity (ft/s/s) |
| $c$ | Speed of exit gases (ft/s) |
| $\rho_a$ | Density of air (slugs/cu ft) |
| $S$ | Cross-sectional area (sq ft) |
| $C_D$ | Drag coefficient |
| $k$ | $\rho_{a0} S C_{D0}/2m$ (1/ft) |
| $T$ | Thrust (lbs) |
| $D$ | Drag (lbs) |
| $W_0$ | Weight at launch (lbs) |
| $MR$ | Mass ratio $m_0/m_b$ |
| $A$ | Area of rocket nozzle (sq ft) |
| $\rho_{rg}$ | Density of rocket exit gases (slug/cu ft) |
| $C_{D0}$ | Subsonic drag coefficient |
| $\rho_{a0}$ | Density of air at sea level (slugs/cu ft) |

Equations for drag coefficient and air density (simplified model for Phase 1 only):

$$C_D = C_{D0}$$

$$\rho_a = \rho_{a0}$$

Constant parameters:

$$g = 32.2\,\text{ft}/\text{s}^2$$

$$\rho_{a0} = 2.38 \times 10^{-3}\ \text{slugs}/\text{ft}^3$$

$$\rho_{rg} = 0.5 \times 10^{-3}\ \text{slugs/ft}^3$$

Parameters changing with rocket design:

$$A = 0.7S$$

$$m_0 = W_0/g$$

$$m_b = m_0/MR$$

Parameters for rocket number one (small model rocket):

$$W_0 = 8 \, \text{lbs}$$

$$MR = 1.15$$

$$S = 0.05 \, \text{ft}^2$$

$$c = 5 \cdot 10^3 \, \text{ft/s}$$

$$C_{D0} = 0.03$$

## PROJECT B: PHASE 1

Project B analyzes the flight performance of a high-performance fighter aircraft. Ultimately (in Phase 2, defined in Chapter 6) we will be interested in answering three fundamental questions: (a) How high can it fly (aircraft ceiling)? (b) How fast can it fly? And (c) How long does it take to go from the ground to the required speed and altitude (minimum time to climb)?

Here in Phase 1 we will have the more limited objective of checking out our program to ensure we have programmed the basic equations of motion and basic thrust and drag equations correctly. To accomplish this we will analyze aircraft dynamics in a special setting: beginning just after the aircraft has taken off from the runway, when it has settled into a constant flight path angle and is gaining speed and altitude. For this special case we will be able to solve the equations analytically, and we can then compare our numerically integrated results with the analytical solutions.

### Definition of Terms

Figure 3.6 illustrates the important forces and kinematic variables and Table 3.2 defines terms used in Phase 1 of Project B.

### Equations

The basic equations for aircraft *translational* motion are

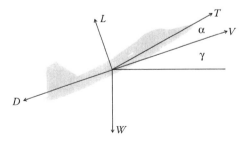

**FIGURE 3.6**  Variables in aircraft longitudinal flight.

---

**TABLE 3.2**

**Definition of Terms for Project B; Phase 1**

| Term | Definition |
|------|------------|
| $h$ | Altitude (ft) |
| $V$ | Speed (ft/s) |
| $\gamma$ | Flight path angle (radians) |
| $m$ | Mass (slugs) |
| $g$ | Acceleration of gravity (ft/s/s) |
| $\rho_a$ | Density of air (slugs/cu ft) |
| $S$ | Cross-sectional area (sq ft) |
| $C_D$ | Drag coefficient |
| $C_L$ | Lift coefficient |
| $T$ | Thrust (lbs) |
| $D$ | Drag (lbs) |
| $W$ | Weight (lbs) |
| $\alpha$ | Angle of attack (radians) |
| $C_{D0}$ | Subsonic drag coefficient |
| $\rho_{a0}$ | Density of air at sea level (slugs/cu ft) |
| $T_0$ | Thrust at sea level (lbs) |
| $k$ | $\rho_{a0}SC_{D0}/2m$ |

---

$$\dot{h} = V\sin\gamma$$

$$\dot{V} = (T\cos\alpha - D - W\sin\gamma)/m$$

$$\dot{\gamma} = (T\sin\alpha + L - W\cos\gamma)/mV$$

The basic lift and drag equations are

$$L = \frac{1}{2}\rho_a SV^2 C_L$$

$$D = \frac{1}{2}\rho_a SV^2 C_D$$

$$C_L = C_{L\alpha}\alpha$$

$$L_\alpha = \frac{\partial L}{\partial \alpha} = \frac{1}{2}\rho_a SV^2 C_{L\alpha}$$

In the special case of low altitude, low-speed flight along a constant flight path, we may approximate

$$T = T_0 = \text{constant}$$

$$C_D = C_{D0} = \text{constant}$$

$$\rho_a = \rho_{a0} = \text{constant}$$

$$\gamma = \gamma_0 = \text{constant}$$

and the equation of motion along the flight path becomes

$$\frac{dV}{dt} = -kV^2 + a$$

where

$$k = \frac{\rho_{a0} S C_{D0}}{2m}$$

$$a = g\left(\frac{T_0}{W} - \sin\gamma_0\right)$$

This is the same form as the equation of motion for the sounding rocket on its descent. (See Section 3.1.1.3.) We apply the results of that analysis here and find that

$$V(t) = \sqrt{\frac{a}{k}}\tanh\left(\sqrt{ka}(t + t_0)\right)$$

$$h(t) = \frac{\sin\gamma_0}{k}\ln\left(\frac{\cosh(\sqrt{ka}(t + t_0))}{\cosh(\sqrt{ka}(t_0))}\right)$$

where $\cosh\theta = (e^\theta + e^{-\theta})/2$ is called the hyperbolic cosine and $t_0$ is found from the initial conditions

$$V(0) = V_0 = \sqrt{\frac{a}{k}}\tanh\left(\sqrt{ka}(t_0)\right)$$

which results in

$$t_0 = \left(\frac{1}{2\sqrt{ka}}\right)\ln\left(\frac{1 + V_0\sqrt{k/a}}{1 - V_0\sqrt{k/a}}\right)$$

In Phase 1 we program the basic equations of aircraft translational motion above with the initial conditions

$$h(0) = 0$$

$$V(0) = 350$$

$$\gamma(0) = 0.25$$

We program the aforementioned basic lift and drag equations, the mass and weight relationship $m = W/g$ and the constant parameters

$$g = 32.2$$

$$\rho_{a0} = 2.38 \times 10^{-3}$$

$$W = 60 \times 10^3$$

$$T_0 = 70 \times 10^3$$

$$C_{L\alpha} = 4.0$$

$$C_{D0} = .0123$$

$$S = 840$$

For Phase 1 only we program

$$C_D = C_{D0}$$

$$\rho_a = \rho_{a0}$$

$$T = T_0$$

$$\alpha = W \cos\gamma / (T + L_\alpha)$$

Also, for Phase 1 only we program the following test parameters:

$$V_C = \sqrt{\frac{a}{k}} \tanh(\theta) \qquad\qquad k = \frac{\rho_{a0} S C_{D0}}{2m}$$

$$h_C = \left(\frac{\sin\gamma_C}{k}\right) \ln\left(\frac{\cosh(\theta)}{\cosh(\theta_0)}\right) \qquad\qquad a = g\left(\frac{T}{W} - \sin\gamma_C\right)$$

$$\theta = \sqrt{ka}(t + t_0) \qquad\qquad \gamma_C = \gamma(0) = 0.25$$

$$t_0 = \left(\frac{1}{2\sqrt{ka}}\right) \ln\left(\frac{1 + V_0\sqrt{k/a}}{1 - V_0\sqrt{k/a}}\right) \qquad\qquad fh_{error} = (h - h_C)/(h + .001)$$

$$V_0 = V(0) = 350 \qquad\qquad fV_{error} = (V - V_C)/V$$

$$\theta_0 = t_0\sqrt{ka} \qquad\qquad f\gamma_{error} = (\gamma - \gamma_C)/\gamma$$

## Desired Output

Run your program for 25 s of flight time. You will know you have programmed the equations correctly when the computed fractional errors $fV_{error}, fh_{error}, f\gamma_{error}$ are on the order of .001 or less. Print out and hand in a copy of

a. Graphs of $h$, $V$, and $\gamma$ as functions of time
b. The computer's instant-by-instant tabulated results at the end of the run for $h, V, \gamma, fh_{\text{error}}, fV_{\text{error}}, f\gamma_{\text{error}}$
c. Your program, including functions you defined.

Save your program for Phase 2.

## Disclaimer

The parameters and models given earlier and the performance computed for the high-performance fighter aircraft are not intended to represent those of any actual aircraft.

# 4 Existence, Uniqueness, and Qualitative Analysis

This chapter explores the conditions under which we can be sure that a single, well defined mathematical solution exists to a given first-order differential equation and also examines analytical methods we can employ when we are unable to solve it in a closed form. It is worth taking a few moments to discuss why engineers should invest some of their time in understanding existence and uniqueness theorems, and the first section of this chapter provides this motivation. Next, we reexamine *successive approximations*, this time in a more general and more theoretically powerful form than we did in Chapter 3. Then we state an existence and uniqueness theorem and use successive approximations as an important element in its proof. Qualitative analysis can often give us important insight into the behavior of the solution to a differential equation when we cannot solve it, and this chapter presents some useful basic concepts. In particular, qualitative analysis reveals that nonlinear ordinary differential equations (ODEs) can have multiple equilibrium points, requiring us to revisit the definition of stability. Finally, with the help of qualitative analysis, we explore a caveat of the existence and uniqueness theorem: that some differential equations may have solutions that apply only for a limited period of time, and present examples where this is the case.

## 4.1 MOTIVATION

We should first note that the theorems and proofs we study in this chapter will also serve us in almost identical form in Chapter 8 for the much more general case of $n$ dimensional systems. The question remains, however, of why engineers should concern themselves with theorems and proofs about existence and uniqueness.

Over the course of a career engineers can find themselves in a variety of roles. In many of these roles the engineer will be expected to solve certain kinds of differential equations and it is the primary objective of this text to enable that capability. In a number of possible roles, however, a deeper understanding of differential equations is required. Differential equations are tools for engineers just as, for example, test equipment is a tool. Engineers must understand conditions and limitations associated with differential equations just as they need to understand conditions and limitations under which test equipment operates correctly.

Frequently the mathematical model of a system is not given to the engineer but must be developed. For complex systems where models derive from empirical data, system equations may take unusual forms. Are the equations reliable? Do they have a single, well defined solution? If we are involved in research, for example investigating leading-edge technological issues, we may find ourselves searching for a solution to a mathematical problem never posed before. How can we be confident that the problem has a solution? How do we prove that a solution exists?

Not all engineers' careers will lead them into the roles just described. However, it is safe to say that all engineers at one point or another in their careers, and probably at many points, will be expected to solve a difficult problem. It is of course essential for engineers to perform everyday engineering tasks reliably, but an engineer's value to his or her organization is greatly enhanced if that engineer is able to step in and find solutions to intractable problems. What enables an engineer to do that? A case can be made for three attributes. First, having a deeper knowledge of fundamental science can often help. For problems in this text, that fundamental science is physics. For other kinds of engineering problems, it may be chemistry or biology. A second valuable attribute for problem-solving is creativity, an ability to think out of the box, to have the knack of visualizing different ways to attack a problem. There are innately creative people, but they hone their talent by voluntarily attacking difficult problems, building skills they can employ later, a practice all of us can benefit from. The Challenge Problems in this text are intended to help with this. But the third attribute is necessary more often than not: an ability to think critically. What is known? What has been proved? Has something been assumed to be true that is not? What, really, is the problem? These are key questions at such times. How does one develop a capability for critical thinking? The discipline involved in understanding the logic in mathematical theorems and proofs can be an important contributor.

In addition, the technique of *successive approximations* can be a powerful tool for the engineer, and this chapter describes another useful version of it.

## 4.2   SUCCESSIVE APPROXIMATIONS (GENERAL FORM)

As mentioned in Section 3.3, the term *successive approximations* applies to a number of methods in which we solve a difficult problem approximately by iteratively solving a sequence of easier problems. Here we use the method to solve a general first-order ODE that is not necessarily *almost linear* in the sense of the RL circuit problem we examined in Section 3.3. We consider an equation of the form

$$\frac{dy}{dt} = f(t, y), \qquad y(t_0) = y_0 \tag{4.1}$$

The successive approximations are the solutions $y_1(t)$, $y_2(t)$,..., $y_n(t)$ to

$$\frac{dy_1}{dt} = f(t, y_0), \qquad y_1(t_0) = y_0$$

$$\frac{dy_2}{dt} = f(t, y_1(t)), \qquad y_2(t_0) = y_0$$

...

$$\frac{dy_n}{dt} = f(t, y_{n-1}(t)), \qquad y_n(t_0) = y_0$$

We are solving the difficult nonlinear ODE by a sequence of equations in which the derivatives $dy_n/dt$ are given as functions of *time* only, and hence can be solved by the methods of integral calculus:

$$y_n(t) = y_0 + \int_{t_0}^{t} f(\tau, y_{n-1}(\tau)) d\tau$$

We will soon see that under certain conditions on $f(t, y)$ the sequential solutions converge to a solution of Equation 4.1.

## 4.2.1   Successive Approximations Example

We consider as an example the ODE

$$\frac{dy}{dt} = 1 + y^2, \qquad y(0) = 0 \tag{4.2}$$

We know that the solution to this equation is $y = \tan t$ and will use this knowledge to determine how quickly the sequence of approximations approaches the true solution. The power series for $\tan t$ is seldom seen because it is written in terms of so-called Bernoulli numbers that are themselves difficult to describe. Its series is

$$\tan t = t + \left(\frac{1}{3}\right)t^3 + \left(\frac{2}{15}\right)t^5 + \left(\frac{17}{315}\right)t^7 + \cdots \tag{4.3}$$

Its successive approximations $y_1(t)$, $y_2(t), \ldots, y_n(t)$ are determined from

$$\frac{dy_n}{dt} = 1 + y_{n-1}^2(t), \qquad y_n(0) = 0 \tag{4.4}$$

so,

$$y_0(t) = 0$$

$$\frac{dy_1}{dt} = 1$$

$$y_1(t) = t$$

$$\frac{dy_2}{dt} = 1 + t^2$$

$$y_2(t) = t + \left(\frac{1}{3}\right)t^3$$

$$\frac{dy_3}{dt} = 1 + \left(t + \left(\frac{1}{3}\right)t^3\right)^2 = 1 + t^2 + \left(\frac{2}{3}\right)t^4 + \left(\frac{1}{9}\right)t^6$$

$$y_3 = t + \left(\frac{1}{3}\right)t^3 + \left(\frac{2}{15}\right)t^5 + \left(\frac{1}{63}\right)t^7$$

$$\frac{dy_4}{dt} = 1 + \left(t + \left(\frac{1}{3}\right)t^3 + \left(\frac{2}{15}\right)t^5 + \left(\frac{1}{63}\right)t^7\right)^2$$

$$= 1 + t^2 + \left(\frac{2}{3}\right)t^4 + \left(\frac{4}{15} + \frac{1}{9}\right)t^6 + \ldots$$

$$y_4 = t + \left(\frac{1}{3}\right)t^3 + \left(\frac{2}{15}\right)t^5 + \left(\frac{1}{7}\right)\left(\frac{4}{15} + \frac{1}{9}\right)t^7 + \dots$$

$$y_4 = t + \left(\frac{1}{3}\right)t^3 + \left(\frac{2}{15}\right)t^5 + \left(\frac{17}{315}\right)t^7 + \dots$$

Comparing the iterations $y_n$ with the expansion of $\tan t$ given in Equation 4.3 we can see that $y_1$ is correct to first order in $t$, $y_2$ is correct to third order, $y_3$ to fifth order, and $y_4$ to seventh order. Hence the successive approximations are converging rapidly to the true solution.

Problems to Solve P4.1 through P4.4 at the end of this chapter involve performing successive approximations on given first-order equations. The equations are linear. From the results of Chapter 2 we know how to solve them analytically and do not require the mechanics of successive approximations to do so. However, the problems exercise the student's ability to execute the process without undue algebraic complication.

## 4.3   EXISTENCE AND UNIQUENESS THEOREM

Before we present the theorem stating the conditions under which we can be sure that a solution exists to a given first-order ODE, and that it is unique, we need to introduce one of the key conditions.

### 4.3.1   LIPSCHITZ CONDITION

Consider the general first-order ODE:

$$\frac{dy}{dt} = f(t, y)$$

$$y(t_0) = y_0$$

(4.5)

where $f(t, y)$ is defined on a rectangle R (Figure 4.1).

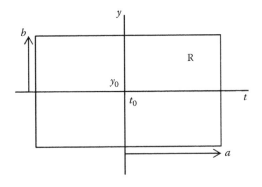

**FIGURE 4.1**   Rectangle R in which $f(t, y)$ satisfies sufficient conditions.

The function $f(t, y)$ is said to satisfy a *Lipschitz condition* with respect to $y$ on R if there exists a constant $K > 0$ such that

$$|f(t, y_1) - f(t, y_2)| \le K |y_1 - y_2| \tag{4.6}$$

for all points $(t, y)$ in R.

Example:

$$\frac{dy}{dt} = g(t) + y^4 \quad \text{R:} \quad \begin{array}{c} 0 \le t \le 10 \\ -2 \le y \le 2 \end{array}$$

$$f(t, y) = g(t) + y^4$$

$$f(t, y_1) - f(t, y_2) = y_1^4 - y_2^4 = (y_1 - y_2)(y_1 + y_2)(y_1^2 + y_2^2)$$

$$\frac{|f(t, y_1) - f(t, y_2)|}{|y_1 - y_2|} = |(y_1 + y_2)(y_1^2 + y_2^2)|$$

This reaches a maximum in R as $y_1$ and $y_2$ together approach ±2. Hence

$$K = (2 + 2)(4 + 4) = 32$$

Note that, in this example, the Lipschitz constant $K$ was given by the point in R where

$$\left| \frac{\partial f(t, y)}{\partial y} \right| = \max$$

This is true more generally:

If $\partial f(t, y) / \partial y$ is continuous in R and

$$\left| \frac{\partial f(t, y)}{\partial y} \right| \le K$$

in R, then $f(t, y)$ satisfies a Lipschitz condition in R with constant $K$.

### 4.3.2   STATEMENT OF THE EXISTENCE AND UNIQUENESS THEOREM

Let the function $f(t, y)$ be defined on the rectangle R: $|t - t_0| \le a$, $|y - y_0| \le b$, as in Figure 4.1, and in R let $f(t, y)$:

1. Be continuous.
2. Be bounded by M.
3. Satisfy a Lipschitz condition with constant K with respect to $y$.

Then a unique solution exists to

$$\frac{dy}{dt} = f(t, y)$$

$$y(t_0) = y_0$$

(4.7)

in the interval

$$I : |t - t_0| \leq \min(a, b / M)$$

(4.8)

### 4.3.3 Successive Approximations Converge to a Solution

Following a line of argument originally developed by Picard, we can prove that a solution exists by demonstrating a process for computing one, that process being successive approximations.

**Theorem:**

The sequence of approximations given by

$$y_0(t) = y_0$$

$$y_1(t) = y_0 + \int_{t_0}^{t} f(\tau, y_0(\tau)) d\tau$$

$$y_2(t) = y_0 + \int_{t_0}^{t} f(\tau, y_1(\tau)) d\tau$$

$$\ldots$$

$$y_n(t) = y_0 + \int_{t_0}^{t} f(\tau, y_{n-1}(\tau)) d\tau$$

(4.9)

converges to a unique solution of Equation 4.7.

**Proof that the Sequence Converges:**

Consider the sequence

$$z_1(t) = y_1(t) - y_0$$

$$z_2(t) = y_2(t) - y_1(t)$$

$$\ldots$$

$$z_n(t) = y_n(t) - y_{n-1}(t)$$

Note that

$$y_n(t) = y_0 + (y_1(t) - y_0(t)) + (y_2(t) - y_1(t)) + \ldots + (y_n(t) - y_{n-1}(t))$$

$$y_n(t) = y_0 + \sum_{i=1}^{n} z_i$$

We must show that the sequence $\sum_{i=1}^{n} z_i$ converges.

Now

$$z_1(t) = y_1(t) - y_0 = \int_{t_0}^{t} f(\tau, y_0) d\tau$$

$$|z_1(t)| = \left| \int_{t_0}^{t} f(\tau, y_0) d\tau \right| \leq \int_{t_0}^{t} |f(\tau, y_0)| d\tau \leq \int_{t_0}^{t} M d\tau = M(t - t_0)$$

$$|z_1(t)| \leq M(t - t_0) \tag{4.10}$$

Next

$$z_2(t) = y_2(t) - y_1(t) = \int_{t_0}^{t} (f(\tau, y_1(\tau)) - f(\tau, y_0(\tau))) d\tau$$

$$|z_2(t)| = \left| \int_{t_0}^{t} (f(\tau, y_1(\tau)) - f(\tau, y_0(\tau))) d\tau \right| \leq \int_{t_0}^{t} |f(\tau, y_1(\tau)) - f(\tau, y_0(\tau))| d\tau \tag{4.11}$$

Using the Lipschitz condition:

$$|f(\tau, y_1(\tau)) - f(\tau, y_0(\tau))| \leq K |y_1(\tau) - y_0(\tau)| = K |z_1(\tau)| \tag{4.12}$$

Using Equations 4.10 and 4.12 in Equation 4.11

$$|z_2(t)| \leq \int_{t_0}^{t} KM(\tau - t_0) d\tau = KM(t - t_0)^2 / 2 \tag{4.13}$$

Similarly,

$$|z_3(t)| = \left| \int_{t_0}^{t} (f(\tau, y_2(\tau)) - f(\tau, y_1(\tau))) d\tau \right| \leq \int_{t_0}^{t} |f(\tau, y_2(\tau)) - f(\tau, y_1(\tau))| d\tau$$

$$|z_3(t)| \leq \int_{t_0}^{t} K |y_2(\tau) - y_1(\tau)| d\tau = \int_{t_0}^{t} K |z_2(\tau)| d\tau \leq K^2 M \int_{t_0}^{t} ((\tau - t_0)^2 / 2) d\tau$$

$$|z_3(t)| \leq K^2 M (t - t_0)^3 / (3 \cdot 2) \tag{4.14}$$

Continuing in this way we find that

$$|z_n(t)| \leq \frac{M}{K} \frac{K^n (t - t_0)^n}{n!} \tag{4.15}$$

From Equation 4.15, each term in the sequence $\sum_{i=1}^{n} z_i$ is smaller than that for a convergent series, namely the series for

$$\frac{M}{K} e^{K(t - t_0)}$$

Hence the series converges, as was to be shown.

We have shown neither that the series limit satisfies the ODE nor that the solution is unique. Challenge Problems 4.1 and 4.2 invite the student to prove this, following proofs given in Coddington [3].

## 4.4  QUALITATIVE ANALYSIS

Consider the ODE

$$\frac{dy}{dt} = (y - 1)(y - 2), \qquad y(0) = y_0 \tag{4.16}$$

We can solve Equation 4.16 by the methods of Chapter 3 (and Solved Problem SP4.4 does so), but even if we could not solve it we could gain useful insights into the behavior of the solution by examining the function $f(t, y) = (y - 1)(y - 2)$. For example, we can create a *direction field*, as in Figure 4.2(a), with vectors drawn for discrete points in the $(y, t)$ plane. The value of $(y - 1)(y - 2)$ is proportional to the tangent of the angle of the vector, the

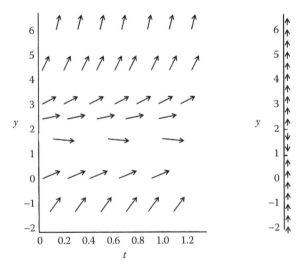

**FIGURE 4.2**  Direction field and direction line for $dy/dt = (y - 1)(y - 2)$ (a) direction field (b) direction line.

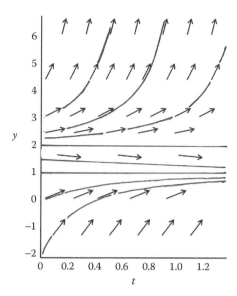

**FIGURE 4.3** Direction field for $dy/dt = (y-1)(y-2)$ overlaid with solution trajectories.

proportionality constant depending on the graph scale chosen. The direction field presents a visual display of the slope of solution trajectories through those points. Figure 4.3 overlays the direction field of Figure 4.2(a) with trajectories drawn from the solution given in Solved Problem 4.4.

For functions $f(t, y)$ that are independent of time, as in Equation 4.16, we can collapse the chart into a *direction line* (sometimes called a *phase line*), which loses the magnitude of the slope but clarifies the ODE solution's equilibrium and stability properties. Figure 4.2(b) is the direction line for Equation 4.16. *Equilibrium points* (sometimes called *critical points*) are points where $f(t, y) = 0$. For Equation 4.16, $y = 1$ and $y = 2$ are equilibrium points. But as the direction line in Figure 4.2(b) shows, $y = 1$ is a *stable* equilibrium point, whereas $y = 2$ is an *unstable* one. Points near $y = 1$ are attracted to $y = 1$, whereas points near $y = 2$ are repelled from that point. From the direction line we can see that if the initial condition $y(0) = y_0$ is less than 2 then the solution eventually approaches 1. If $y_0 > 2$ then the solution $y(t)$ increases without bound. Solved Problem 4.4 explores this behavior further.

## 4.5 STABILITY REVISITED

In Chapter 2 we defined stability for a linear system in terms of its response to initial conditions, where the only possible equilibrium point was the origin. As we have seen in the previous Section 4.4, nonlinear systems, with their potential for multiple equilibrium points, require us to refine that definition.

Loosely speaking, an equilibrium point $y_E$ is said to be *stable* if $y$ ultimately stays near $y_E$ when $y$ comes sufficiently close to $y_E$. An equilibrium point $y_E$ is said to *asymptotically stable* if $\lim_{t \to \infty} y = y_E$ when $y$ comes sufficiently close to $y_E$.

More rigorously, an equilibrium point $y_E$ is said to be stable for $t > t_0$ if for every $\varepsilon > 0$ there is a $\delta > 0$ (which may be a function of $t_0$ and $\varepsilon$) such that if $|y(t_0) - y_E| < \delta$ then $|y(t) - y_E| < \varepsilon$ for all $t > t_0$. If, in addition, $\lim_{t \to \infty} y(t) = y_E$, then the equilibrium point is asymptotically stable.

Equilibrium points can also be stable only when approached from above or only when approached from below, in which case they are called *semistable*. See Word Problem WP4.1.

## 4.6   LOCAL SOLUTIONS

The existence and uniqueness theorem provides only a lower bound on the length of time $t_E$ in which a solution is certain to exist:

$$|t - t_0| \le t_E = \min\,(a, b/M)$$

This restriction on the time interval ensures that $y(t)$ stays within the rectangle where $f(t, y)$ is defined:

$$y(t) = y_0 + \int_{t_0}^t f(\tau, y(\tau))d\tau$$

$$y(t) - y_0 = \int_{t_0}^t f(\tau, y(\tau))d\tau$$

$$|y(t) - y_0| \le \left| \int_{t_0}^t f(\tau, y(\tau))d\tau \right| \le \int_{t_0}^t |f(\tau, y(\tau))|\, d\tau \le M|t - t_0|$$

$$|y(t) - y_0| \le M(b/M) = b$$

But the caveat leaves the possibility that the solution may indeed only be local; that its duration is in fact limited. We present examples below in which this is the case.

### Example 1

Consider the ODE

$$\frac{dy}{dt} = \frac{1}{4 - y}, \qquad y(0) = 0 \tag{4.17}$$

Using a direction line we see that the solution moves toward the value 4 independent of the initial condition. For clarity we consider here the initial condition $y(0) = 0$. The solution $y(t)$ begins at zero and moves in the $+y$ direction. The function

$$f(t, y) = \frac{1}{4 - y}$$

is not a function of $t$, so we may set any positive number for the bound $a$. However, $f(t, y) = 1/(4 - y)$ is not continuous at $y = 4$. To apply the theorem we must bound $y$ by a number less than 4. We could choose any number $0 < b < 4$.

We choose $a = 10$ (arbitrarily) and let $b$ be unspecified for the moment. Then

$$f(t,y) \le \frac{1}{4-b}$$

That is, the bound $M$ on $f(t,y)$ as required in the theorem is $M = 1/(4-b)$. Provided $y < 4$,

$$\frac{\partial f}{\partial y} = \frac{1}{(4-y)^2}$$

is continuous in R, so we are assured that $f(t,y)$ satisfies a Lipschitz condition. Therefore $f(t,y)$ satisfies the conditions of the theorem and a solution is certain to exist in the interval

$$|t - t_0| = |t| \le t_E = \min(a, b/M) = \min(10, (4-b)b).$$

It is easy to determine that the function $b(4-b)$ is always less than 10 for $0 < b < 4$, hence

$$t_E = \min(10, b(4-b)) = b(4-b) \qquad (4.18)$$

Note that, via Equation 4.18, the length of time the theorem states a solution is certain to exist is a function of the bound $b$ *we set* on the variable $y$. By setting

$$\frac{dt_E}{db} = 0$$

we can find the bound $b$ maximizing $t_E$. We have

$$\frac{dt_E}{db} = \frac{d}{db}(b(4-b)) = \frac{d}{db}(4b - b^2) = 4 - 2b = 0.$$

Hence the bound maximizing $t_E$ is $b = 2$ and the theorem's highest bound on the time a solution is certain to exist is $t_E = 4$.

Solved Problem SP4.5 solves Equation 4.17 and shows that the solution is indeed local, but that it actually exists for $t < 8$. We can determine the theorem's highest bound on $t_E$, but the theorem uses only rough measures (bounds) as conditions and so develops only crude, conservative estimates of it.

**Example 2**

Consider the three ODEs

$$\frac{dy_A}{dt} = f_A(t, y_A) = 1 + y_A^2, \qquad y_A(0) = 0$$

$$\frac{dy_B}{dt} = f_B(t, y_B) = (1 + y_B)^2, \qquad y_B(0) = 0$$

$$\frac{dy_C}{dt} = f_C(t, y_C) = (1 + y_C)(2 + y_C), \qquad y_C(0) = 0$$

Using direction lines we see that the solution of each ODE starts at zero and moves in the $+y$ direction. Clearly, over the trajectories of the solutions, for a given $t$ and $y$,

$$f_C(t, y) > f_B(t, y) > f_A(t, y) > 0$$

The solution to the first is familiar to us: $y_A(t) = \tan t$. The solution is defined for $0 \le t < \pi/2$; $y_A(t)$ goes to infinity as $t$ goes to $\pi/2$. Since $f_B(t, y) > f_A(t, y)$ we expect that $y_B(t)$ goes to infinity as well, and in a shorter period of time than does $y_A(t)$, and since $f_C(t, y) > f_B(t, y)$ we expect that $y_C(t)$ goes to infinity also and in even a shorter period. This is indeed so. Solved Problem SP4.6 studies $y_B(t)$ and the study of $y_C(t)$ is left as an exercise for the student in Problem to Solve P4.12.

## 4.7   SUMMARY

If, in a defined rectangle around the point $(t_0, y_0)$ the function $f(t, y)$ is sufficiently smooth (is continuous and $|f(t, y_1) - f(t, y_2)| \le K|y_1 - y_2|$ for some $K$) then a unique solution exists to

$$\frac{dy}{dt} = f(t, y)$$

$$y(t_0) = y_0$$

in a certain interval around $t_0$.

We know that a solution exists in that interval because we can demonstrate a process for computing one, that process being successive approximations:

$$y_0(t) = y_0$$

$$y_1(t) = y_0 + \int_{t_0}^{t} f(\tau, y_0(\tau)) d\tau$$

$$y_2(t) = y_0 + \int_{t_0}^{t} f(\tau, y_1(\tau)) d\tau$$

$$\cdots$$

$$y_n(t) = y_0 + \int_{t_0}^{t} f(\tau, y_{n-1}(\tau)) d\tau$$

A direction field and direction line can provide useful insight into the behavior of the solution to an ODE when we cannot solve it. A direction field is a graph of vectors at discrete points in the $(y, t)$ plane with the tangent of the angle of the vector drawn proportional to the value of $f(t, y)$.

For functions that are independent of time, the direction line is created by drawing arrows at discrete points along the $y$ axis, the direction of the arrow up if $f(y) > 0$, and down if $f(y) < 0$. The direction line immediately reveals the ODE's equilibrium points, that is, where the derivative is zero. Nonlinear ODEs can have multiple equilibrium points, where linear equations can have at most one. The direction line also immediately identifies whether the point is stable or unstable. Loosely speaking, an equilibrium point $y_E$ is said to be *stable* if $y$ ultimately stays near $y_E$ when $y$ comes sufficiently close to $y_E$. An equilibrium point $y_E$ is said to be *asymptotically stable* if $\lim_{t \to \infty} y = y_E$ when $y$ comes sufficiently close to $y_E$.

The existence and uniqueness theorem provides only a lower bound on the length of time in which a solution exists, leaving the possibility that solutions may only be local. This chapter gave a number of examples in which this is the case. The theorem's lower bound is typically very conservative.

## Solved Problems

**SP4.1** Compute the first four iterations $(y_0(t), y_1(t), y_2(t), y_3(t))$ in successive approximation to solution of

$$\frac{dy}{dt} - 4y = 1, \qquad y(0) = 2$$

*Solution*: The equation for successive approximations for this problem is

$$\frac{dy_{n+1}}{dt} = 1 + 4y_n, \qquad y_{n+1}(0) = 2$$

$n = 0$:

The zeroth approximation $y_0(t)$ is the initial condition $y(0) = 2$. Hence

$$\frac{dy_1}{dt} = 1 + 4y_0 = 1 + 4 \cdot 2 = 9$$

$$\int_{y_1(0)}^{y_1(t)} \left( \frac{dy_1}{d\tau} \right) d\tau = y_1(t) - y_1(0) = \int_0^t (9) d\tau = 9t$$

$$y_1(t) = y_1(0) + 9t = 2 + 9t$$

$n = 1$:

$$\frac{dy_2}{dt} = 1 + 4y_1, \qquad y_2(0) = 2$$

$$\frac{dy_2}{dt} = 1 + 4y_1 = 1 + 4 \cdot (2 + 9t) = 9 + 36t$$

$$\int_{y_2(0)}^{y_2(t)} \left( \frac{dy_2}{d\tau} \right) d\tau = y_2(t) - y_2(0) = \int_0^t (9 + 36\tau) d\tau = 9t + 18t^2$$

$$y_2(t) = y_2(0) + 9t + 18t^2 = 2 + 9t + 18t^2$$

$n = 2$:

$$\frac{dy_3}{dt} = 1 + 4y_2, \qquad y_3(0) = 2$$

$$\frac{dy_3}{dt} = 1 + 4y_2 = 1 + 4 \cdot (2 + 9t + 18t^2) = 9 + 36t + 72t^2$$

$$\int_{y_3(0)}^{y_3(t)} \left(\frac{dy_3}{d\tau}\right) d\tau = y_3(t) - y_3(0) = \int_0^t (9 + 36\tau + 72\tau^2) d\tau = 9t + 18t^2 + 24t^3$$

$$y_3(t) = y_3(0) + 9t + 18t^2 + 24t^3 = 2 + 9t + 18t^2 + 24t^3$$

**SP4.2** Compute the first four iterations $(y_0(t), y_1(t), y_2(t), y_3(t))$ in successive approximation to solution of

$$\frac{dy}{dt} + 3y = 2t, \qquad y(0) = 4$$

*Solution*: The equation for successive approximations for this problem is

$$\frac{dy_{n+1}}{dt} = 2t - 3y_n, \qquad y_{n+1}(0) = 4$$

$n = 0$:

The zeroth approximation $y_0(t)$ is the initial condition $y(0) = 4$. Hence

$$\frac{dy_1}{dt} = 2t - 3y_0 = 2t - 3 \cdot 4 = -12 + 2t$$

$$\int_{y_1(0)}^{y_1(t)} \left(\frac{dy_1}{d\tau}\right) d\tau = y_1(t) - y_1(0) = \int_0^t (-12 + 2t) d\tau = -12t + t^2$$

$$y_1(t) = y_1(0) - 12t + t^2 = 4 - 12t + t^2$$

$n = 1$:

$$\frac{dy_2}{dt} = 2t - 3y_1, \qquad y_2(0) = 2$$

$$\frac{dy_2}{dt} = 2t - 3y_1 = 2t - 3(4 - 12t + t^2) = -12 + 38t - 3t^2$$

$$\int_{y_2(0)}^{y_2(t)} \left(\frac{dy_2}{d\tau}\right) d\tau = y_2(t) - y_2(0) = \int_0^t (-12 + 38\tau - 3\tau^2) d\tau = -12t + 19t^2 - t^3$$

$$y_2(t) = y_2(0) - 12t + 19t^2 - t^3 = 4 - 12t + 19t^2 - t^3$$

$n = 2$:

$$\frac{dy_3}{dt} = 2t - 3y_2, \qquad y_3(0) = 2$$

$$\frac{dy_3}{dt} = 2t - 3y_2 = 2t - 3(4 - 12t + 19t^2 - t^3) = -12 + 38t - 57t^2 + 3t^3$$

$$\int_{y_3(0)}^{y_3(t)} \left(\frac{dy_3}{d\tau}\right) d\tau = y_3(t) - y_3(0)$$

$$= \int_0^t (-12 + 38\tau - 57\tau^2 + 3\tau^3) d\tau = -12t + 19t^2 - 19t^3 + (3/4)t^4$$

$$y_3(t) = 4 - 12t + 19t^2 - 19t^3 + (3/4)t^4$$

**SP4.3** Without solving the equation, determine the equilibrium points for

$$\frac{dy}{dt} = 2y^3 - 24y^2 + 88y - 96 \tag{4.19}$$

Use a direction line to determine which equilibrium points are stable and which are unstable.

*Solution*:

We can simplify Equation 4.19 somewhat:

$$\frac{dy}{dt} = 2(y^3 - 12y^2 + 44y - 48) \tag{4.20}$$

There is an algebraic formula for finding roots to cubic polynomials but it is cumbersome to use. Chapter 6 will give us some practice in finding them by an iterative process, which is useful for engineers to know how to do, but here we opt for the convenience of going out on the Internet and finding a "root calculator" to tell us that the polynomial in Equation 4.20 has roots 2, 4 and 6, which are the equilibrium points for this system. As a check, we multiply out $(y-2)(y-4)(y-6)$ to be sure that it equals $(y^3 - 12y^2 + 44y - 48)$. It does, and Equation 4.20 can then be written as

$$\frac{dy}{dt} = 2(y-2)(y-4)(y-6) \tag{4.21}$$

We can substitute numbers for $y$ into Equation 4.21 to help us see that

$$\frac{dy}{dt} < 0 \qquad \text{for } y < 2$$

$$\frac{dy}{dt} > 0 \qquad \text{for } 2 < y < 4$$

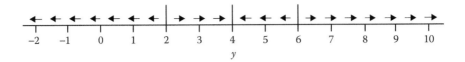

**FIGURE 4.4**   Direction line for $dy/dt = 2(y-2)(y-4)(y-6)$.

$$\frac{dy}{dt} < 0 \qquad \text{for } 4 < y < 6$$

$$\frac{dy}{dt} > 0 \qquad \text{for } y > 6$$

From these inequalities we can create the direction line in Figure 4.4, and from that diagram determine that the points $y = 2$ and $y = 6$ are unstable equilibrium points and $y = 4$ is a stable equilibrium point.

For printing convenience we have turned the direction line in Figure 4.4 on its side.

**SP4.4** Solve

$$\frac{dy}{dt} = (y-1)(y-2), \qquad y(0) = y_0 \tag{4.22}$$

and examine the behavior of the solution as it depends on $y_0$.

*Solution*:

Separating variables:

$$\frac{dy}{(y-1)(y-2)} = dt$$

Expanding in partial fractions:

$$\frac{1}{(y-1)(y-2)} = \frac{c_1}{y-1} + \frac{c_2}{y-2} = \frac{c_1(y-2)+c_2(y-1)}{(y-1)(y-2)} = \frac{(c_1+c_2)y-2c_1-c_2}{(y-1)(y-2)}$$

Equating coefficients of similar powers of $y$ in the numerator:

$$c_1 + c_2 = 0, \qquad -2c_1 - c_2 = 1$$

which have solution $c_1 = -1, c_2 = 1$. Then

$$\frac{dy}{(y-1)(y-2)} = \frac{-dy}{y-1} + \frac{dy}{y-2} = dt$$

$$-\int_{y_0}^{y} \frac{dy}{y-1} + \int_{y_0}^{y} \frac{dy}{y-2} = t$$

$$-\ln(y-1) + \ln(y_0-1) + \ln(y-2) - \ln(y_0-2) = t$$

$$\ln\left[\left(\frac{y_0-1}{y-1}\right)\left(\frac{y-2}{y_0-2}\right)\right]=t$$

$$\left(\frac{y_0-1}{y-1}\right)\left(\frac{y-2}{y_0-2}\right)=e^t$$

$$(y-2)(y_0-1)=(y-1)(y_0-2)e^t$$

$$y\left((y_0-1)-(y_0-2)e^t\right)=2(y_0-1)-(y_0-2)e^t$$

$$y(t)=\frac{e^t-2\left(\dfrac{y_0-1}{y_0-2}\right)}{e^t-\left(\dfrac{y_0-1}{y_0-2}\right)}$$

For clarity we substitute specific values for $y_0$.
Let $y_0=0$. Then

$$y(t)=\frac{e^t-1}{e^t-(1/2)}$$

from which we see that a solution exists for all $t>0$ and furthermore that

$$\lim_{t\to\infty}y(t)=1$$

It can be shown that this is the case for all $y_0<2$.
Let $y_0=3$. Then

$$y(t)=\frac{e^t-4}{e^t-2}$$

The denominator goes to zero as $e^t$ approaches 2, or $t$ approaches $\ln 2=0.693$. Hence a solution exists only for a finite time. It can be shown that this is the case for all $y_0>2$. Figure 4.5 graphs the solution of Equation 4.22 for various initial conditions.

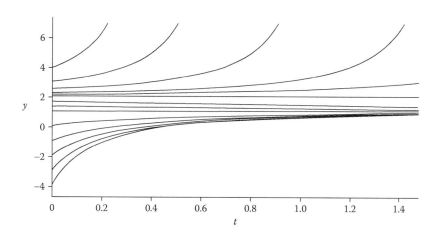

**FIGURE 4.5**  Graph of the solution of $dy/dt=(y-1)(y-2)$ for various initial conditions.

**SP4.5** Solve

$$\frac{dy}{dt} = \frac{1}{4-y}, \qquad y(0) = 0 \tag{4.23}$$

and determine the maximum time a solution exists. Compare with the results of Example 1 in Section 4.6.

*Solution*:
Separating variables:

$$(4-y)dy = dt$$

Integrating:

$$\int_{y(0)}^{y} (4-y)dy = t$$

$$4y - y^2/2 = t$$

$$y^2 - 8y + 2t = 0$$

$$(y-4)^2 - 16 + 2t = 0$$

$$y = 4 \pm \sqrt{16 - 2t}$$

To meet the initial condition the negative root must apply:

$$y(t) = 4 - \sqrt{16 - 2t} \tag{4.24}$$

Example 1 of Section 4.6 used only the terms and conditions of the existence and uniqueness theorem to show that a solution must exist for at least $t_E = 4$. We based the estimate on the time to reach a bound $y = 2$. Using the true solution in Equation 4.24 we see that $y = 2$ at $t = 6$ and that the ODE has a valid solution for $t \leq 8$.

**SP4.6** Without solving the equation, determine the highest estimate of the time, according to the terms and conditions of the existence and uniqueness theorem in Section 4.3.1, in which a solution to the following equation is certain to exist:

$$\frac{dy}{dt} = (1+y)^2, \qquad y(0) = 0 \tag{4.25}$$

Then solve Equation 4.25 and determine the maximum time a solution actually exists. Compare with the results of Example 2 in Section 4.6.

*Solution*:
Because $f(t,y) = (1+y)^2$ is unbounded as $y$ goes to infinity we must set a bound on $y$ to have the theorem apply. We set a bound $y \leq b$, which we leave unspecified for the moment. Using a direction line we see that, from an initial value at the origin, $y(t)$ will move in the $+y$ direction, so $(1+y)^2$ is bounded by $M = (1+b)^2$. Also,

$$\frac{\partial f}{\partial y} = 2(1+y) \le 2(1+b)$$

so, the function $f(t, y)$ satisfies a Lipschitz condition and the conditions of the theorem are met. In this problem the function $f(t, y)$ is independent of $t$ so we can set an arbitrarily large value on the bound $a$. The theorem states that a solution is certain to exist for a time

$$t_E = \min(a, b/M) = \min\left(a, b/(1+b)^2\right)$$

We can set $a$ large enough that $a > b/(1+b)^2$ and then $t_E = b/(1+b)^2$. Hence $t_E$ depends on the bound we set. We can find a bound maximizing $t_E$ by setting

$$\frac{dt_E}{db} = 0$$

Now

$$\frac{dt_E}{db} = \frac{d}{db}\left(\frac{b}{(1+b)^2}\right) = \frac{1}{(1+b)^2} - \frac{2b}{(1+b)^3} = \frac{1-b}{(1+b)^3}$$

so, the bound maximizing $t_E$ is $b_m = 1$. Then the theorem's highest bound on the time in which a solution is certain to exist is

$$t_E = \frac{b_m}{(1+b_m)^2} = \frac{1}{4} \tag{4.26}$$

To work the second part of the problem, actually solving Equation 4.25, we separate variables:

$$\frac{dy}{(1+y)^2} = dt$$

Integrating:

$$\int_{y(0)}^{y} \frac{dy}{(1+y)^2} = t$$

$$1 - \frac{1}{(1+y)} = t$$

$$\frac{y}{(1+y)} = t$$

$$y = (1+y)t$$

$$y(t) = \frac{t}{1-t} \tag{4.27}$$

The solution is valid for $t < 1$, which is less than $\pi/2$, so the first of the expectations expressed in Example 2 of Section 4.6 is borne out. For the second see Problem 4.12 below.

Note that, again, the actual value for the time in which a solution is valid is considerably greater than the lower bound provided by the existence and uniqueness theorem, which, from Equation 4.26, is $t_E = 1/4$. We based that estimate on $y(t)$ reaching the bound $y = 1$. The actual solution reaches $y = 1$ at $t = 1/2$, and a solution is valid for twice that time. We repeat: because the theorem uses only rough measures (bounds), we can derive only crude, conservative lower bounds on $t_E$ from it.

## Problems to Solve

**P4.1** Compute the first four iterations $(y_0(t), y_1(t), y_2(t), y_3(t))$ in successive approximation to solution of

$$\frac{dy}{dt} + 2y = 3, \qquad y(0) = 1$$

**P4.2** Compute the first four iterations $(y_0(t), y_1(t), y_2(t), y_3(t))$ in successive approximation to solution of

$$\frac{dy}{dt} - y = 4t, \qquad y(0) = -2$$

**P4.3** Compute the first four iterations $(y_0(t), y_1(t), y_2(t), y_3(t))$ in successive approximation to solution of

$$\frac{dy}{dt} + 4y = 2 - t, \qquad y(0) = 3$$

**P4.4** Compute the first four iterations $(y_0(t), y_1(t), y_2(t), y_3(t))$ in successive approximation to solution of

$$\frac{dy}{dt} - 3y = 2t^2, \qquad y(0) = -1$$

**P4.5** Without solving the equation, determine the equilibrium points for

$$\frac{dy}{dt} = 3y^2 - 33y + 84$$

Use a direction line to determine which equilibrium points are stable and which are unstable.

**P4.6** Without solving the equation, determine the equilibrium points for

$$\frac{dy}{dt} = -2y^2 + 6y + 36$$

Use a direction line to determine which equilibrium points are stable and which are unstable.

**P4.7** Without solving the equation, determine the equilibrium points for

$$\frac{dy}{dt} = y^4 - 20y^2 + 64$$

Use a direction line to determine which equilibrium points are stable and which are unstable.

**P4.8** Without solving the equation, determine the equilibrium points for

$$\frac{dy}{dt} = 4\sin 2y$$

Use a direction line to determine which equilibrium points are stable and which are unstable.

**P4.9** Without solving the equation, determine the equilibrium points for

$$\frac{dy}{dt} = 3(e^{-y} - y)(e^y + y)$$

Use a direction line to determine which equilibrium points are stable and which are unstable.

**P4.10** Solve

$$\frac{dy}{dt} = 4y^3, \qquad y(0) = 1$$

and determine the maximum time a solution exists.

**P4.11** Solve

$$\frac{dy}{dt} = \frac{1}{\sqrt{1-y}}, \qquad y(0) = 0$$

and determine the maximum time a solution exists.

**P4.12** Without solving the equation, determine the highest lower bound, according to the terms and conditions of the existence and uniqueness theorem in Section 4.3.1, on the time in which a solution to the following equation is certain to exist:

$$\frac{dy}{dt} = (1+y)(2+y), \qquad y(0) = 0$$

Then solve the equation and determine the maximum time a solution actually exists. Compare with the results of Solved Problem SP4.6 and the expectations of Example 2 in Section 4.6.

**P4.13** Without solving the equation, determine the highest lower bound, according to the terms and conditions of the existence and uniqueness theorem in Section 4.3.1, on the time in which a solution to the following equation is certain to exist:

$$\frac{dy}{dt} = \frac{1}{5 - y^4}, \qquad y(0) = 0$$

**P4.14** Without solving the equation, determine the highest lower bound, according to the terms and conditions of the existence and uniqueness theorem in Section 4.3.1, on the time in which a solution to the following equation is certain to exist:

$$\frac{dy}{dt} = 48 + y^4, \qquad y(0) = 0$$

## Word Problems

**WP4.1** Consider the ODE

$$\frac{dy}{dt} = 1 + \sin y, \qquad y(0) = y_0 \tag{4.28}$$

Part (a): Examine Equation 4.28 and show that the solution has an infinity of equilibrium points, depending on $y_0$.

Part (b): Using a direction line, show that each of the equilibrium points is stable only when approached from below. Such a system is said to be *semistable*. Show that, for $y_0 = 0$, $y = 3\pi/2$ is a stable equilibrium point.

Part (c): Solve Equation 4.1 and verify that, for $y_0 = 0$,

$$\lim_{t \to \infty} y(t) = 3\pi/2$$

Hint: Show that $1 + \sin y = 2\cos^2(y/2 - \pi/4)$.

**WP4.2** Consider a hypothetical spacecraft that controls its spin rate $\omega$ by changing its moment of inertia, as shown in Figure 4.6. The spacecraft has arms on opposite sides of its central body that can be extended or retracted. Each arm has a large mass at its end. Neglecting the mass of the arms themselves, the spacecraft's moment of inertia can be approximated by $I = I_0 + 2mr^2$, where $m$ is the mass at the end of each arm, $r$ is the distance of each mass from the center line of the body, and $I_0$ is the moment of inertia of the central body. The spacecraft experiences no torque from external sources, so its angular momentum is constant. Then

$$\frac{d}{dt}(I\omega) = 0$$

$$I\dot{\omega} + \dot{I}\omega = 0$$

$$\dot{\omega} = -\omega\dot{I}/I$$

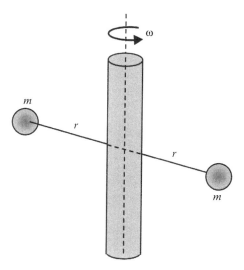

**FIGURE 4.6**    Spacecraft controlling spin rate by changing moment of inertia.

The spacecraft's computer is programmed to control the arms so that

$$\frac{\dot{I}}{I} = k(\omega - \omega_D)$$

where $\omega_D > 0$ is the desired spin rate for the spacecraft. This means the arms move in or out under computer control at a rate

$$\dot{r} = \left(\frac{I_0 + 2mr^2}{4mr}\right) k(\omega - \omega_D)$$

Write the equation governing the spacecraft's spin rate as a function of time. Show that the programmed control law produces more than one equilibrium point. Use a direction line to determine those that are stable. Determine requirements on the initial spin rate $\omega_0$ for the spacecraft spin rate $\omega(t)$ to eventually reach the desired spin rate $\omega_D$. Assuming that $\omega_0$ satisfies those requirements, solve the ODE and verify that $\lim_{t \to \infty} \omega(t) = \omega_D$.

Note: Your analysis should reveal the reason this specific approach is not used to stabilize spacecraft.

## Challenge Problems

**CP4.1** Section 4.3.3 proved that the sequence of successive approximations

$$y_{n+1}(t) = y_0 + \int_{t_0}^{t} f(\tau, y_n(\tau)) d\tau$$

converges. We invite you to show here that the sequence converges to a solution of the integral equation

$$y(t) = y_0 + \int_{t_0}^t f(\tau, y(\tau))d\tau$$

and hence the differential equation in Equation 4.5. You must show that, with $y_n(t)$ and its limit $y(t)$ as defined as in Section 4.3.3,

$$\lim_{n\to\infty} \int_{t_0}^t f(\tau, y_n(\tau))d\tau = \int_{t_0}^t f(\tau, y(\tau))d\tau \tag{4.29}$$

Part (a): Using the Lipschitz condition, show that

$$\left| \int_{t_0}^t f(\tau, y_n(\tau))d\tau - \int_{t_0}^t f(\tau, y(\tau))d\tau \right| \le K \int_{t_0}^t \left| y_n(\tau) - y(\tau) \right| d\tau \tag{4.30}$$

Part (b): Using the results of Section 4.3.3, show that

$$\left| y_n(\tau) - y(\tau) \right| = \left| \sum_{m=n}^\infty (y_{m+1}(\tau) - y_m(\tau)) \right| = \left| \sum_{m=n}^\infty z_{m+1} \right| \le \frac{M}{K} \sum_{m=n}^\infty \frac{K^{m+1}(\tau - t_0)^{m+1}}{(m+1)!} \tag{4.31}$$

Part (c): Recalling the series expansion of the exponential function and that the solution region of the ODE was defined in part by $|t - t_0| \le a$, show that

$$\sum_{m=n}^\infty \frac{K^{m+1}(\tau - t_0)^{m+1}}{(m+1)!} = \frac{K^{n+1}(\tau - t_0)^{n+1}}{(n+1)!} e^{K(\tau - t_0)} \le \frac{(Ka)^{n+1}}{(n+1)!} e^{Ka} \tag{4.32}$$

Part (d): Given that $(Ka)^{n+1}/(n+1)!$ is the $(n + 1)$th term in the series expansion of $e^{Ka}$ and hence must go to zero as n goes to infinity, use Equations 4.30 through 4.32 to prove Equation 4.29.

**CP4.2** Section 4.3.3 proved that the sequence of successive approximations

$$y_{n+1}(t) = y_0 + \int_{t_0}^t f(\tau, y_n(\tau))d\tau$$

converges and the preceding problem, CP4.1, outlined a proof that the sequence converges to a solution of the ODE in Equation 4.5. We invite you here to show that the solution so constructed is unique. The proof outlined below is based on demonstrating that small changes in the function $f(t,y)$ and initial condition $y_0$ lead to small changes in the solution $y(t)$; changes bounded in such a way that if two functions $\phi(t)$ and $\psi(t)$ each satisfy ODEs with the same $f(t,y)$ and $y_0$ they must be identical.

Let $\phi(t)$ and $\psi(t)$ be the solutions to

$$\frac{dy}{dt} = f(t,y) \qquad \frac{dy}{dt} = g(t,y)$$

$$y(t_0) = y_0 \qquad y(t_0) = w_0$$

where:
  $f(t,y)$ is as described in Section 4.3
  $g(t,y)$ is continuous in R

Then

$$\phi(t) = y_0 + \int_{t_0}^{t} f(\tau, \phi(\tau)) d\tau$$

$$\psi(t) = w_0 + \int_{t_0}^{t} g(\tau, \psi(\tau)) d\tau$$

Let $|f(t,y) - g(t,y)| < \varepsilon$ and $|y_0 - w_0| < \delta$.

Part (a): Show that

$$|\phi(t) - \psi(t)| \le \delta + \varepsilon |t - t_0| + \int_{t_0}^{t} |f(\tau, \phi(\tau)) - f(\tau, \psi(\tau))| d\tau$$

Part (b): Use the Lipschitz condition on $f(t,y)$ and define $E(t) = \int_{t_0}^{t} |\phi(\tau) - \psi(\tau)| d\tau$. Show that $E(t)$ satisfies the differential inequality

$$\frac{dE}{dt} - KE \le \delta + \varepsilon |t - t_0|, \qquad E(t_0) = 0$$

Part(c): Use an integrating factor and show that, for $t > t_0$,

$$|\phi(t) - \psi(t)| \le \delta e^{K(t-t_0)} + \varepsilon(e^{K(t-t_0)} - 1) / K^2$$

and hence if $\delta$ and $\varepsilon$ are zero, (i.e, the functions $f(t,y)$ and $g(t,y)$ are the same and the initial conditions $y_0$ and $w_0$ are the same), then $\phi(t)$ and $\psi(t)$ must be identical, and therefore the solution to the ODE is unique.

# 5 Second-Order Linear Ordinary Differential Equations

Chapters 2 through 4 focused on first-order ordinary differential equations (ODEs). In this chapter, we begin our march to higher and higher dimensional systems.

A second-order linear ODE has the form

$$\frac{d^2y}{dt^2} + p(t)\frac{dy}{dt} + q(t)y = g(t) \tag{5.1}$$

Second-order linear ODEs probably command the attention of more engineers more of the time than any other type of differential equation.

## 5.1 SECOND-ORDER LINEAR TIME-INVARIANT HOMOGENEOUS ORDINARY DIFFERENTIAL EQUATIONS

In a *time-invariant* linear second-order ODE, the coefficients $p(t)$ and $q(t)$ in Equation 5.1 are constant. A *homogeneous* equation has input $g(t) = 0$.

Consider the electrical circuit shown in Figure 5.1.

Recall that the circuits we analyzed in Chapter 2 included a resistor accompanied by either a capacitor or an inductor. The circuit in Figure 5.1 combines all three. As discussed in Chapter 2, Kirchhoff's voltage law says that the sum of the voltage drops around the circuit equals zero. In review, the voltage drop over a resistor is $IR$, where $I$ is the current through the resistor, the voltage drop over an inductor is $LdI/dt$, and the voltage drop over a capacitor is $q/C$, where $q$ is the electrical charge on the capacitor. The terms $R, L,$ and $C$ are known constants. We assume that the quantity of interest is the voltage over the capacitor, $V$, that a voltage $V_0$ exists on the capacitor at time $t = 0$ and then a switch is thrown, setting a current in motion. Conservation of charge requires that $I = dq/dt$, so Kirchhoff's voltage law can be written as

$$L\frac{dI}{dt} + RI + \frac{q}{C} = 0$$

$$LC\frac{d^2V}{dt^2} + RC\frac{dV}{dt} + V = 0$$

$$\frac{d^2V}{dt^2} + \frac{R}{L}\frac{dV}{dt} + \frac{1}{LC}V = 0 \tag{5.2}$$

This is a linear, time-invariant (LTI) second-order homogeneous ODE. In this case the initial conditions are

$$V(0) = V_0$$

$$\frac{dV}{dt}(0) = \frac{1}{C}I(0) = 0 \tag{5.3}$$

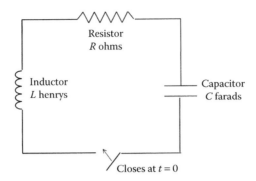

**FIGURE 5.1**   Second-order electrical circuit, an RLC circuit.

Second-order ODEs posed as initial value problems must have *two* stated initial conditions to yield a unique solution. How do we solve Equations 5.2 and 5.3? Based on our experience with first-order LTI ODEs we try the solution

$$V(t) = e^{rt} \tag{5.4}$$

Then

$$\frac{d}{dt}(e^{rt}) = re^{rt} \tag{5.5}$$

$$\frac{d^2}{dt^2}(e^{rt}) = r^2 e^{rt} \tag{5.6}$$

Substituting Equations 5.4, 5.5, and 5.6 into Equation 5.2 we have

$$\left( r^2 + \frac{R}{L}r + \frac{1}{LC} \right)e^{rt} = 0 \tag{5.7}$$

Since $e^{rt} > 0$ we can divide through by $e^{rt}$ and we are left with

$$r^2 + \frac{R}{L}r + \frac{1}{LC} = 0 \tag{5.8}$$

Equation 5.8, a second-order polynomial equation in $r$, is called the *characteristic equation* for the ODE in Equation 5.2. Its solutions are

$$r = -\frac{R}{2L} \pm \sqrt{\left( \frac{R}{2L} \right)^2 - \frac{1}{LC}} \tag{5.9}$$

Solutions to the ODE differ in form and behavior depending on the value of the expression under the radical. If

$$\left( \frac{R}{2L} \right)^2 - \frac{1}{LC} > 0$$

the characteristic equation has two real roots. If

$$\left(\frac{R}{2L}\right)^2 - \frac{1}{LC} = 0$$

it has repeated roots, and if

$$\left(\frac{R}{2L}\right)^2 - \frac{1}{LC} < 0$$

it has complex conjugate roots.

## 5.1.1 TWO REAL ROOTS

If

$$\left(\frac{R}{2L}\right)^2 - \frac{1}{LC} > 0$$

the roots of the characteristic equation can be written

$$
\begin{aligned}
r_1 &= -\alpha - \beta \\
r_2 &= -\alpha + \beta
\end{aligned}
\tag{5.10}
$$

where:

$$\alpha = \frac{R}{2L} \tag{5.11}$$

$$\beta = \sqrt{\left(\frac{R}{2L}\right)^2 - \frac{1}{LC}} \tag{5.12}$$

and we see that $\alpha > \beta > 0$.

As Section 5.2 will prove, any solution to the ODE in Equation 5.2 can be written as

$$V(t) = c_1 e^{r_1 t} + c_2 e^{r_2 t} \tag{5.13}$$

where the coefficients $c_1$ and $c_2$ are constants that depend on initial conditions. Here, to satisfy Equation 5.3, we must have

$$
\begin{aligned}
V(0) &= c_1 + c_2 = V_0 \\
\frac{dV}{dt}(0) &= c_1(-\alpha - \beta) + c_2(-\alpha + \beta) = 0
\end{aligned}
\tag{5.14}
$$

Equations 5.14 form two linear simultaneous equations in two unknowns. The solutions are

$$
\begin{aligned}
c_1 &= -\frac{(\alpha - \beta)}{2\beta} V_0 \\
c_2 &= \frac{(\alpha + \beta)}{2\beta} V_0
\end{aligned}
\tag{5.15}
$$

Hence the full solution to Equations 5.2 and 5.3, in the case where the characteristic equation has two real roots, is

$$V(t) = V_0 \left( -\frac{(\alpha - \beta)}{2\beta} e^{(-\alpha - \beta)t} + \frac{(\alpha + \beta)}{2\beta} e^{(-\alpha + \beta)t} \right) \tag{5.16}$$

Note that, in terms of $\alpha$ and $\beta$, Equation 5.2 can be rewritten as

$$\frac{d^2V}{dt^2} + 2\alpha \frac{dV}{dt} + (\alpha^2 - \beta^2)V = 0 \tag{5.17}$$

Any equation of the form in Equation 5.17 has the general solution

$$V(t) = c_1 e^{-(\alpha + \beta)t} + c_2 e^{-(\alpha - \beta)t} \tag{5.18}$$

where $c_1$ and $c_2$ are constants determined by initial conditions. Again, Section 5.2 provides a general proof of this.

Engineers call a system modeled by Equation 5.17 *overdamped*, a term we can explain more clearly in Section 5.1.4. If the ODE is of the form

$$\frac{d^2V}{dt^2} + a\frac{dV}{dt} + bV = 0 \tag{5.19}$$

the system is overdamped if $a > 0$, $b > 0$, and

$$b - a^2/4 < 0 \tag{5.20}$$

## 5.1.2  REPEATED ROOTS

If the electrical circuit in Figure 5.1 satisfies

$$\left( \frac{R}{2L} \right)^2 - \frac{1}{LC} = 0$$

then $\beta = 0$ in Equation 5.16 and $V(t)$ is undefined. We determine the solution by taking the limit of Equation 5.16 as the two roots

$$r_1 = -\alpha - \beta$$

$$r_2 = -\alpha + \beta$$

approach one another; that is, as $\beta$ goes to zero:

$$V(t) = \lim_{\beta \to 0} V_0 \left( -\frac{(\alpha - \beta)}{2\beta} e^{(-\alpha - \beta)t} + \frac{(\alpha + \beta)}{2\beta} e^{(-\alpha + \beta)t} \right) \tag{5.21}$$

We begin by dividing through by $V_0 e^{-\alpha t}$:

$$\frac{V(t)}{V_0 e^{-\alpha t}} = -\frac{(\alpha - \beta)}{2\beta} e^{-\beta t} + \frac{(\alpha + \beta)}{2\beta} e^{\beta t} \tag{5.22}$$

Next we expand $e^{-\beta t}$ and $e^{\beta t}$ in power series:

$$e^{-\beta t} = 1 - \beta t + \frac{(\beta t)^2}{2} + \dots \tag{5.23}$$

$$e^{\beta t} = 1 + \beta t + \frac{(\beta t)^2}{2} + \dots \tag{5.24}$$

Substituting Equations 5.23 and 5.24 into Equation 5.22:

$$\frac{V(t)}{V_0 e^{-\alpha t}} = -\frac{(\alpha - \beta)}{2\beta}\left(1 - \beta t + \frac{(\beta t)^2}{2} + \dots\right) + \frac{(\alpha + \beta)}{2\beta}\left(1 + \beta t + \frac{(\beta t)^2}{2} + \dots\right)$$

$$\frac{V(t)}{V_0 e^{-\alpha t}} = \frac{1}{2\beta}\left(-\alpha + \beta + (\alpha - \beta)\beta t - \frac{(\alpha - \beta)(\beta t)^2}{2} + \dots\right)$$

$$+ \frac{1}{2\beta}\left(\alpha + \beta + (\alpha + \beta)\beta t + \frac{(\alpha + \beta)(\beta t)^2}{2} + \dots\right)$$

$$= \frac{1}{2\beta}\left(2\beta + 2\alpha\beta t + \beta^3 t^2 + \dots\right) = 1 + \alpha t + \frac{\beta^2 t^2}{2} + \dots$$

Hence

$$\lim_{\beta \to 0} \frac{V(t)}{V_0 e^{-\alpha t}} = 1 + \alpha t$$

$$V(t) = V_0 e^{-\alpha t}(1 + \alpha t) \tag{5.25}$$

Equation 5.25 is the solution to Equation 5.2 with the specific initial conditions given in Equation 5.3. Note that when $\beta = 0$ Equation 5.2 can be written as

$$\frac{d^2 V}{dt^2} + 2\alpha \frac{dV}{dt} + \alpha^2 V = 0 \tag{5.26}$$

One can verify by direct substitution into Equation 5.26 that $V_1 = e^{-\alpha t}$ and $V_2 = te^{-\alpha t}$ are solutions to Equation 5.26. Any solution to 5.26 can be written as

$$V(t) = c_1 e^{-\alpha t} + c_2 t e^{-\alpha t} \tag{5.27}$$

where coefficients $c_1$ and $c_2$ are constants that depend on initial conditions.

Engineers call a system modeled by Equation 5.26 *critically damped*, another term we will defer explaining until in a later section. If the ODE is of the form

$$\frac{d^2V}{dt^2} + a\frac{dV}{dt} + bV = 0 \tag{5.28}$$

the system is critically damped if $a > 0$, $b > 0$, and

$$b - a^2/4 = 0 \tag{5.29}$$

### 5.1.3  COMPLEX CONJUGATE ROOTS

If the electrical circuit in Figure 5.1 satisfies

$$\left(\frac{R}{2L}\right)^2 - \frac{1}{LC} < 0$$

then according to Equation 5.9 the roots of the characteristic equation can be written as

$$r_1 = -\lambda + i\omega \tag{5.30}$$
$$r_2 = -\lambda - i\omega$$

where:

$$\lambda = \frac{R}{2L} \tag{5.31}$$

$$\omega = \sqrt{\frac{1}{LC} - \left(\frac{R}{2L}\right)^2} \tag{5.32}$$

and $i = \sqrt{-1}$, an imaginary number.

To summarize the situation, in seeking to solve Equation 5.2 we proposed in Equation 5.4 the trial solution $e^{rt}$ and Equations 5.30 through 5.32 resulted. A general solution would be of the form

$$V(t) = c_1 e^{r_1 t} + c_2 e^{r_2 t} = c_1 e^{(-\lambda+i\omega)t} + c_2 e^{(-\lambda-i\omega)t} = c_1 e^{-\lambda t} e^{i\omega t} + c_2 e^{-\lambda t} e^{-i\omega t} \tag{5.33}$$

What does Equation 5.33 mean? What are $e^{i\omega t}$ and $e^{-i\omega t}$?

Consider the series expansion

$$e^x = 1 + x + \left(\frac{1}{2}\right)x^2 + \left(\frac{1}{3\cdot2}\right)x^3 + \left(\frac{1}{4\cdot3\cdot2}\right)x^4 + \dots + \left(\frac{1}{n!}\right)x^n + \dots \tag{5.34}$$

Let $x = i\omega t$ in Equation 5.34, remembering that

$$i^2 = -1$$
$$i^3 = -i$$
$$i^4 = 1$$
$$i^5 = i$$

Then Equation 5.34 becomes

$$e^{i\omega t} = 1 + i\omega t - \left(\frac{1}{2}\right)(\omega t)^2 - i\left(\frac{1}{3\cdot 2}\right)(\omega t)^3 + \left(\frac{1}{4\cdot 3\cdot 2}\right)(\omega t)^4 + i\left(\frac{1}{5\cdot 4\cdot 3\cdot 2}\right)(\omega t)^5 \dots$$

$$= 1 - \left(\frac{1}{2}\right)(\omega t)^2 + \left(\frac{1}{4\cdot 3\cdot 2}\right)(\omega t)^4 + \dots \tag{5.35}$$

$$+ i\omega t - i\left(\frac{1}{3\cdot 2}\right)(\omega t)^3 + i\left(\frac{1}{5\cdot 4\cdot 3\cdot 2}\right)(\omega t)^5 - \dots$$

We recognize the real part of Equation 5.35 as

$$\cos(\omega t) = 1 - \left(\frac{1}{2}\right)(\omega t)^2 + \left(\frac{1}{4\cdot 3\cdot 2}\right)(\omega t)^4 + \dots + \frac{(-1)^n}{(2n)!}(\omega t)^{2n} + \dots$$

and the imaginary part as

$$i\sin(\omega t) = i(\omega t - \left(\frac{1}{3\cdot 2}\right)(\omega t)^3 + \left(\frac{1}{5\cdot 3\cdot 2}\right)(\omega t)^5 - \dots + \frac{(-1)^n}{(2n+1)!}(\omega t)^{2n+1} + \dots)$$

Hence we can write

$$e^{i\omega t} = \cos(\omega t) + i\sin(\omega t) \tag{5.36}$$

Equation 5.36 is known as Euler's Equation.

We have, from Equation 5.33,

$$V(t) = c_1 e^{-\lambda t} e^{i\omega t} + c_2 e^{-\lambda t} e^{-i\omega t}$$

and using Euler's Equation

$$V(t) = c_1 e^{-\lambda t}(\cos(\omega t) + i\sin(\omega t)) + c_2 e^{-\lambda t}(\cos(\omega t) - i\sin(\omega t))$$

$$V(t) = (c_1 + c_2)e^{-\lambda t}\cos(\omega t) + i(c_1 - c_2)e^{-\lambda t}\sin(\omega t) \tag{5.37}$$

$$V(t) = c_3 e^{-\lambda t}\cos(\omega t) + c_4 e^{-\lambda t}\sin(\omega t)$$

Now if $V(t)$ is a real solution then the coefficients $c_3 = c_1 + c_2$ and $c_4 = i(c_1 - c_2)$ must be real. Since the coefficients $c_1$ and $c_2$ are arbitrary, we see that they must be complex numbers. If we let $c_1 = (c_3 - ic_4)/2$, $c_2 = (c_3 + ic_4)/2$, then

$$V(t) = c_1 e^{-\lambda t} e^{i\omega t} + c_2 e^{-\lambda t} e^{-i\omega t}$$

is equivalent to

$$V(t) = c_3 e^{-\lambda t}\cos\omega t + c_4 e^{-\lambda t}\sin\omega t \tag{5.38}$$

where all the constants in Equation 5.38 are real valued. We can verify by direct substitution that $V_1(t) = e^{-\lambda t}\cos\omega t$ and $V_2(t) = e^{-\lambda t}\sin\omega t$ are solutions to

$$\frac{d^2V}{dt^2} + \frac{R}{L}\frac{dV}{dt} + \frac{1}{LC}V = 0 \tag{5.39}$$

where, again,

$$\lambda = \frac{R}{2L}$$

and

$$\omega = \sqrt{\frac{1}{LC} - \left(\frac{R}{2L}\right)^2}$$

For the initial conditions $V(0) = V_0$ and $dV/dt = 0$, we have

$$\begin{aligned} V(0) &= c_3 = V_0 \\ \frac{dV}{dt}(0) &= -\lambda c_3 + \omega c_4 = 0 \end{aligned} \tag{5.40}$$

so, the solution to Equations 5.2 and 5.3 is

$$V(t) = V_0\left(e^{-\lambda t}\cos\omega t + (\lambda/\omega)e^{-\lambda t}\sin\omega t\right) \tag{5.41}$$

Figure 5.2 presents a graph of an example of this function, showing the decreasing envelope due to the damping factor $\lambda$ and the oscillatory behavior with frequency $\omega$.

Note that, in terms of $\lambda$ and $\omega$, Equation 5.39 can be rewritten as

$$\frac{d^2V}{dt^2} + 2\lambda\frac{dV}{dt} + (\lambda^2 + \omega^2)V = 0 \tag{5.42}$$

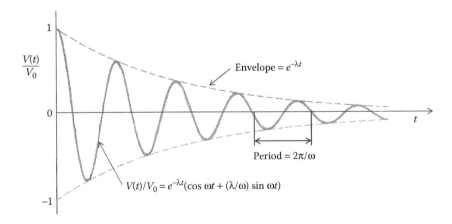

**FIGURE 5.2**  Graph of example solution of a second-order LTI ODE when the characteristic equation has complex roots.

Any equation of the form in Equation 5.42 has the general solution

$$V(t) = c_1 e^{-\lambda t} \cos \omega t + c_2 e^{-\lambda t} \sin \omega t \tag{5.43}$$

where $c_1$ and $c_2$ are constants determined by initial conditions.

It is important to note that in solving a new problem we do not have to go through the intermediate steps represented here by Equation 5.33 and invocation of Euler's equation. We can go directly from identification of a complex root of the characteristic equation to the real solution represented by Equation 5.43. See, for example, Solved Problem SP5.3 at the end of this chapter.

Engineers call a system modeled by Equation 5.42 *underdamped,* and call $\omega$ the circuit's *natural frequency.* If the ODE is of the form

$$\frac{d^2 V}{dt^2} + a \frac{dV}{dt} + bV = 0 \tag{5.44}$$

the system is underdamped if $a > 0$, $b > 0$, and

$$b - a^2/4 > 0 \tag{5.45}$$

## 5.1.4 Under-, Over-, and Critically Damped Systems

We are now in a position to explain why engineers use the terminology under-, over-, and critically damped systems. You will recall that engineers say an LTI system is damped if, with no external input, it ultimately rests at the origin independent of initial conditions.

Consider now a general second-order LTI homogeneous ODE with $a > 0$, $b > 0$, and specific initial conditions:

$$\frac{d^2 y}{dt^2} + a \frac{dy}{dt} + by = 0$$
$$y(0) = 1 \tag{5.46}$$
$$\dot{y}(0) = 0$$

The solution in all detail depends on the two parameters $a$ and $b$ but the qualitative character of the solution, that is, the shape of the curve $y(t)$, is determined by just one parameter. We can scale (nondimensionalize) time with the substitution $t_* = t\sqrt{b}$ and Equations 5.46 are converted to a form that can be written as

$$\ddot{y} + 2\zeta\dot{y} + y = 0$$
$$y(0) = 1 \tag{5.47}$$
$$\dot{y}(0) = 0$$

where the derivatives are understood to be with respect to the nondimensionalized time and $\zeta = a/(2\sqrt{b})$ is called the *damping ratio.* This is the form employed by control system engineers. Communication system engineers prefer the form

$$\ddot{y} + \left(\frac{1}{Q}\right)\dot{y} + y = 0$$

where $Q = \sqrt{b}/a$ is called the *quality factor*. The point is that the solution as a function of scaled time is governed by just one parameter.

Figure 5.3 graphs the solution of Equation 5.47 as a function of scaled time and damping ratio. We see that solutions for $\zeta < 1$, the underdamped systems, overshoot the origin and oscillate about it before finally settling in. This behavior is more prominent the lower the damping ratio.

Solutions for $\zeta > 1$, the overdamped systems, do not overshoot and oscillate but progress relatively slowly toward the origin, the higher the damping ratio the slower the motion. The system with $\zeta = 1$, the critically damped system, is the fastest to reach any given distance from the origin among the nonoscillating systems. Because it is fast and nonoscillating, critical damping is often sought as a design goal. But perhaps more often, systems with $\zeta \cong 0.75$ are implemented because they are faster yet and their overshoot and oscillation are minimal.

### 5.1.5  STABILITY OF SECOND-ORDER SYSTEMS

From the forms the fundamental solutions can take and their dependence on the coefficients $a$ and $b$ in Equations 5.19, 5.28, and 5.44, we can easily see that second-order time-invariant systems are stable if $a \geq 0$ and $b > 0$ and damped if $a > 0$ and $b > 0$. If either $a$ or $b$ is negative the system is unstable. Any second-order electrical circuit consisting only of inductors and capacitors is stable and any consisting only of resistors, inductors, and capacitors is stable and damped.

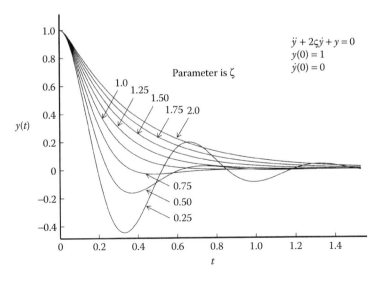

**FIGURE 5.3**   Graph of solution of second-order LTI ODE as a function of time and damping ratio.

## 5.2   FUNDAMENTAL SOLUTIONS

Consider the general homogeneous linear second-order ODE

$$\frac{d^2 y}{dt^2} + p(t)\frac{dy}{dt} + q(t)y = 0 \tag{5.48}$$

with initial conditions

$$y(0) = y_0$$
$$\frac{dy}{dt}(0) = \dot{y}_0 \tag{5.49}$$

Solutions $y_1(t)$ and $y_2(t)$ to Equation 5.48 are called *fundamental solutions* if any solution to Equation 5.48 can be written as

$$y(t) = c_1 y_1(t) + c_2 y_2(t) \tag{5.50}$$

What makes a solution *fundamental*? It is easy to prove that a function in the form Equation 5.50 satisfies Equation 5.48 (see Solved Problem SP5.4) but that is insufficient. A function in the form Equation 5.50 must also be able to satisfy the initial conditions, Equation 5.49. That is, we must be able to find coefficients $c_1$ and $c_2$ satisfying

$$c_1 y_1(0) + c_2 y_2(0) = y_0$$
$$c_1 \dot{y}_1(0) + c_2 \dot{y}_2(0) = \dot{y}_0 \tag{5.51}$$

From algebra we recall that we will be able to do this provided the coefficients of $c_1$ and $c_2$ have a nonzero determinant:

$$\det\begin{pmatrix} y_1(0) & y_2(0) \\ \dot{y}_1(0) & \dot{y}_2(0) \end{pmatrix} = y_1(0)\dot{y}_2(0) - y_2(0)\dot{y}_1(0) \neq 0 \tag{5.52}$$

Therefore, functions $y_1(t)$ and $y_2(t)$ are *fundamental solutions* provided that they satisfy Equations 5.48 and 5.52. The determinant

$$W(t) = \det\begin{pmatrix} y_1(t) & y_2(t) \\ \dot{y}_1(t) & \dot{y}_2(t) \end{pmatrix} = y_1(t)\dot{y}_2(t) - y_2(t)\dot{y}_1(t) \tag{5.53}$$

is called the *Wronskian*. Word Problem 5.7 asks the student to show that the Wronskian satisfies the differential equation

$$\frac{dW}{dt} + p(t)W = 0 \tag{5.54}$$

which we know from results of Chapter 2 has solution

$$W(t) = c \exp\left(-\int_0^t p(\tau)d\tau\right) \tag{5.55}$$

where $c$ is a constant. From Equation 5.55 it follows that if $W(0) \neq 0$ then $W(t) \neq 0$ for all $t$. This is an important result because it implies that if $y_1(t)$ and $y_2(t)$ are found to be fundamental solutions at a specific point in time then they are fundamental solutions for all time.

The Wronskian also proves important in the study of nonhomogeneous linear ODEs, as Section 5.3.2 will discuss.

## 5.3 SECOND-ORDER LINEAR NONHOMOGENEOUS ORDINARY DIFFERENTIAL EQUATIONS

A nonhomogeneous second-order linear ODE has the form

$$\frac{d^2 y}{dt^2} + p(t)\frac{dy}{dt} + q(t)y = g(t) \tag{5.56}$$

where $g(t)$ is nonzero.

### 5.3.1 EXAMPLE: AUTOMOBILE CRUISE CONTROL

Consider an automobile under cruise (speed)-control transitioning from a level roadway to an ascending grade (Figure 5.4). We will invest some time to derive the equations governing this process to demonstrate an important application but also to illustrate the technique of *linearization* so vital to engineers.

As discussed in Section 3.1.1.2, Newton's second law of motion is

$$F = ma$$

where:
  $F$ is the total force on the vehicle
  $m$ is its mass
  $a$ is its acceleration

We assume that the vehicle is in steady state (i.e., at a constant speed $V_0$) on the level roadway before transitioning to the ascending grade at time $t = 0$. Prior to $t = 0$, on an axis parallel to the level roadway, the car's constant engine thrust $T_0$ balances the air resistance (drag) $D_0$ and horizontal road resistance $R_0$:

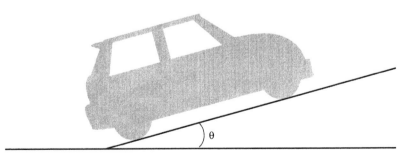

**FIGURE 5.4**  Schematic for automobile cruise control.

$$T_0 = D_0 + R_0$$

$$T_0 = \frac{\rho S C_D}{2} V_0^2 + R_0 \tag{5.57}$$

where, as given in Section 3.1.1.2
  $\rho$ is the air density
  $S$ the vehicle's cross-sectional area
  $C_D$ is its drag coefficient, assumed constant

When the car transitions to the ascending grade, in an axis parallel to the ascending grade a component of the car's weight is added to the forces on the car. Then the equation of motion on that axis becomes

$$m\frac{dV}{dt} = T - \frac{\rho S C_D}{2} V^2 - R - mg\sin\theta \tag{5.58}$$

Now we assume that $R = R_0$ and let

$$V = V_0 + u$$

$$T = T_0 + \Delta T \tag{5.59}$$

From Equation 5.59,

$$\frac{dV}{dt} = \frac{du}{dt} \tag{5.60}$$

We assume that $u$ is small relative to $V_0$, so that we can approximate

$$V^2 = V_0^2 + 2V_0 u + u^2 \cong V_0^2 + 2V_0 u \tag{5.61}$$

That is, we *linearize* Equation 5.58. We are examining perturbations about the nominal condition. Our approximations will be valid as long as the perturbations remain small, which will be the case if the angle $\theta$ of the grade is not too large and cruise control is effective.

What is the change in thrust $\Delta T$? We assume that the cruise-control system is based on *feedback* as shown in Figure 5.5; that is, that a change in thrust results when the speedometer measures that the speed deviates from that desired.

In Figure 5.5 we model the response of "Controller + Actuator" as

$$\Delta T = -k_1 u - k_2 \int^t u(\tau)d\tau \tag{5.62}$$

In doing so we are assuming a *proportional plus integral control law* and that the time required for the vehicle's engine to deliver increased horsepower in response to the controller's command for added gasoline flow is small. We should note here that our analysis is a considerable simplification of reality. A realistic control-system design would take into account engine dynamics and be digitally implemented in the form of difference equations. We are presenting here an introductory, top-level view.

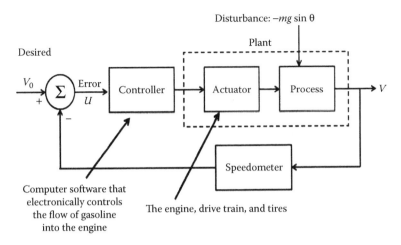

**FIGURE 5.5**   Block diagram for feedback control of vehicle speed.

Combining Equations 5.58 through 5.62 results in our model for the "Process" in Figure 5.5:

$$m\frac{du}{dt} = \left(T_0 - \frac{\rho SC_D}{2}V_0^2 - R_0\right) - \left(\rho SC_D V_0 + k_1\right)u - k_2\int^t u(\tau)d\tau - mg\sin\theta \qquad (5.63)$$

From Equation 5.57 the first term in parentheses on the right-hand side equals zero, so Equation 5.63 becomes

$$m\frac{du}{dt} = -\left(\rho SC_D V_0 + k_1\right)u - k_2\int^t u(\tau)d\tau - mg\sin\theta \qquad (5.64)$$

Defining a new variable $y(t) = \int^t u(\tau)d\tau$ and rearranging, Equation 5.64 becomes

$$\frac{d^2y}{dt^2} + a\frac{dy}{dt} + by = -g\sin\theta \qquad (5.65)$$

where:

$$a = (\rho SC_D V_0 + k_1)/m$$
$$b = k_2/m \qquad (5.66)$$

With knowledge of vehicle characteristics and density of air the designer of the vehicle's cruise-control system can achieve a desired response by *scheduling gains* $k_1$ and $k_2$ as a function of the speed $V_0$ set by the driver. One would expect the control system designer to seek a rapid, smooth transion between power states: rapid so as not to cause traffic problems, smooth (without overshoot and oscillation) so as not to cause driver and passenger discomfort. Consistent with these design goals, we assume here that the designer selects a *critically damped* response (recall the discussion in Section 5.1.4) with a given time constant, which can be accomplished by setting

$$a = 2\alpha$$
$$b = \alpha^2$$

(5.67)

From Equations 5.65 and 5.67, the equation of motion of the vehicle under cruise control is

$$\frac{d^2 y}{dt^2} + 2\alpha \frac{dy}{dt} + \alpha^2 y = -g \sin\theta$$

(5.68)

From Section 5.1.2 we know that the fundamental solutions to the homogeneous form of Equation 5.68 are

$$y_1(t) = e^{-\alpha t}$$
$$y_2(t) = te^{-\alpha t}$$

(5.69)

Section 5.3.2 together with Solved Problem SP5.5 will prove that the solution to Equation 5.68 can be written as the convolution

$$y(t) = \int_0^t K(t-\tau)(-g\sin\theta)d\tau$$

(5.70)

where the kernel $K(t-\tau)$ is given in this case by

$$K(t-\tau) = (t-\tau)e^{-\alpha(t-\tau)}$$

(5.71)

Substituting Equation 5.71 into Equation 5.70, we find

$$y(t) = \int_0^t (t-\tau)e^{-\alpha(t-\tau)}(-g\sin\theta)d\tau = -g\sin\theta \cdot e^{-\alpha t}\left(t\int_0^t e^{\alpha\tau}d\tau - \int_0^t \tau e^{\alpha\tau}d\tau\right)$$

(5.72)

Now

$$\int_0^t e^{\alpha\tau}d\tau = \frac{1}{\alpha}(e^{\alpha\tau} - 1)$$

and, using integration by parts,

$$\int_0^t \tau e^{\alpha\tau}d\tau = \int udv = uv - \int vdu$$

Let $u = \tau$, which means that $du = d\tau$, $dv = e^{\alpha\tau}d\tau$, $v = e^{\alpha\tau}/\alpha$, and so

$$\int_0^t \tau e^{\alpha\tau}d\tau = \frac{1}{\alpha}\tau e^{\alpha\tau}\Big|_0^t - \frac{1}{\alpha}\int_0^t e^{\alpha\tau}d\tau = \frac{1}{\alpha^2}(\alpha te^{\alpha t} - e^{\alpha t} + 1)$$

and then Equation 5.72 becomes

$$y(t) = -g\sin\theta \cdot e^{-\alpha t}\left( t\int_0^t e^{\alpha\tau}d\tau - \int_0^t \tau e^{\alpha\tau}d\tau \right)$$

$$= -g\sin\theta\left[ te^{-\alpha t}(e^{\alpha t}-1)/\alpha - e^{-\alpha t}\left(\alpha t e^{\alpha t} - e^{\alpha t}+1\right)/\alpha^2 \right]$$

$$y(t) = -g\sin\theta \cdot \left(1-(1+\alpha t)e^{-\alpha t}\right)/\alpha^2 \tag{5.73}$$

Recall now that the speed error $u(t) = dy/dt$:

$$u(t) = -gt\sin\theta \cdot e^{-\alpha t} \tag{5.74}$$

Figure 5.6a graphs this result for an incline of 5° and a time constant $1/\alpha$ as a parameter. Under the same conditions, Figure 5.6b graphs the second derivative of the speed error (the derivative of the acceleration), which is frequently used as a measure of rider discomfort. It is often called "jerk". The lower the value of the time constant the smaller the speed error, but the greater the jerk. The control system engineer must make a decision on the design value of $\alpha$ balancing speed error with comfort of the ride.

In summary, an automobile in cruise control mode transitioning from a level roadway to an ascending grade experiences a speed error given by Equation 5.74 if it utilizes a proportional plus integral control law set for critically damped response. Note that the steady-state

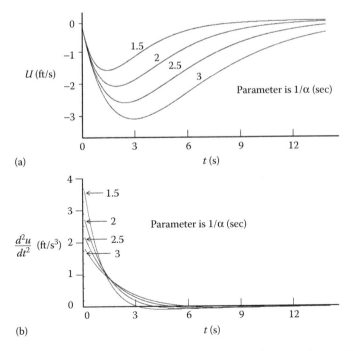

**FIGURE 5.6**  Speed error and "jerk" due to an abrupt 5° change in grade for a vehicle under cruise control: (a) speed error and (b) derivative of acceleration (a measure of rider discomfort.)

error is zero. It is left as an exercise for the reader to show that the steady-state error is not zero if the control law is merely proportional (i.e., if $k_2 = 0$ in Equation 5.62).

## 5.3.2 VARIATION OF PARAMETERS (KERNEL) METHOD

**Theorem:**

If $y_1(t)$ and $y_2(t)$ are fundamental solutions to

$$\frac{d^2 y}{dt^2} + p(t)\frac{dy}{dt} + q(t)y = 0 \tag{5.75}$$

then the solution to

$$\frac{d^2 y}{dt^2} + p(t)\frac{dy}{dt} + q(t)y = g(t) \tag{5.76}$$

$$y(t_0) = 0 \tag{5.77}$$
$$\dot{y}(t_0) = 0$$

is

$$y(t) = \int_{t_0}^{t} K(t,\tau)g(\tau)d\tau \tag{5.78}$$

where:

$$K(t,\tau) = (-y_1(t)y_2(\tau) + y_2(t)y_1(\tau))/W(\tau) \tag{5.79}$$

and

$$W(\tau) = y_1(\tau)\dot{y}_2(\tau) - y_2(\tau)\dot{y}_1(\tau) \tag{5.80}$$

is the Wronskian of $y_1(\tau)$ and $y_2(\tau)$.

**Proof:**

We try a solution in the form

$$y(t) = u_1(t)y_1(t) + u_2(t)y_2(t) \tag{5.81}$$

Then

$$\dot{y} = \dot{u}_1 y_1 + u_1 \dot{y}_1 + \dot{u}_2 y_2 + u_2 \dot{y}_2 \tag{5.82}$$

We *set*

$$\dot{u}_1 y_1 + \dot{u}_2 y_2 = 0 \tag{5.83}$$

and are left with

$$\dot{y} = u_1 \dot{y}_1 + u_2 \dot{y}_2 \tag{5.84}$$

Then

$$\ddot{y} = \dot{u}_1 \dot{y}_1 + u_1 \ddot{y}_1 + \dot{u}_2 \dot{y}_2 + u_2 \ddot{y}_2 \tag{5.85}$$

Substituting equations for $y(t)$ (Equation 5.81), $\dot{y}(t)$ (Equation 5.84) and $\ddot{y}(t)$(Equation 5.85) into the ODE

$$\frac{d^2 y}{dt^2} + p(t)\frac{dy}{dt} + q(t)y = g(t) \tag{5.86}$$

we have, after some rearranging,

$$u_1 \left[\ddot{y}_1 + p(t)\dot{y}_1 + q(t)y_1\right] + u_2 \left[\ddot{y}_2 + p(t)\dot{y}_2 + q(t)y_2\right] + \dot{u}_1 \dot{y}_1 + \dot{u}_2 \dot{y}_2 = g(t) \tag{5.87}$$

Now

$$\ddot{y}_1 + p(t)\dot{y}_1 + q(t)y_1 = 0$$
$$\ddot{y}_2 + p(t)\dot{y}_2 + q(t)y_2 = 0 \tag{5.88}$$

so

$$\dot{u}_1 \dot{y}_1 + \dot{u}_2 \dot{y}_2 = g(t) \tag{5.89}$$

Equations 5.83 and 5.89 constitute two linear equations in two unknowns, and their solution is

$$\dot{u}_1 = -y_2 g / W \tag{5.90}$$

$$\dot{u}_2 = y_1 g / W \tag{5.91}$$

where

$$W = y_1 \dot{y}_2 - y_2 \dot{y}_1 \tag{5.92}$$

Integrating Equations 5.90 and 5.91 we have

$$u_1(t) = -\int_{t_0}^{t} \frac{y_2(\tau)g(\tau)}{W(\tau)} d\tau$$

$$u_2(t) = \int_{t_0}^{t} \frac{y_1(\tau)g(\tau)}{W(\tau)} d\tau \tag{5.93}$$

Incorporating Equation 5.93 into Equation 5.81 results in

$$y(t) = -y_1(t)\int_{t_0}^{t} \frac{y_2(\tau)g(\tau)}{W(\tau)} d\tau + y_2(t)\int_{t_0}^{t} \frac{y_1(\tau)g(\tau)}{W(\tau)} d\tau \tag{5.94}$$

Defining

$$K(t,\tau) = (-y_1(t)y_2(\tau) + y_2(t)y_1(\tau))/W(\tau) \tag{5.95}$$

we arrive at

$$y(t) = \int_{t_0}^{t} K(t,\tau)g(\tau)d\tau \tag{5.96}$$

It remains to show that $y(t)$ defined by Equation 5.96 satisfies the initial conditions of Equation 5.77:

$$y(t_0) = 0$$

$$\dot{y}(t_0) = 0$$

It is clear that

$$y(t_0) = \int_{t_0}^{t_0} K(t_0,\tau)g(\tau)d\tau = 0$$

Now

$$\dot{y}(t) = K(t,t)g(t) + \int_{t_0}^{t} \frac{\partial K}{\partial t}(t,\tau)g(\tau)d\tau \tag{5.97}$$

By its definition (Equation 5.95),

$$K(t,t) = 0$$

so Equation 5.97 becomes

$$\dot{y}(t) = \int_{t_0}^{t} \frac{\partial K}{\partial t}(t,\tau)g(\tau)d\tau \tag{5.98}$$

Then

$$\dot{y}(t_0) = \int_{t_0}^{t_0} \frac{\partial K}{\partial t}(t_0,\tau)g(\tau)d\tau = 0 \tag{5.99}$$

and so $y(t)$ as defined by Equations 5.96, 5.95, and 5.92 satisfies both the ODE Equation 5.76 and initial conditions Equation 5.77, as was to be shown.

For LTI nonhomogeneous second-order ODEs Equation 5.96 becomes the convolution

$$y(t) = \int_{t_0}^{t} K(t-\tau)g(\tau)d\tau$$

Solved Problem SP5.5 derives the forms the kernel $K(t-\tau)$ can take for overdamped, critically damped, and underdamped second-order systems.

When a linear nonhomogeneous ODE has nonzero initial conditions its solution will be of the form

$$y(t) = c_1 y_1(t) + c_2 y_2(t) + y_p(t) \tag{5.100}$$

where $y_p(t)$ is the *particular solution* (or, recalling Section 2.2.3, the *forced response*) given by Equations 5.92, 5.95, and 5.96 and the coefficients $c_1$ and $c_2$ are determined by the initial conditions.

Equation 5.100 follows from the ODE's linearity. The proof utilizes logic similar to that employed in Solved Problem SP5.4.

The variation of parameters method is the most powerful of the three methods we will examine for the solution of second- and higher-order linear nonhomogeneous ODEs. It works for time-varying equations with continuous coefficients and for all continuous inputs $g(t)$, whereas the other methods do not. Its disadvantage compared to the other methods is the time and energy required to execute it. We will meet the second solution method in Section 5.3.3 and the third in Chapter 7.

### 5.3.3  Undetermined Coefficients Method

The simplest of the three approaches we will meet for solving LTI nonhomogeneous ODEs, but the most limited in its power, is the undetermined coefficients method. In this method we let the form of the input determine a trial solution. Table 5.1 lists solutions to try, given a specific input function.

We substitute the trial solution into the ODE and solve for the undetermined coefficients.

#### 5.3.3.1  Steady-State Solutions

Recall the definition of *steady-state solution* in Section 2.2.3: in a stable, damped system, the steady-state solution remains when transients have died away. The undetermined coefficient method is especially attractive when we seek only the steady-state solution, because we are then spared the effort of factoring the characteristic equation and determining fundamental solutions. We must determine that the system is stable and damped, but recalling Section 5.1.5 we can do this for second-order ODEs by examining whether the ODE's coefficients are greater than zero.

---

**TABLE 5.1**

**Trial Solutions for Given Input Functions Using the Undetermined Coefficients Method**

| Input | Trial Solution |
|---|---|
| $e^{-kt}$ | $Ce^{-kt}$ |
| $t$ | $Ct + D$ |
| $te^{-kt}$ | $(Ct + D)e^{-kt}$ |
| $t^2$ | $Ct^2 + Dt + E$ |
| $\sin \omega t$ | $C \sin \omega t + D \cos \omega t$ |
| $e^{-kt} \sin \omega t$ | $(C \sin \omega t + D \cos \omega t)e^{-kt}$ |
| $t \sin \omega t$ | $(Ct + D) \sin \omega t + (Et + F) \cos \omega t$ |
| $te^{-kt} \sin \omega t$ | $((Ct + D) \sin \omega t + (Et + F) \cos \omega t)e^{-kt}$ |
| $t^2 e^{-kt} \sin \omega t$ | $((Ct^2 + Dt + E) \sin \omega t + (Ft^2 + Gt + H) \cos \omega t)e^{-kt}$ |

---

**Example:**

Problem: Find the steady-state solution to the ODE

$$\frac{d^2y}{dt^2} + 3\frac{dy}{dt} + 2y = 12t \qquad (5.101)$$

Solution: The ODE's coefficients are greater than zero, so the system is stable and damped. Then, following Table 5.1, we try the solution

$$y(t) = Ct + D \qquad (5.102)$$

and solve for $C$ and $D$. Now, from Equation 5.102,

$$\frac{dy}{dt} = C \qquad (5.103)$$

and

$$\frac{d^2y}{dt^2} = 0 \qquad (5.104)$$

Substituting Equations 5.102, 5.103, and 5.104 into Equation 5.101 we have

$$0 + 3C + 2(Ct + D) = 12t \qquad (5.105)$$

Collecting coefficients of powers of $t$, we have

$$(3C + 2D) + (2C)t = 12t \qquad (5.106)$$

Equating coefficients of powers of $t$,

$$3C + 2D = 0$$
$$\qquad (5.107)$$
$$2C = 12$$

Equation 5.107 are two simultaneous linear equations in two unknowns. The solutions are

$$C = 6$$
$$D = -3C/2 = -9$$

and the steady-state solution to Equation 5.101 is

$$y(t) = 6t - 9$$

### 5.3.3.2   Example: RLC Circuit with Sine-Wave Source

Consider the electrical circuit shown in Figure 5.7:

We are interested in the steady-state voltage $V(t)$ over the capacitor when the voltage source has the form $V_0 \sin \omega t$. Following the results in Section 5.1 we can write the circuit equation as

$$\frac{d^2V}{dt^2} + \frac{R}{L}\frac{dV}{dt} + \frac{1}{LC}V = \frac{1}{LC}V_0 \sin \omega t \tag{5.108}$$

For simplicity let

$$a = \frac{R}{L}$$
$$b = \frac{1}{LC} \tag{5.109}$$

Then Equation 5.108 becomes

$$\frac{d^2V}{dt^2} + a\frac{dV}{dt} + bV = bV_0 \sin \omega t \tag{5.110}$$

Using Table 5.1 we try the solution

$$V(t) = C\sin \omega t + D\cos \omega t \tag{5.111}$$

Then

$$\frac{dV}{dt} = \omega C \cos \omega t - \omega D \sin \omega t \tag{5.112}$$

$$\frac{d^2V}{dt^2} = -\omega^2 C \sin \omega t - \omega^2 D \cos \omega t \tag{5.113}$$

Substituting Equations 5.111, 5.112, and 5.113 into Equation 5.110 we have, after some rearranging,

$$\left[(b-\omega^2)C - a\omega D\right]\sin \omega t + \left[a\omega C + (b-\omega^2)D\right]\cos \omega t = bV_0 \sin \omega t \tag{5.114}$$

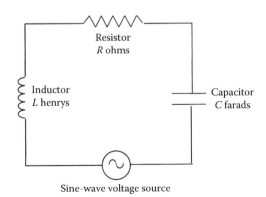

**FIGURE 5.7**   Second-order RLC electrical circuit with sine-wave voltage source.

Equation 5.114 must be true for all $t$. When $\omega t = 2n\pi$, $\sin \omega t = 0$, and $\cos \omega t = 1$ and we see that Equation 5.114 implies that

$$a\omega C + (b - \omega^2)D = 0 \tag{5.115}$$

When $\omega t = 2n\pi + \pi/2$, $\sin \omega t = 1$, and $\cos \omega t = 0$, in which case Equation 5.114 implies that

$$(b - \omega^2)C - a\omega D = bV_0 \tag{5.116}$$

Equations 5.115 and 5.116 are two simultaneous linear equations in the two unknowns $C$ and $D$. Their solutions are

$$C = \frac{b - \omega^2}{(b - \omega^2)^2 + (a\omega)^2} bV_0$$

$$\tag{5.117}$$

$$D = \frac{-a\omega}{(b - \omega^2)^2 + (a\omega)^2} bV_0$$

We can simplify the form of the answer. Following Section 2.2.3, instead of writing

$$V(t) = C \sin \omega t + D \cos \omega t$$

we can define

$$\cos \theta = \frac{C}{\sqrt{C^2 + D^2}}$$

$$\sin \theta = \frac{-D}{\sqrt{C^2 + D^2}}$$

and write

$$V(t) = \sqrt{C^2 + D^2} (\cos \theta \sin \omega t - \sin \theta \cos \omega t)$$

$$\tag{5.118}$$

$$V(t) = \sqrt{C^2 + D^2} \sin(\omega t - \theta)$$

The steady-state solution to Equation 5.110, then, is

$$V(t) = GV_0 \sin(\omega t - \theta) \tag{5.119}$$

where

$$G = \frac{b}{\sqrt{(b - \omega^2)^2 + (a\omega)^2}} \tag{5.120}$$

$$\theta = \tan^{-1}(a\omega, b - \omega^2) \tag{5.121}$$

As discussed in Section 2.2.3 for a water pipe temperature problem, Equations 5.120 and 5.121 provide the *gain* and *phase lag* of the frequency response of the RLC circuit in Figure 5.7.

Note that we have derived these results without knowing the magnitude of $a = R/L$ and $b = 1/LC$. We have neither factored the ODE's characteristic equation nor determined the ODE's fundamental solutions.

### 5.3.3.3 Resonance with a Sine-Wave Source

Let us examine the behavior of the gain $G$ in Equation 5.120 as a function of the input frequency $\omega$. Recall that Section 2.2.3 determined that the gain of the frequency response for a first-order LTI system decreases monotonically with input frequency. We examine here the possibility that, for second-order LTI systems, a certain input frequency may maximize the gain.

The gain is maximized when the denominator

$$F(\omega) = (b - \omega^2)^2 + (a\omega)^2 \tag{5.122}$$

is minimized. We let $x = \omega^2$ and expand Equation 5.122:

$$F(x) = b^2 - 2bx + x^2 + a^2 x = b^2 + (a^2 - 2b)x + x^2$$

To minimize $F(x)$ we set its derivative equal to zero:

$$\frac{dF}{dx} = (a^2 - 2b) + 2x = 0$$

resulting in

$$x = b - a^2/2 \tag{5.123}$$

Recalling that $x = \omega^2$ we see that we must have

$$b - a^2/2 > 0$$

to have a real input frequency maximizing the gain. Looking back at Section 5.1.3 we see from Equations 5.44 and 5.45 that the gain has a maximum only when the system is underdamped. For this case, that is, when $\omega_m = \sqrt{b - a^2/2}$ is real, the maximum gain is given by

$$G_m = \frac{b}{\sqrt{(b - \omega_m^2)^2 + (a\omega_m)^2}} = \frac{b}{\sqrt{(a^2/2)^2 + a^2(b - a^2/2)}}$$

$$G_m = \frac{b}{a\sqrt{b - a^2/4}} \tag{5.124}$$

When $a^2/4 \ll b$ Equation 5.124 becomes

$$G_m = \frac{\sqrt{b}}{a} \gg \frac{1}{2} \tag{5.125}$$

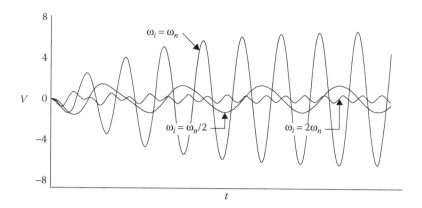

**FIGURE 5.8**  Output of an example underdamped second-order system of natural frequency $\omega_n$ when excited by sine-wave inputs of unit magnitude but different frequencies $\omega_i$.

and we see that the gain can become very large. That is, when a system is only lightly damped, an input frequency near the natural frequency of the system results in a very large magnitude of the output, a phenomenon known as *resonance*.

Figure 5.8 presents the output of the underdamped second-order system whose natural response is shown in Figure 5.2 when that system is excited by three sine-wave inputs, each with unit magnitude but with different frequencies. When the input frequency $\omega_i$ equals the system's natural frequency $\omega_n$ the magnitude of the output in steady state is much greater than it is when the input frequency is a factor of two greater or less than the natural frequency.

In summary, when driven by a sine-wave source, an underdamped second-order system can experience a sharp rise in its gain when the input frequency approaches its natural frequency, in contrast to the monotonically decreasing gain in the frequency response of a first-order system.

### 5.3.3.4  Resonance with a Periodic Input Other than a Sine Wave

The quantitative results of Sections 5.3.3.2 and 5.3.3.3 apply to a sine-wave input, but the qualitative conclusions apply much more generally. We will see in Chapter 7 that the steady-state response of a stable, damped nth order LTI system to any periodic input is periodic and that an underdamped system can experience resonance when the input frequency approaches the system's natural frequency, independent of the input's waveform.

### 5.3.3.5  Satisfying Initial Conditions Using the Undetermined Coefficients Method

Suppose that the values of the solution $y(t)$ and its derivative $\dot{y}(t)$ are specified at time $t_0$ as $y_0$ and $\dot{y}_0$. Let us call the solution we derive using the undetermined coefficients method $y_C(t)$. In contrast to the particular solution generated by the variation of parameters method, which has zero initial value and zero initial derivative, typically $y_C(t_0) \neq 0$ and $\dot{y}_C(t_0) \neq 0$. To satisfy given initial conditions we must find the ODE's fundamental solutions $y_1(t)$ and $y_2(t)$ and then choose coefficients $c_1$ and $c_2$ to satisfy

$$y_0 = y_C(t_0) + c_1 y_1(t_0) + c_2 y_2(t_0)$$

$$\dot{y}_0 = \dot{y}_C(t_0) + c_1 \dot{y}_1(t_0) + c_2 \dot{y}_2(t_0)$$

### 5.3.3.6 Limitations of the Undetermined Coefficients Method

The undetermined coefficients method applies only to LTI ODEs. It also applies only to a certain class of input functions, namely those that are themselves fundamental solutions to the class of LTI ODEs. One cannot successfully apply the undetermined coefficients method when the input function is, for example, $\ln(1+t)$ or $1/(1+t)$ or $\tan \omega t$.

Finally, the undetermined coefficients method is, in practice, a trial and error method. Even if one recalls Table 5.1 faithfully (e.g., try $Ct + D$, not $Ct$, when the input is a multiple of $t$) the trial solutions in Table 5.1 fail when the input is itself a fundamental solution of the particular ODE. For second-order equations the solution will then involve either $t$ or $t^2$ times the trial solution suggested in Table 5.1.

## 5.4 EXISTENCE AND UNIQUENESS OF SOLUTIONS TO SECOND-ORDER LINEAR ORDINARY DIFFERENTIAL EQUATIONS

**Theorem:**

Consider the second-order linear ODE

$$\frac{d^2y}{dt^2} + p(t)\frac{dy}{dt} + q(t)y = g(t) \tag{5.126}$$

Suppose that the values of $y(t)$ and $dy/dt$ are specified to be $y_0$ and $\dot{y}_0$ at time $t_0$. If the functions $p(t)$, $q(t)$, and $g(t)$ are continuous on an open interval $I$ that contains the point $t_0$ then a solution to Equation 5.126 exists throughout the interval $I$ and the solution is unique.

(Stated without proof. See [3].)

**Corollary:**

If the functions $p(t)$, $q(t)$, and $g(t)$ are continuous everywhere then Equation 5.126 has a unique solution at every instant of time. In particular, LTI equations have unique solutions at every instant of time.

## 5.5 SUMMARY

We have considered in this chapter equations of the form

$$\frac{d^2y}{dt^2} + p(t)\frac{dy}{dt} + q(t)y = g(t)$$

Two solutions $y_1(t)$ and $y_2(t)$ to the homogeneous ODE (i.e., when $g(t)=0$) are called fundamental if any solution to the ODE can be written in the form

$$y(t) = c_1 y_1(t) + c_2 y_2(t)$$

Fundamental solutions must satisfy the ODE and have nonzero Wronskian

$$W(t) = \det \begin{pmatrix} y_1(t) & y_2(t) \\ \dot{y}_1(t) & \dot{y}_2(t) \end{pmatrix} = y_1(t)\dot{y}_2(t) - y_2(t)\dot{y}_1(t)$$

so that constants $c_1$ and $c_2$ can be found to satisfy the given initial conditions

$$y(0) = c_1 y_1(0) + c_2 y_2(0) = y_0$$

$$\dot{y}(0) = c_1 \dot{y}_1(0) + c_2 \dot{y}_2(0) = \dot{y}_0$$

If the Wronskian is nonzero at one point in time it is nonzero everywhere and if it is zero at one point it is zero everywhere.

Solutions to the time-invariant homogeneous ODE

$$\frac{d^2 y}{dt^2} + a\frac{dy}{dt} + by = 0$$

are found by trying $y(t) = e^{rt}$, which leads to the characteristic equation

$$r^2 + ar + b = 0$$

which has solutions

$$r_{1,2} = -(a/2) \pm \sqrt{(a/2)^2 - b}$$

If $b - a^2/4 < 0$ then the two fundamental solutions to the ODE are

$$y_1(t) = \exp(r_1 t) \quad y_2(t) = \exp(r_2 t)$$

If $b - a^2/4 = 0$ then the two solutions are

$$y_1(t) = \exp(r_1 t) \quad y_2(t) = t\exp(r_1 t)$$

If $b - a^2/4 > 0$ then the two solutions are

$$y_1(t) = e^{\lambda t}\cos\omega t \quad y_2(t) = e^{\lambda t}\sin\omega t$$

where $\lambda = -(a/2)$ and $\omega = \sqrt{b - a^2/4}$.

A time-invariant system in the aforementioned form is stable if $a \geq 0$ and $b > 0$ and asymptotically stable (damped) if $a > 0$ and $b > 0$. It is unstable if either $a$ or $b$ is less than zero.

If $y_1(t)$ and $y_2(t)$ are fundamental solutions to the homogeneous ODE

$$\frac{d^2 y}{dt^2} + p(t)\frac{dy}{dt} + q(t)y = 0$$

then the solution to

$$\frac{d^2 y}{dt^2} + p(t)\frac{dy}{dt} + q(t)y = g(t)$$

with initial conditions $y(t_0) = 0$ and $\dot{y}(t_0) = 0$ is

$$y_P(t) = \int_{t_0}^{t} K(t,\tau) g(\tau) d\tau$$

where

$$K(t,\tau) = (-y_1(t) y_2(\tau) + y_2(t) y_1(\tau)) / W(\tau)$$

is the kernel of the ODE and

$$W(\tau) = y_1(\tau) \dot{y}_2(\tau) - y_2(\tau) \dot{y}_1(\tau)$$

is the Wronskian of $y_1(\tau)$ and $y_2(\tau)$. Any solution to the nonhomogeneous ODE can be written as

$$y(t) = y_P(t) + c_1 y_1(t) + c_2 y_2(t)$$

where the constants $c_1$ and $c_2$ are chosen to satisfy initial conditions.

A sometimes quicker approach to solving nonhomogeneous time-invariant ODEs is the method of undetermined coefficients. In this method one chooses a trial solution with unknown coefficients based on the form of the input (e.g., if $g(t) = \sin \omega_0 t$, try $y(t) = C \sin \omega_0 t + D \cos \omega_0 t$), substitutes the trial solution in the ODE, and solves for the unknown coefficients. This chapter provided a table of trial solutions for specific inputs and discussed some limitations of the method.

When the input to a significantly underdamped time-invariant system oscillates at a frequency near the natural frequency of the system, the magnitude of the output significantly increases. This is called resonance.

If the coefficients $p(t)$ and $q(t)$ in a second-order linear ODE are continuous in an open interval, and initial conditions are specified at a point in that interval then the ODE has a solution in that interval and the solution is unique. In particular, LTI ODEs have solutions everywhere.

## Solved Problems

**SP5.1** Solve:

$$\frac{d^2 y}{dt^2} + 13 \frac{dy}{dt} + 12 y = 0, \qquad y(0) = 1, \qquad \dot{y}(0) = -2 \qquad (5.127)$$

*Solution*: We try $y(t) = e^{rt}$. Now $dy/dt = re^{rt}$ and $d^2 y/dt^2 = r^2 e^{rt}$. Substituting these into the ODE yields

$$(r^2 + 13r + 12) e^{rt} = 0$$

Since $e^{rt} > 0$ we can divide through by it and obtain the characteristic equation

$$r^2 + 13r + 12 = 0$$

This factors into

$$(r+1)(r+12) = 0$$

and so the roots of the characteristic equation are $r_1 = -1$ and $r_2 = -12$. Therefore, the fundamental solutions are $y_1(t) = e^{-t}$ and $y_2(t) = e^{-12t}$. The solution $y(t)$ to Equation 5.127 must be a linear combination of these: $y(t) = c_1 e^{-t} + c_2 e^{-12t}$. We find the coefficients $c_1$ and $c_2$ from the initial conditions:

$$y(0) = c_1 e^{(0)} + c_2 e^{(0)} = c_1 + c_2 = 1$$

$$\frac{dy}{dt} = \frac{d}{dt}(c_1 e^{-t} + c_2 e^{-12t}) = -c_1 e^{-t} - 12 c_2 e^{-12t}$$

$$\frac{dy}{dt}(0) = -c_1 e^{(0)} - 12 c_2 e^{(0)} = -c_1 - 12 c_2 = -2$$

We have two linear equations in two unknowns:

$$c_1 + c_2 = 1$$

$$-c_1 - 12 c_2 = -2$$

Solving:

$$c_2 = 1 - c_1$$

$$-c_1 - 12(1 - c_1) = -2$$

$$11 c_1 = 10$$

$$c_1 = 10/11$$

$$c_2 = 1/11$$

Therefore, the solution to Equation 5.127 is

$$y(t) = (10/11) e^{-t} + (1/11) e^{-12t}$$

**SP5.2** Solve:

$$\frac{d^2 y}{dt^2} - 12 \frac{dy}{dt} + 36 y = 0, \qquad y(0) = 2, \qquad \dot{y}(0) = -3 \tag{5.128}$$

*Solution*: We try $y(t) = e^{rt}$. Now $dy/dt = r e^{rt}$ and $d^2 y/dt^2 = r^2 e^{rt}$. Substituting these into the ODE yields

$$(r^2 - 12r + 36) e^{rt} = 0.$$

Since $e^{rt} > 0$ we can divide through by it and obtain the characteristic equation

$$r^2 - 12r + 36 = 0$$

This factors into

$$(r-6)^2 - 0$$

and so the two roots of the characteristic equation are $r_1 = r_2 = 6$. Therefore, the fundamental solutions are $y_1(t) = e^{6t}$ and $y_2(t) = te^{6t}$. (Recall the results of Section 5.1.2.) The solution $y(t)$ to Equation 5.128 must be a linear combination of these: $y(t) = c_1 e^{6t} + c_2 te^{6t}$. We find the coefficients $c_1$ and $c_2$ from the initial conditions:

$$y(0) = c_1 e^{(0)} + c_2(0)e^{(0)} = c_1 = 2$$

$$\frac{dy}{dt} = \frac{d}{dt}(c_1 e^{6t} + c_2 te^{6t}) = 6c_1 e^{6t} + c_2 e^{6t} + 6tc_2 e^{6t}$$

$$\frac{dy}{dt}(0) = 6c_1 e^{(0)} + c_2 e^{(0)} + 6(0)c_2 e^{(0)} = 6c_1 + c_2 = -3$$

We have

$$c_1 = 2$$

$$6c_1 + c_2 = -3$$

so, $c_2 = -15$ and the solution to Equation 5.128 is

$$y(t) = 2e^{6t} - 15te^{6t}$$

**SP5.3** Solve:

$$\frac{d^2 y}{dt^2} + 4\frac{dy}{dt} + 20y = 0, \qquad y(0) = 3, \qquad \dot{y}(0) = 2 \qquad\qquad (5.129)$$

*Solution*: Trying $y(t) = e^{rt}$, remembering that $dy/dt = re^{rt}$ and $d^2y/dt^2 = r^2 e^{rt}$, substituting these into the ODE and dividing through by $e^{rt}$ (which is always nonzero) yields the characteristic equation

$$r^2 + 4r + 20 = 0$$

Completing the square (recall Section 1.1.2.1):

$$r^2 + 4r + 20 = (r+2)^2 - 4 + 20 = (r+2)^2 + 16 = 0$$

$$(r+2)^2 = -16$$

$$r + 2 = \pm\sqrt{-16} = \pm 4\sqrt{-1} = \pm 4i$$

$$r = -2 \pm 4i$$

and so the roots of the characteristic equation are $r_1 = -2 + 4i$ and $r_2 = -2 - 4i$. Therefore, the fundamental solutions are $y_1(t) = e^{-2t} \cos 4t$ and $y_2(t) = e^{-2t} \sin 4t$. (Recall Equation 5.38 in Section 5.1.3.) The solution $y(t)$ to Equation 5.129 must be a linear combination of these: $y(t) = c_1 e^{-2t} \cos 4t + c_2 e^{-2t} \sin 4t$. We find the coefficients $c_1$ and $c_2$ from the initial conditions:

$$y(0) = c_1 e^{(0)} \cos(0) + c_2 e^{(0)} \sin(0) = c_1 \cdot 1 + c_2 \cdot (0) = c_1 = 3$$

$$\frac{dy}{dt} = \frac{d}{dt}(c_1 e^{-2t} \cos 4t + c_2 e^{-2t} \sin 4t) =$$

$$-2c_1 e^{-2t} \cos 4t - 4c_1 e^{-2t} \sin 4t - 2c_2 e^{-2t} \sin 4t + 4c_2 e^{-2t} \cos 4t$$

$$\frac{dy}{dt}(0) = -2c_1 e^{(0)} \cos(0) - 4c_1 e^{(0)} \sin(0) - 2c_2 e^{(0)} \sin(0) + 4c_2 e^{(0)} \cos(0)$$

$$\frac{dy}{dt}(0) = -2c_1 \cdot 1 - 4c_1 \cdot (0) - 2c_2 \cdot (0) + 4c_2 \cdot (1) = -2c_1 + 4c_2 = 2$$

We have

$$c_1 = 3$$

$$-2c_1 + 4c_2 = 2$$

so, $c_2 = 2$ and the solution to Equation 5.129 is

$$y(t) = 3e^{-2t} \cos 4t + 2e^{-2t} \sin 4t$$

**SP5.4** Show that if $y_1(t)$ and $y_2(t)$ satisfy

$$\frac{d^2 y}{dt^2} + p(t)\frac{dy}{dt} + q(t)y = 0 \qquad (5.130)$$

then so does $y(t) = c_1 y_1(t) + c_2 y_2(t)$

*Solution*:
The function $y(t) = c_1 y_1(t) + c_2 y_2(t)$ satisfies Equation 5.130 because it is a linear combination of solutions to the ODE and because the ODE is linear:

$$\frac{d^2 y}{dt^2} + p(t)\frac{dy}{dt} + q(t)y = \frac{d^2}{dt^2}(c_1 y_1 + c_2 y_2) + p(t)\frac{d}{dt}(c_1 y_1 + c_2 y_2) + q(t)(c_1 y_1 + c_2 y_2)$$

$$= c_1\left(\frac{d^2 y_1}{dt^2} + p(t)\frac{dy_1}{dt} + q(t)y_1\right) + c_2\left(\frac{d^2 y_2}{dt^2} + p(t)\frac{dy_2}{dt} + q(t)y_2\right)$$

$$= c_1 \cdot (0) + c_2 \cdot (0) = 0$$

**SP5.5** Derive the form of the Wronskian and convolution kernel for an (a) overdamped, (b) critically damped, and (c) underdamped LTI second-order system.

*Solution*:
The Wronskian is defined as

$$W(t) = \det \begin{pmatrix} y_1(t) & y_2(t) \\ \dot{y}_1(t) & \dot{y}_2(t) \end{pmatrix} = y_1(t)\dot{y}_2(t) - y_2(t)\dot{y}_1(t)$$

and the kernel is defined as

$$K(t-\tau) = (-y_1(t)y_2(\tau) + y_2(t)y_1(\tau)) / W(\tau)$$

(a) Fundamental solutions for overdamped systems have the form $y_1(t) = e^{-\lambda_1 t}$ and $y_2(t) = e^{-\lambda_2 t}$ where $\lambda_1 > 0$, $\lambda_2 > 0$, and $\lambda_1 \neq \lambda_2$. Hence, for an overdamped system,

$$W(t) = e^{-\lambda_1 t}(-\lambda_2 e^{-\lambda_2 t}) - e^{-\lambda_2 t}(-\lambda_1 e^{-\lambda_1 t}) = (\lambda_1 - \lambda_2)e^{-(\lambda_1 + \lambda_2)t}$$

$$K(t-\tau) = ((-e^{-\lambda_1 t})(e^{-\lambda_2 \tau}) + (e^{-\lambda_2 t})(e^{-\lambda_1 \tau})) / ((\lambda_1 - \lambda_2)e^{-(\lambda_1 + \lambda_2)\tau})$$

$$= ((-e^{-\lambda_1 t})(e^{-\lambda_2 \tau})(e^{+(\lambda_1 + \lambda_2)\tau}) + (e^{-\lambda_2 t})(e^{-\lambda_1 \tau})(e^{+(\lambda_1 + \lambda_2)\tau})) / (\lambda_1 - \lambda_2)$$

$$= ((-e^{-\lambda_1 t})(e^{+\lambda_1 \tau}) + (e^{-\lambda_2 t})(e^{+\lambda_2 \tau})) / (\lambda_1 - \lambda_2)$$

$$= (e^{-\lambda_1(t-\tau)} - e^{-\lambda_2(t-\tau)}) / (\lambda_2 - \lambda_1)$$

(b) Fundamental solutions for critically damped systems have the form $y_1(t) = e^{-\lambda t}$ and $y_2(t) = te^{-\lambda t}$. Hence, for a critically damped system,

$$W(t) = e^{-\lambda t}(-\lambda te^{-\lambda t} + e^{-\lambda t}) - te^{-\lambda t}(-\lambda e^{-\lambda t}) = e^{-2\lambda t}$$

$$K(t-\tau) = ((-e^{-\lambda t})(\tau e^{-\lambda \tau}) + (te^{-\lambda t})(e^{-\lambda \tau})) / (e^{-2\lambda \tau})$$

$$= (-e^{-\lambda t})(\tau e^{-\lambda \tau})(e^{+2\lambda \tau}) + (te^{-\lambda t})(e^{-\lambda \tau})(e^{+2\lambda \tau})$$

$$= (-e^{-\lambda t})(\tau e^{+\lambda \tau}) + (te^{-\lambda t})(e^{+\lambda \tau})$$

$$= (t-\tau)e^{-\lambda(t-\tau)}$$

(c)  Fundamental solutions for underdamped systems have the form $y_1(t) = e^{-\lambda t} \cos \omega t$ and $y_2(t) = e^{-\lambda t} \sin \omega t$. Hence, for an underdamped system,

$$W(t) = (e^{-\lambda t} \cos \omega t)(-\lambda e^{-\lambda t} \sin \omega t + \omega e^{-\lambda t} \cos \omega t)$$

$$- (e^{-\lambda t} \sin \omega t)(-\lambda e^{-\lambda t} \cos \omega t - \omega e^{-\lambda t} \sin \omega t)$$

$$= \omega e^{-2\lambda t}(\cos^2 \omega t + \sin^2 \omega t) = \omega e^{-2\lambda t}$$

$$K(t - \tau) = ((-e^{-\lambda t} \cos \omega t)(e^{-\lambda \tau} \sin \omega \tau) + (e^{-\lambda t} \sin \omega t)(e^{-\lambda \tau} \cos \omega \tau)) / (\omega e^{-2\lambda \tau})$$

$$= ((-e^{-\lambda t} \cos \omega t)(e^{-\lambda \tau} \sin \omega \tau)(e^{+2\lambda \tau}) + (e^{-\lambda t} \sin \omega t)(e^{-\lambda \tau} \cos \omega \tau)(e^{+2\lambda \tau})) / \omega$$

$$= ((-e^{-\lambda t} \cos \omega t)(e^{+\lambda \tau} \sin \omega \tau) + (e^{-\lambda t} \sin \omega t)(e^{+\lambda \tau} \cos \omega \tau)) / \omega$$

$$= e^{-\lambda(t-\tau)}(\sin \omega t \cos \omega \tau - \cos \omega t \sin \omega \tau) / \omega$$

$$= e^{-\lambda(t-\tau)} \sin \omega(t - \tau) / \omega$$

**SP5.6** Solve:

$$\frac{d^2 y}{dt^2} + 5\frac{dy}{dt} + 6y = 4e^{-t}, \qquad y(0) = 0, \qquad \dot{y}(0) = 0 \tag{5.131}$$

*Solution*:
*Step 1*: Find the fundamental solutions to the homogeneous equation

$$\frac{d^2 y}{dt^2} + 5\frac{dy}{dt} + 6y = 0$$

Trying $y(t) = e^{rt}$ we find the characteristic equation

$$r^2 + 5r + 6 = 0$$

which factors into

$$(r + 2)(r + 3) = 0$$

and hence has roots $r_1 = -2$, $r_2 = -3$. Therefore the fundamental solutions are

$$y_1(t) = e^{-2t}, y_2(t) = e^{-3t}$$

*Step 2*: Calculate the Wronskian:

$$W(\tau) = y_1(\tau)\dot{y}_2(\tau) - y_2(\tau)\dot{y}_1(\tau)$$

$$-e^{-2\tau}(-3e^{-3\tau}) - e^{-3\tau}(-2e^{-2\tau})$$

$$= (-3+2)e^{-5\tau} = -e^{-5\tau}$$

*Step 3*: Calculate the kernel:

$$K(t,\tau) = (-y_1(t)y_2(\tau) + y_2(t)y_1(\tau))/W(\tau) =$$

$$(-e^{-2t}(e^{-3\tau}) + e^{-3t}(e^{-2\tau}))/(-e^{-5\tau}) = e^{-2t}e^{+2\tau} - e^{-3t}e^{+3\tau} =$$

$$e^{-2(t-\tau)} - e^{-3(t-\tau)}$$

*Step 4*: Perform the integrations:

$$y(t) = \int_0^t K(t,\tau)g(\tau)d\tau = \int_0^t (e^{-2(t-\tau)} - e^{-3(t-\tau)})(4e^{-\tau})d\tau$$

$$= \int_0^t e^{-2(t-\tau)}(4e^{-\tau})d\tau - \int_0^t e^{-3(t-\tau)}(4e^{-\tau})d\tau$$

$$= 4e^{-2t}\int_0^t e^{(+2-1)\tau}d\tau - 4e^{-3t}\int_0^t e^{(+3-1)\tau}d\tau$$

$$= 4e^{-2t}\int_0^t e^{\tau}d\tau - 4e^{-3t}\int_0^t e^{2\tau}d\tau$$

$$= 4e^{-2t}(e^t - 1) - 4e^{-3t}(e^{2t} - 1)/2$$

*Step 5*: Calculate the particular solution

$$y_p(t) = 4e^{-2t}(e^t - 1) - 4e^{-3t}(e^{2t} - 1)/2$$

$$= 4e^{-t} - 4e^{-2t} - 2e^{-t} + 2e^{-3t}$$

$$y_p(t) = 2e^{-t} - 4e^{-2t} + 2e^{-3t} \qquad (5.132)$$

*Step 6*: Add in the response due to initial conditions:
In this problem the initial conditions are zero, so the particular solution is the complete
    solution.

*Step 7*: Check the solution by computing its value at $t = 0$. Does it match the initial conditions? From Equation 5.6.2,

$$y_p(0) = 2e^{(0)} - 4e^{(0)} + 2e^{(0)} = 2 - 4 + 2 = 0$$

$$\dot{y}_p(t) = -2e^{-t} + 8e^{-2t} - 6e^{-3t}$$

$$\dot{y}_p(0) = -2e^{(0)} + 8e^{(0)} - 6e^{(0)} = -2 + 8 - 6 = 0$$

If we have the time we can also substitute the calculated solution into the ODE and see if the equation is satisfied.

**SP5.7** Find the steady-state solution to

$$\frac{d^2y}{dt^2} + 3\frac{dy}{dt} + 4y = 5te^{2t} \tag{5.133}$$

*Solution*: Referring to Table 5.1, we try

$$y(t) = (Ct + D)e^{2t} \tag{5.134}$$

Then

$$\frac{dy}{dt} = 2(Ct + D)e^{2t} + Ce^{2t} = 2Cte^{2t} + (2D + C)e^{2t} \tag{5.135}$$

$$\frac{d^2y}{dt^2} = 4Cte^{2t} + 2Ce^{2t} + 2(2D + C)e^{2t} = 4Cte^{2t} + (4D + 4C)e^{2t} \tag{5.136}$$

Substituting Equations 5.134, 5.135, and 5.136 into Equation 5.133:

$$\left[4Cte^{2t} + (4D + 4C)e^{2t}\right] + 3\left[2Cte^{2t} + (2D + C)e^{2t}\right] + 4\left[(Ct + D)e^{2t}\right] = 5te^{2t}$$

Dividing through by $e^{2t}$ and collecting coefficients in like powers of $t$:

$$((4 + 3 \cdot 2 + 4)C)t + ((4 + 3 \cdot 2 + 4)D + (4 + 3)C) = 5t$$

$$(14C)t + (14D + 7C) = 5t$$

Matching coefficients of similar powers of $t$:

$$14C = 5$$

$$14D + 7C = 0$$

The solutions are

$$C = 5/14$$

$$D = -5/28$$

Hence the steady-state solution to Equation 5.131 is

$$y(t) = (5/14)(t - 1/2)e^{2t}$$

## Problems to Solve

In Problems 5.1 through 5.12, find $y(t)$.

**P5.1** $\dfrac{d^2 y}{dt^2} + 8\dfrac{dy}{dt} + 15y = 0,$ $\qquad y(0) = 3$ , $\quad \dot{y}(0) = -2$

**P5.2** $\dfrac{d^2 y}{dt^2} - \dfrac{dy}{dt} - 20y = 0,$ $\qquad y(0) = -2,$ $\qquad \dot{y}(0) = 1$

**P5.3** $\dfrac{d^2 y}{dt^2} - 10\dfrac{dy}{dt} + 21y = 0,$ $\qquad y(0) = 1,$ $\qquad \dot{y}(0) = 3$

**P5.4** $\dfrac{d^2 y}{dt^2} - 16y = 0,$ $\qquad y(0) = 2,$ $\qquad \dot{y}(0) = 0$

**P5.5** $\dfrac{d^2 y}{dt^2} + 6\dfrac{dy}{dt} + 9y = 0,$ $\qquad y(0) = 4,$ $\qquad \dot{y}(0) = 2$

**P5.6** $\dfrac{d^2 y}{dt^2} - 14\dfrac{dy}{dt} + 49y = 0,$ $\qquad y(0) = 5,$ $\qquad \dot{y}(0) = -1$

**P5.7** $\dfrac{d^2 y}{dt^2} + 18\dfrac{dy}{dt} + 81y = 0,$ $\qquad y(0) = -3,$ $\qquad \dot{y}(0) = 4$

**P5.8** $\dfrac{d^2 y}{dt^2} - 2\dfrac{dy}{dt} + y = 0,$ $\qquad y(0) = -1,$ $\qquad \dot{y}(0) = -3$

**P5.9** $\dfrac{d^2 y}{dt^2} - 8\dfrac{dy}{dt} + 41y = 0,$ $\qquad y(0) = 6,$ $\qquad \dot{y}(0) = -2$

**P5.10** $\dfrac{d^2 y}{dt^2} + 16\dfrac{dy}{dt} + 80y = 0,$ $\qquad y(0) = 2,$ $\qquad \dot{y}(0) = 1$

**P5.11** $\dfrac{d^2 y}{dt^2} + 36y = 0,$ $\qquad y(0) = 1,$ $\qquad \dot{y}(0) = 5$

**P5.12** $\dfrac{d^2 y}{dt^2} + 14\dfrac{dy}{dt} + 53y = 0,$ $\qquad y(0) = -4,$ $\qquad \dot{y}(0) = 3$

For Problems 5.13 through 5.22, determine the Wronskian $W(\tau)$ and the convolution kernel $K(t - \tau)$.

**P5.13** $\dfrac{d^2 y}{dt^2} + 7\dfrac{dy}{dt} - 18y = 0$

**P5.14** $\dfrac{d^2 y}{dt^2} + 13\dfrac{dy}{dt} + 40y = 0$

**P5.15** $\dfrac{d^2 y}{dt^2} - 12\dfrac{dy}{dt} + 35y = 0$

**P5.16** $\dfrac{d^2 y}{dt^2} + 10\dfrac{dy}{dt} + 25y = 0$

**P5.17** $\dfrac{d^2 y}{dt^2} - 4\dfrac{dy}{dt} + 4y = 0$

**P5.18** $\dfrac{d^2 y}{dt^2} + 16\dfrac{dy}{dt} + 64y = 0$

**P5.19** $\dfrac{d^2 y}{dt^2} + 6\dfrac{dy}{dt} + 90y = 0$

**P5.20** $\dfrac{d^2 y}{dt^2} + 10\dfrac{dy}{dt} + 34y = 0$

**P5.21** $\dfrac{d^2 y}{dt^2} + 49y = 0$

**P5.22** $\dfrac{d^2 y}{dt^2} - 4\dfrac{dy}{dt} + 5y = 0$

For Problems 5.23 through 5.26, use the variation of parameters method to find $y(t)$. Assume zero initial conditions.

**P5.23** $\dfrac{d^2 y}{dt^2} + 10\dfrac{dy}{dt} + 24y = 6e^{-2t}$

**P5.24** $\dfrac{d^2 y}{dt^2} + 16y = 2\sin 4t$

**P5.25** $\dfrac{d^2 y}{dt^2} + 6\dfrac{dy}{dt} + 8y = 3t$

**P5.26** $\dfrac{d^2 y}{dt^2} + 4\dfrac{dy}{dt} + 3y = 4te^{-3t}$

For Problems 5.27 through 5.31, use the undetermined coefficient method to find the steady-state solution $y(t)$.

**P5.27** $\dfrac{d^2y}{dt^2} + 5\dfrac{dy}{dt} + 4y = 2t$

**P5.28** $\dfrac{d^2y}{dt^2} + 3\dfrac{dy}{dt} + 4y = 3e^{2t}$

**P5.29** $\dfrac{d^2y}{dt^2} + 2\dfrac{dy}{dt} + 3y = 2\cos 3t$

**P5.30** $\dfrac{d^2y}{dt^2} + 3\dfrac{dy}{dt} + 2y = 6te^{4t}$

**P5.31** $\dfrac{d^2y}{dt^2} + 4\dfrac{dy}{dt} + 2y = 5e^{3t}\sin 2t$

For Problems 5.32 through 5.36, use the undetermined coefficients method to find $y(t)$.

**P5.32** $\dfrac{d^2y}{dt^2} + 4\dfrac{dy}{dt} + 3y = 2t^2 + 3t, \qquad y(0) = 2, \qquad \dot{y}(0) = -2$

**P5.33** $\dfrac{d^2y}{dt^2} + 16y = 9\sin 5t, \qquad y(0) = 3, \qquad \dot{y}(0) = 2$

**P5.34** $\dfrac{d^2y}{dt^2} + 4\dfrac{dy}{dt} + 8y = 4te^{-3t}, \qquad y(0) = -2, \qquad \dot{y}(0) = 1$

**P5.35** $\dfrac{d^2y}{dt^2} + 4\dfrac{dy}{dt} + 4y = 8e^{-4t}\sin 2t, \qquad y(0) = 0, \qquad \dot{y}(0) = 0$

**P5.36** $\dfrac{d^2y}{dt^2} + 5\dfrac{dy}{dt} + 6y = 7t\cos 2t, \qquad y(0) = 1, \qquad \dot{y}(0) = -1$

## Word Problems

**WP5.1** Figure 5.9 depicts a weathervane mounted on a post, initially pointed into the wind. The wind shifts instantaneously by $\theta_0$ degrees at time $t = 0$ and stays constant thereafter at speed $V$. For small angles $\theta$, the aerodynamic force of the wind on the weathervane can be modeled as having magnitude $F_W = kV^2\theta$ and a distribution across the weathervane as if all the forces were concentrated at the center of pressure at a distance $d$ behind the post. The angular momentum of the weathervane is $I\dot{\theta}$, where $I$ is the moment of inertia of the weathervane about the post. The torque due to the wind is $-F_W d$. The weathervane's movement about the post is resisted by a frictional torque that can be approximately modeled as $-\gamma\dot{\theta}$. Newton's law applied to rotational motion states that the derivative of the angular momentum equals the torque on the body. With these facts in hand: (a) Write the equation of motion of the weathervane, (b) given the numerical values in Table 5.2, solve the equation, and (c) show that the weathervane eventually points into the wind once again.

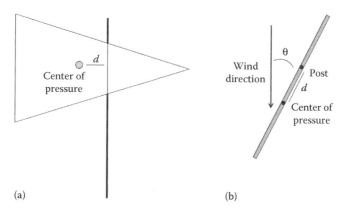

(a)    (b)

**FIGURE 5.9**    Weathervane: (a) side view and (b) top view.

**TABLE 5.2**
**Weathervane Parameter Values**

| Parameter | Values (English units) | Values (Metric units) |
|---|---|---|
| $V$ | 20 ft / s | 6.1 m/s |
| $d$ | 0.725 ft | 0.22 m |
| $k$ | 0.1 slugs/ft | 4.82 kg/m |
| $I$ | 1.0 slugs/ft$^2$ | 1.36 kg·m$^2$ |
| $\gamma$ | 4.0 slugs·ft$^2$/s | 5.44 kg·m$^2$/s |

**WP5.2** Figure 5.10 depicts a spring hanging vertically. With no mass attached the end of the spring is at position $y_0$. When a mass $m$ is attached, in steady state the end of the spring is extended to $y_E$, to counter the force of gravity. If the mass is pulled to a new position $y_I$ beneath $y_E$ and released, the mass experiences three forces: (1) gravity: $F_1 = mg$, (2) the spring force, according to Hooke's law: $F_2 = -k(y - y_0)$, and (3) an energy-diminishing force due to air resistance and heat lost through internal friction in the bending of the spring that can be approximately modeled as $F_3 = -\beta(dy/dt)$, where $\beta$ is a constant. Applying Newton's second law, $F = ma$,

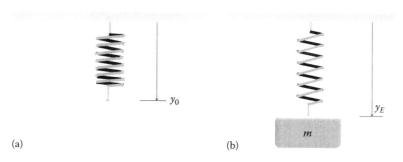

(a)    (b)

**FIGURE 5.10**    Mass-spring system hanging vertically: (a) unextended spring and (b) spring extended due to mass under gravity.

remembering that $a = d^2 y/dt^2$, write the equation of motion of the mass. Consider now a new variable $u = y - y_E$, representing the mass' deviation from its steady-state position. Substitute $y = u + y_E$ in the equation of motion and search for terms that cancel. What equation does $u$ satisfy?

**WP5.3** Consider the ODE whose solutions were graphed in Figure 5.3

$$\ddot{y} + 2\zeta\dot{y} + y = 0$$

$$y(0) = 1$$

$$\dot{y}(0) = 0$$

Part (a): If $\zeta < 1$, determine, as a function of $\zeta$, the maximum amount by which $y(t)$ overshoots the origin. Hint: at the instant of maximum overshoot, what is $\dot{y}(t)$?

Part (b): If $\zeta = 1$, show that $y(t)$ does not overshoot the origin.

Part (c): If $\zeta > 1$, show that $y(t)$ does not overshoot the origin.

Hints for Parts (b) and (c): Show that

$$\lim_{t \to \infty} y(t) = 0$$

$$\dot{y}(t) < 0 \quad \text{for all } t$$

and that these two facts prove the proposition.

**WP5.4** Figure 5.11 diagrams a circuit in which a resistor, inductor, and capacitor are wired in parallel. Prior to time $t = 0$ the capacitor is isolated from the other circuit elements and carries a voltage $V_0$. Also prior to $t = 0$ no current flows in the circuit formed by the resistor and inductor. At $t = 0$ the switch is thrown to join the capacitor with the resistor and inductor.

By the conservation of charge, the current flowing from the capacitor equals the sum of the current flowing into the resistor and the current flowing into the inductor. Because the circuit elements are wired in parallel, each has the same voltage drop $v$ across it. The current flowing into the resistor can be expressed as $v/R$. The current flowing into the inductor can be expressed as $(\int^t v(\tau)d\tau)/L$.

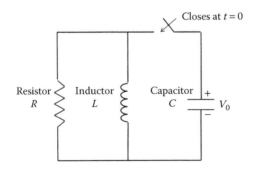

**FIGURE 5.11**   Resistor–Inductor–Capacitor (RLC) circuit wired in parallel.

The current flowing from the capacitor can be expressed as $-C(dv/dt)$. With these facts in hand, derive an equation for the voltage $v$ as a function of time. Differentiate your equation with respect to time to arrive at a second-order LTI equation for the circuit. Because voltage cannot change instantaneously across a capacitor $v(0) = V_0$. Hence the current flowing through the resistor at $t = 0^+$, that is, immediately after the switch is thrown, must be $V_0/R$. Because current flowing into an inductor cannot change instantaneously the current initially flowing through the inductor must be zero. Therefore all the current initially flowing from the capacitor must flow into the resistor, from which we determine that $\dot{v}(0) = -V_0/RC$. With these initial conditions in hand, and given that $R = 10^3$, $L = 10^{-2}$, and $C = 10^{-8}$ in compatible units, solve the equation.

**WP5.5** Current sources that deliver a specified amount of current independent of the rest of the circuit are not as commonly used as voltage sources (think of the ubiquitous battery), but they are easily constructed and useful in many applications. Figure 5.12 diagrams a current source connected to a resistor, inductor, and capacitor wired in parallel. From the conservation of charge, the current flowing from the current source must equal the sum of the current flowing into the resistor, the inductor, and the capacitor. WP5.4 showed how the current flowing into each of these elements may be described in terms of the voltage, which must be the same across each element. Assume that prior to $t = 0$ there is no current flowing in the circuit and no voltage drop across the elements. At $t = 0$ the current source is activated. Part (a): Derive an equation for the variable $w(t) = \int_0^t v(\tau)d\tau$ as a function of time. Part (b): Suppose the current source produces a constant current $I_0$. One initial condition is $w(0) = 0$. Because voltage cannot change instantaneously across a capacitor $v(0) = \dot{w}(0) = 0$. Assume that the resistor, inductor, and capacitor have the same values as in WP5.4. Solve the circuit equation and determine the voltage $v$ as a function of time.

**WP5.6** Gravity-gradient stabilization is a passive means of controlling a satellite to point toward the Earth. The moon is gravity-gradient stabilized due to a bulge in the direction of the Earth, which is why we only ever see one face of it. The technique is based on the diminution of gravity with increased distance from the Earth;

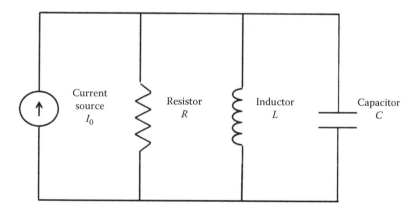

**FIGURE 5.12** Resistor, inductor, and capacitor wired in parallel with current source.

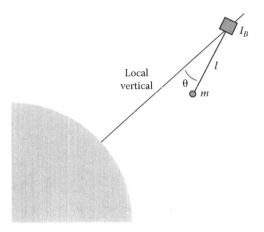

**FIGURE 5.13**   Gravity–gradient–stabilized satellite.

parts of the satellite nearer Earth experience greater force than those parts farther away. This produces a stabilizing torque. Figure 5.13 illustrates how a long boom extended toward Earth enhances this effect. For the particular case where the body of the spacecraft can be modeled as a uniformly distributed cube or sphere, the equation of motion in the plane of the orbit is

$$(I_B + ml^2)\ddot{\theta} + 3(ml^2)\omega_0^2\theta = \tau$$

where:
  $\theta$ is the angle between the satellite's orientation and the local vertical
  $I_B$ is the moment of inertia of the satellite's main body
  $l$ the length of the boom
  $m$ is the mass of the object at the end of the boom
  $\omega_0$ is the orbital radian frequency
  $\tau$ any disturbance torques

A residual magnetic moment in the spacecraft in the direction of the boom can interact with the Earth's magnetic field to produce such a disturbance torque, as shown in Figure 5.14. Let $M_S$ be the magnitude of the satellite's magnetic moment and $B_E$ the magnitude of the Earth's magnetic field at the Equator. If (a) the satellite is in a polar orbit, (b) we use the approximate model of the Earth's magnetic field as that of a dipole magnet in the center of the Earth oriented due North–South, and (c) $M_S B_E \ll ml^2 \omega_0^2$, then the torque is approximated by $\tau = M_S B_E \sin \omega_0 t$. Find the steady-state motion of the satellite about the local vertical. Show that an undesirable resonance occurs if $I_B = 2ml^2$, and determine the behavior of $\theta$ as a function of time in that case, assuming zero initial conditions.

**WP5.7** Show that if two solutions of the ODE

$$\ddot{y} + p(t)\dot{y} + q(t)y = 0$$

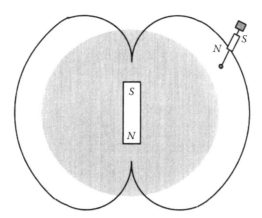

**FIGURE 5.14** Earth's magnetic field interacting with a residual satellite magnetic moment. (Note that the Earth's dipole must have its South Pole pointed toward the Earth's North Pole.)

are $y_1(t)$ and $y_2(t)$ then the Wronskian

$$W(t) = y_1(t)\dot{y}_2(t) - y_2(t)\dot{y}_1(t)$$

satisfies

$$\dot{W} + p(t)W = 0$$

## Challenge Problems

**CP5.1** Figure 5.15 depicts a thin-walled, hollow cylinder of mass $m$ and radius $r$ rolling without slipping (i.e., without skidding) on a curved platform in the form of an arc with constant radius of curvature $R$. Determine the motion of the cylinder

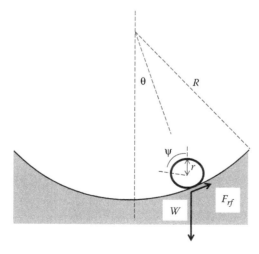

**FIGURE 5.15** Cylinder rolling on a curved platform under the influence of gravity and rolling friction.

under two circumstances: Case (a): the only moment (torque) on the cylinder is that due to the tangential rolling friction $F_{rf}$; Case (b): The cylinder also experiences a moment due to an offset in the normal force. Show that the motion in Case (a) is perpetual, whereas the motion in Case (b) eventually ceases.

*Guidance*: The cylinder's motion is governed by Newton's laws of translational and rotational motion. Recognizing that the cylinder's distance along the arc is given by $R\theta$, we have

$$ma = m\frac{d^2}{dt^2}(R\theta) = mR\ddot{\theta} = \sum \text{forces} \tag{5.137}$$

$$I\ddot{\psi} = \sum \text{moments} \tag{5.138}$$

where $I$ is the cylinder's moment of inertia. For a hollow, thin-walled cylinder,

$$I = mr^2 \tag{5.139}$$

In Case (a), the only moment on the cylinder is that due to rolling friction, which is proportional to the normal force $N$ in the direction opposite to the cylinder's motion:

$$\tau = rF_{rf} = -r\mu N \,\text{sgn}(\dot{\theta}) = -r\mu mg\cos\theta\,\text{sgn}(\dot{\theta}) \tag{5.140}$$

Here $\mu$ is the coefficient of friction and $\text{sgn}(\dot{\theta})$ is the sign of $\dot{\theta}$; that is,

$$\text{sgn}(\dot{\theta}) = 1 \quad \text{if} \quad \dot{\theta} > 0$$

$$\text{sgn}(\dot{\theta}) = -1 \quad \text{if} \quad \dot{\theta} < 0$$

The forces parallel to the arc are the rolling friction and the tangential component of gravity:

$$mR\ddot{\theta} = -mg\sin\theta - \mu mg\cos\theta\,\text{sgn}(\dot{\theta}) \tag{5.141}$$

Since the cylinder rolls without slipping,

$$r\ddot{\psi} = -R\dot{\theta} \tag{5.142}$$

From Equations 5.138, 5.139, 5.140, and 5.142,

$$-\mu mg\cos\theta\,\text{sgn}(\dot{\theta}) = \left(\frac{I}{r}\right)\ddot{\psi} = mr\ddot{\psi} = mr\left(\frac{-R\ddot{\theta}}{r}\right) = -mR\ddot{\theta} \tag{5.143}$$

Substituting Equation 5.143 into Equation 5.141, we have

$$mR\ddot{\theta} = -mg\sin\theta - mR\ddot{\theta}$$

$$\ddot{\theta} + \left(\frac{g}{2R}\right)\sin\theta = 0$$

For small angles, this becomes

$$\ddot{\theta} + \left(\frac{g}{2R}\right)\theta = 0 \tag{5.144}$$

Use Equation 5.144 to show that the motion continues indefinitely.

Case (b). A number of nearly invisible effects eventually cause the motion pictured in Figure 5.15 to cease. One possibility is skidding, but we will analyze here another potential contributor. Inelastic deformation of the cylinder or platform surface, or the wearing down of surface irregularities, as shown in Figure 5.16, can cause an additional moment on the cylinder [4].

Considering these effects, on average after rolling over many small irregularities in only an instant of time, the moment is given by

$$\tau = rF_{rf} - \rho N \, \text{sgn}(\dot{\psi}) = -r\mu mg \cos\theta \, \text{sgn}(\dot{\theta}) - \rho mg \cos\theta \, \text{sgn}(\dot{\psi}) \tag{5.145}$$

where:

$\rho$ is the average moment arm for the offset normal force
$\text{sgn}(\alpha)$ denotes "sign of $\alpha$"

Show that Equations 5.138, 5.139, 5.141, 5.142, and 5.145 lead to

$$\ddot{\theta} + \left(\frac{g}{2R}\right)\sin\theta = -\left(\frac{g}{2R}\right)\left(\frac{\rho}{r}\right)\cos\theta \, \text{sgn}(\dot{\theta}) \tag{5.146}$$

Let the initial conditions be $\theta(0) = \theta_0 > 0$ and $\dot{\theta}(0) = 0$. Assuming small angles and also that $\rho / r \ll 1$, show that, at the end of half-cycle $n$,

$$\theta(t_n) = (-1)^n \left[\theta_0 - 2n\left(\frac{\rho}{r}\right)\right] \tag{5.147}$$

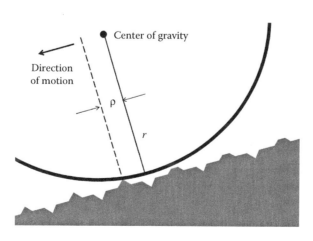

**FIGURE 5.16** Illustrating the mechanism for an offset normal force on a cylinder rolling to the left. (Vertical scale of surface irregularities exaggerated.)

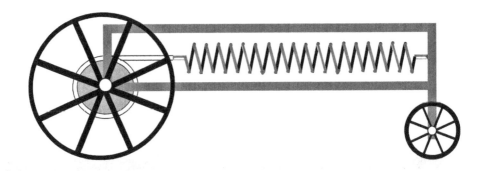

**FIGURE 5.17**   Spring-powered vehicle.

Use Equations 5.141 and 5.147 to show that after half-cycle $m$, where

$$\tan|\theta(t_m)| \leq \mu$$

gravity is unable to overcome friction and motion stops.

**CP5.2** Consider a vehicle on flat ground powered by an internal spring with spring constant $k$, as shown in Figure 5.17. One end of the spring is fixed to the front of the vehicle and the other tied to a cord that spools around a drum fixed to the rear axle. The rear wheels are wound backward to extend the spring an amount $z_0$ from its equilibrium position. When the wheels are released the spring pulls on the cord, which pulls on the drum and applies a moment (torque) to the rear wheels. The length of the cord is such that it breaks free of the drum as the spring reaches its equilibrium position.

Assume that (1) each wheel rolls without slipping and experiences rolling friction given by $F = \mu N$, where $\mu$ is the coefficient of friction and $N$ is the normal force that wheel bears; (2) vehicle velocity is low enough that air resistance can be ignored; and (3) the cord wound around the rear axle drum does not slip. The challenge: In terms of parameter definitions in Table 5.3, determine the distance $x$ traveled by the vehicle as a function of time after release. Also, show that, after the cord has broken free of the drum, the vehicle continues indefinitely at constant speed.

*Guidance*: The vehicle's motion along the ground is governed by Newton's law:

$$m\ddot{x} = \sum \text{forces} \tag{5.148}$$

The horizontal forces on the vehicle are provided by the ground. We can write

$$\sum \text{forces} = F_R + F_F \tag{5.149}$$

The wheels' motion is governed by the rotational version of Newton's law:

$$I_R\ddot{\psi}_R = \tau_R = -r_D k(z - z_0) - r_R F_R$$
$$I_F\ddot{\psi}_F = \tau_F = -r_F F_F \tag{5.150}$$

---

**TABLE 5.3**

**Parameter Definitions**

| | |
|---|---|
| $I_R$ | Moment of inertia of rear wheels, axle, and drum |
| $r_R$ | Radius of rear wheels |
| $r_D$ | Radius of drum about which the cord is wrapped |
| $I_F$ | Moment of inertia of front wheels and axle |
| $r_F$ | Radius of front wheels |
| $F_R$ | Force on rear wheels |
| $\tau_R$ | Torque on rear wheels, axle, and drum |
| $F_F$ | Force on front wheels |
| $\tau_F$ | Torque on front wheels and axle |
| $\psi_R$ | Angle of rotation in rear wheels |
| $\psi_F$ | Angle of rotation in front wheels |
| $z$ | Change in extension of spring after release |

---

Now, since the wheels are rolling without slipping,

$$r_R \psi_R = x$$
$$r_F \psi_F = x \tag{5.151}$$

Finally, since the cord does not slip on the rear axle drum:

$$z = r_D \psi_R \tag{5.152}$$

Use Equations 5.150, 5.151, and 5.152 to substitute for the forces $F_R$ and $F_F$ in Equations 5.148 and 5.149 and arrive at

$$m_E \ddot{x} + \left(\frac{r_D}{r_R}\right)^2 kx = \left(\frac{r_D}{r_R}\right) kz_0 \tag{5.153}$$

where:

$$m_E = m + I_R/r_R^2 + I_F/r_F^2$$

Solve Equation 5.153 to determine the distance traveled as a function of time, while the spring is engaged. Recalling that the cord breaks free when $z = z_0$, show that, at that moment,

$$\dot{x} = z_0 \sqrt{\frac{k}{m_E}}$$

To show that the vehicle continues indefinitely at this speed, set $k = 0$ in Equation 5.153.

**CP5.3** Figure 5.18 depicts a simple frictionless pendulum, a mass at the end of a taut string or rod, which we model as without mass. The pendulum's equation of motion can be derived from Newton's second law of motion. The pendulum bob

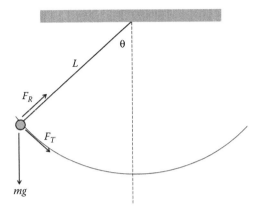

**FIGURE 5.18**   Pendulum at large angles.

experiences forces due to gravity and due to the tension of the string. The force
due to the string is entirely radial. The only tangential force component is that due
to gravity: $F_T = -mg \sin\theta$. The bob's position along its arc of travel is given by $L\theta$.
From force equals mass times acceleration:

$$m\frac{d^2}{dt^2}(L\theta) = -mg\sin\theta$$

$$\ddot{\theta} + \left(\frac{g}{L}\right)\sin\theta = 0 \qquad\qquad (5.154)$$

For small angles one can approximate $\sin\theta \cong \theta$, and the equation becomes

$$\ddot{\theta} + \left(\frac{g}{L}\right)\theta = 0$$

which is easily solved. For the initial conditions $\theta(0) = \theta_0$ and $\dot{\theta}(0) = 0$ the solution is

$$\theta(t) = \theta_0 \cos\omega t \qquad\qquad (5.155)$$

where $\omega = \sqrt{g/L}$.

In this problem we endeavor to construct an approximate solution to Equation 5.154
that is more accurate than Equation 5.155 for larger angles. We will employ a slight
adaptation of the method of successive approximations for almost linear systems
introduced in Chapter 3. We can rewrite Equation 5.154 as

$$\ddot{\theta} + \omega^2\theta + \omega^2(\sin\theta - \theta) = 0 \qquad\qquad (5.156)$$

Our first approximation is from Equation 5.155, the linear solution $\theta_1(t) = \theta_0 \cos\omega t$.
The second will derive from first approximating $\sin\theta \cong \theta - \theta^3/6$ in Equation 5.156
and then approximating the resulting equation with

$$\ddot{\theta}_2 + \omega^2\theta_2 = \omega^2\theta_1^3/6 \qquad\qquad (5.157)$$

where $\theta_1(t) = \theta_0 \cos \omega t$ is obviously a known function of time. With the initial conditions $\theta(0) = \theta_0$ and $\dot{\theta}(0) = 0$ the solution of Equation 5.157 can be written as

$$\theta_2(t) = \theta_1(t) + \int_0^t K(t-\tau) \left( \frac{\omega^2 \theta_1^3(\tau)}{6} \right) d\tau \qquad (5.158)$$

where $K(t-\tau)$ is the kernel for Equation 5.157.

Part(a): Show that the solution of Equation 5.158 is

$$\theta_2(t) = \theta_1(t) + \psi(t) \qquad (5.159)$$

where:

$$\psi(t) = \frac{\theta_0^3}{48}(3\omega t + \cos \omega t \sin \omega t) \sin \omega t \qquad (5.160)$$

The recursive formula

$$\int_0^t \cos^n(\tau)d\tau = \frac{1}{n}\cos^{n-1}(\tau)\sin(\tau)\Big|_0^t + \left(\frac{n-1}{n}\right)\int_0^t \cos^{n-2}(\tau)d\tau \qquad (5.161)$$

is useful.

Part(b): For the third approximation we use $\sin \theta \cong \theta - \theta^3/6 + \theta^5/120$ in Equation 5.156 and then write

$$\theta_3(t) = \theta_1(t) + \int_0^t K(t-\tau)\left(\frac{\omega^2 \theta_2^3(\tau)}{6} - \frac{\omega^2 \theta_2^5(\tau)}{120}\right)d\tau \qquad (5.162)$$

Show that, if we combine terms involving $\theta_0^n$ for similar values of the power $n$ and retain only powers of $n \le 5$, Equation 5.162 reduces to

$$\theta_3(t) = \theta_2(t) + \int_0^t K(t-\tau)\left(\frac{\omega^2 \theta_1^2(\tau)\psi(\tau)}{2} - \frac{\omega^2 \theta_1^5(\tau)}{120}\right)d\tau \qquad (5.163)$$

Integration of Equation 5.163 and subsequent simplification is straightforward but tedious. You may prefer to simply accept that the result is

$$\theta_3(t) = \theta_2(t) + \Delta_3(t) \qquad (5.164)$$

where:

$$\Delta_3(t) = \left(\frac{\theta_0^5}{768}\right)(\varphi_A(t) + \varphi_B(t))$$

$$\varphi_A(t) = 3\omega t\left(-\left(\frac{1}{2}\right)\omega t \cos \omega t + \sin \omega t\left(\left(\frac{1}{4}\right) - \cos^2 \omega t\right)\right) \qquad (5.165)$$

$$\varphi_B(t) = \cos \omega t\left(\left(\frac{23}{20}\right) - \left(\frac{7}{4}\right)\cos^2 \omega t + \left(\frac{3}{5}\right)\cos^4 \omega t\right)$$

Part(c): Perform a numerical analysis of the error involved in each of these approximations. Use a numerical integration routine to compute accurately the exact solution of Equation 5.154. Call that solution $\theta_C(t)$. For simplicity let $\omega = 1$. For $\theta_0 = \pi/6$, $\pi/4$, and $\pi/3$, calculate the following measures of accuracy for each approximation:

1. Over one full cycle of the motion of $\theta_C(t)$, that is, over its true period $T_C$, the average absolute error:

$$\frac{1}{T_C} \int_0^{T_C} \left| \theta_n(t) - \theta_C(t) \right| dt.$$

2. Over one full cycle of the motion of $\theta_C(t)$, that is, over its true period $T_C$, the maximum absolute error:

$$\max \left| \theta_n(t) - \theta_C(t) \right| \quad 0 < t < T_C$$

3. The error in the period for the first cycle: $\left| T_n - T_C \right|$.
4. The error in the value of angular position at the end of the first cycle: $\left| \theta_n(T_n) - \theta_C(T_C) \right|$.

Show that, for each of the three initial values $\theta_0$ considered, (a) no approximation is superior in all measures and each approximation is superior in at least one measure; (b) performance in the first two measures improves markedly as $n$ increases and degrades as the initial value $\theta_0$ increases.
Note that $\theta_2(t)$ and $\theta_3(t)$ are not as accurate for later cycles as they are for the first. It is an artifact of the approach that these approximate solutions are not periodic. In practice, given that one can determine from physical considerations that the motion must be periodic, one could simply repeat first-cycle values in later cycles. Or, better yet, to avoid discontinuities, for

$$kT_n < t < (k+1)T_n,$$

$$\theta(t) = \theta(T_n - \delta t) \text{ for } k \text{ odd}$$

$$\theta(t) = \theta(\delta t) \text{ for } k \text{ even, where}$$

$$\delta t = t - kT_n$$

# 6 Higher-Order Linear Ordinary Differential Equations

This chapter discusses linear ordinary differential equations (ODEs) of order higher than two. The general form is

$$\frac{d^n y}{dt^n} + a_1(t)\frac{d^{n-1}y}{dt^{n-1}} + \ldots + a_{n-1}(t)\frac{dy}{dt} + a_n(t)y = g(t)$$

The chapter begins by examining an important application and then presents general results. We will find many similarities to second-order equations and so will summarize these results succinctly. Chapter 8 introduces another, more powerful, approach to representing higher-order linear systems and we will devote much more attention to that methodology.

## 6.1 EXAMPLE: SATELLITE ORBIT DECAY

Figure 6.1 depicts a satellite in low Earth orbit (roughly 200–400 miles or 320–640 km altitude). The primary force on the satellite, of course, is Earth's gravity. However, although the atmosphere is very thin at these altitudes, the long-term effect of air resistance (drag) on the satellite cannot be ignored.

The satellite's equations of motion are more easily written in Cartesian $(x, y)$ coordinates and more easily solved in polar $(R, \theta)$ coordinates. From Newton's second law of motion, $F = ma$, or $a = F/m$, the equations of motion in Cartesian coordinates are

$$\ddot{x} = -\frac{GM}{(x^2 + y^2)^{3/2}}x - \frac{\rho S C_D}{2m}\left(\dot{x}^2 + \dot{y}^2\right)^{1/2}\dot{x}$$

$$\ddot{y} = -\frac{GM}{(x^2 + y^2)^{3/2}}y - \frac{\rho S C_D}{2m}\left(\dot{x}^2 + \dot{y}^2\right)^{1/2}\dot{y}$$

(6.1)

where:
  $G$ is the gravitational constant
  $M$ is the mass of the earth
  $\rho$ is the air density
  $S$ is the satellite's cross-sectional area
  $C_D$ is the satellite's drag coefficient
  $m$ is the satellite's mass

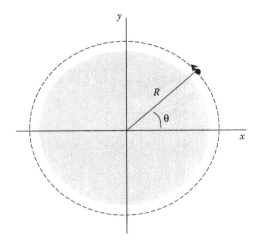

**FIGURE 6.1** Satellite in low Earth orbit.

Challenge Problem CP6.1 guides the student through the transformation of Equations 6.1 into their polar form:

$$\ddot{R} = -\frac{GM}{R^2} + R\dot{\theta}^2 - \frac{\rho S C_D \dot{R}}{2m}\sqrt{\left(R\dot{\theta}\right)^2 + \dot{R}^2}$$

$$R\ddot{\theta} = -2\dot{R}\dot{\theta} - \frac{\rho S C_D R\dot{\theta}}{2m}\sqrt{\left(R\dot{\theta}\right)^2 + \dot{R}^2}$$

(6.2)

We now assume that the air density $\rho$ is small enough that the effects of drag can be considered a perturbation about a nominally circular zero-drag trajectory. Setting $\rho = 0$ in Equation 6.2, the nominal trajectory is thus defined by the equations

$$0 = -\frac{GM}{R_0^2} + R_0\dot{\theta}_0^2$$

$$0 = -2\dot{R}_0\dot{\theta}_0$$

(6.3)

From Equation 6.3,

$$\dot{\theta}_0^2 = \frac{GM}{R_0^3}$$

$$\dot{R}_0 = 0$$

(6.4)

For simplicity let $\dot{\theta} = \omega$. Now consider small motions about the nominal trajectory:

$$\omega = \omega_0 + \nu$$

$$R = R_0 + r$$

(6.5)

How does air density vary with altitude? Measurement and theory show that it varies roughly exponentially about a given altitude:

$$\rho(R) = \rho(R_0)\exp(-Z(R - R_0))$$ (6.6)

References [5–7] enable us to determine typical values for the key quantities

$$\rho_0 = \rho(R_0)$$ (6.7)

$$Z = -\frac{1}{\rho_0}\left(\frac{d\rho}{dR}\right)_0$$ (6.8)

and

$$k = \frac{\rho_0 S C_D}{2m}$$ (6.9)

The next steps are not conceptually difficult but they take time and involve equations that are somewhat lengthy. Given the exceptionally simple conclusion we reach, it seems preferable to sidestep details here and move quickly to the bottom line. The details are available in Challenge Problems CP6.2, 6.3, and 6.4. We satisfy ourselves here with a summary of the steps.

First, we linearize Equations 6.2, yielding two linear time-invariant equations in the variables $r$ and $v$. (Recall Equations 6.5.) We then use these two equations to eliminate $v$ and obtain a third-order ODE for $r$, the variation in altitude, alone. The ODE is nonhomogeneous but the input is a constant, which allows us to transform the ODE into a homogeneous one by simply defining a new dependent variable $r_* = r - 1/Z$ where $Z$ is as given in Equation 6.8. For notational simplicity we next define a new independent variable $t_* = \omega_0 t$. Finally, we calculate values for the equation's parameters and determine that a number of terms in the equation can safely be neglected.

The bottom line is

$$\dddot{r}_* + \dot{r}_* - \gamma r_* = 0$$ (6.10)

where:

$$\gamma = 2kZR_0^2$$ (6.11)

and the derivatives are understood to be with respect to the nondimensional time $t_* = \omega_0 t$. Initial conditions are

$$r_*(0) = -1/Z$$

$$\dot{r}_*(0) = 0$$ (6.12)

$$\ddot{r}_*(0) = 0$$

To solve Equation 6.10 we try $r_*(t_*) = \exp(pt_*)$ and find the characteristic equation

$$p^3 + p - \gamma = 0 \tag{6.13}$$

Because $\gamma \ll 1$, we can quickly obtain an approximate factorization of the cubic polynomial in Equation 6.13:

$$(p - \gamma)(p^2 + \gamma p + 1) = p^3 + (1 - \gamma^2)p - \gamma \cong p^3 + p - \gamma \tag{6.14}$$

Hence the roots of the characteristic equation are approximately

$$p_1 = \gamma \qquad p_2 = -\frac{\gamma}{2} + i \qquad p_3 = -\frac{\gamma}{2} - i \tag{6.15}$$

and the corresponding fundamental solutions are

$$r_{*1}(t_*) = \exp(\gamma t_*)$$
$$r_{*2}(t_*) = \exp(-\gamma t_*/2)\cos t_* \tag{6.16}$$
$$r_{*3}(t_*) = \exp(-\gamma t_*/2)\sin t_*$$

Expressing

$$r_* = c_1 r_{*1} + c_2 r_{*2} + c_3 r_{*3} \tag{6.17}$$

and selecting the coefficients $c_1, c_2$, and $c_3$ to satisfy Equation 6.12, we find that, approximately,

$$r_*(t_*) = -\frac{1}{Z}\left(\exp(\gamma t_*) - \gamma \exp(-\gamma t_*/2)\sin t_* + 2\gamma^2 \exp(-\gamma t_*/2)\cos t_*\right) \tag{6.18}$$

Returning to dimensional time, the satellite's deviation in altitude from a circular orbit is approximated by

$$r(t) = -\frac{1}{Z}\left(\exp(\gamma\omega_0 t) - 1 - \gamma \exp(-\gamma\omega_0 t/2)\sin \omega_0 t + 2\gamma^2 \exp(-\gamma\omega_0 t/2)\cos \omega_0 t\right) \tag{6.19}$$

where, again,

$$Z = -\frac{1}{\rho_0}\left(\frac{d\rho}{dR}\right)_0$$
$$\gamma = 2kZR_0^2$$
$$\omega_0 = \sqrt{GM/R^3} \tag{6.20}$$
$$k = \frac{\rho_0 SC_D}{2m}$$

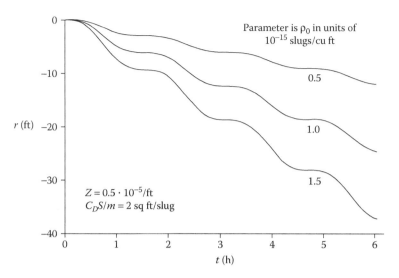

**FIGURE 6.2** Deviation in altitude from a circular orbit for a typical small satellite launched into an initial 300 statute mile orbit, with atmospheric density at the initial altitude as a parameter.

The terms involving $\sin\omega_0 t$ and $\cos\omega_0 t$ are needed to satisfy initial conditions but they are transients that become negligible in short order with respect to the first term, and then the satellite's deviation in altitude from a circular orbit is approximated by

$$r(t) = -\frac{1}{Z}\left(\exp(\gamma\omega_0 t)-1\right) \tag{6.21}$$

Equation 6.21 shows that a low earth orbit is unstable; the satellite will begin to fall at an accelerating rate. Figure 6.2 considers the case of a typical satellite launched into an initial orbit of 300 statute miles above the Earth's surface. The figure graphs the deviation in altitude as a function of time, with atmospheric density at the initial altitude as a parameter. Due to solar variations, space weather is much more variable than weather on Earth's surface. The values selected for density are representative, but higher and lower densities can occur.

Our analysis has revealed the satellite's initial motions. To determine its lifetime in orbit we would have to solve the full nonlinear equations, which must be done numerically, and input a more complete model of air density dependence on altitude and its variation in time.

## 6.2 HIGHER-ORDER HOMOGENEOUS LINEAR EQUATIONS

Consider the general nth order *homogeneous* linear ODE:

$$\frac{d^n y}{dt^n} + a_1(t)\frac{d^{n-1}y}{dt^{n-1}} + \ldots + a_n(t)y = 0 \tag{6.22}$$

If the functions $y_1(t), y_2(t), \ldots, y_{n-1}(t), y_n(t)$ are solutions to Equation 6.22 and their Wronskian

$$W(t) = \begin{vmatrix} y_1 & y_2 & .. & y_{n-1} & y_n \\ \dot{y}_1 & \dot{y}_2 & .. & \dot{y}_{n-1} & \dot{y}_n \\ .. & .. & .. & .. & .. \\ y_1^{(n-2)} & y_2^{(n-2)} & .. & y_{n-1}^{(n-2)} & y_n^{(n-2)} \\ y_1^{(n-1)} & y_2^{(n-1)} & .. & y_{n-1}^{(n-1)} & y_n^{(n-1)} \end{vmatrix} \tag{6.23}$$

is nonzero then any solution to Equation 6.22 has the form

$$y(t) = c_1 y_1(t) + c_2 y_2(t) + ... + c_{n-1} y_{n-1}(t) + c_n y_n(t) \tag{6.24}$$

Consider the general nth order linear *time-invariant* homogeneous ODE

$$\frac{d^n y}{dt^n} + a_1 \frac{d^{n-1} y}{dt^{n-1}} + a_2 \frac{d^{n-2} y}{dt^{n-2}} + ... + a_n y = 0 \tag{6.25}$$

Trying the solution $y(t) = e^{rt}$ we derive the characteristic equation for Equation 6.25:

$$r^n + a_1 r^{n-1} + a_2 r^{n-2} + ... + a_n = 0 \tag{6.26}$$

Roots to Equation 6.26 can have four different patterns:

1. Distinct real roots (appearing only once)
2. Repeated real roots
3. Distinct conjugate pairs
4. Repeated conjugate pairs

Each of these patterns of roots corresponds to a different form of solution. A distinct real root corresponds to a solution of the form $y(t) = c_1 e^{rt}$. If a real root $r_i$ appears m times it represents the solution $y(t) = (c_1 + c_2 t + ... + c_m t^{m-1}) e^{rt}$. If the conjugate pair $r = \lambda \pm i\omega$ appears only once it represents the solution $c_1 e^{\lambda t} \cos \omega t + c_2 e^{\lambda t} \sin \omega t$. If the conjugate pair $r = \lambda \pm i\omega$ appears m times it corresponds to the solution

$$y(t) = \left( c_1 + c_2 t + ... + c_m t^{m-1} \right) e^{\lambda t} \cos \omega t + \left( c_{m+1} + c_{m+2} t + ... + c_{2m} t^{m-1} \right) e^{\lambda t} \sin \omega t$$

Coefficients are determined from equations for the initial conditions: $n$ linear equations in $n$ unknowns, providing a unique solution because the determinant of the equations (the Wronskian) is nonzero:

$$c_1 y_1(t_0) + c_2 y_2(t_0) + ... + c_n y_n(t_0) = y_0$$

$$c_1 \dot{y}_1(t_0) + c_2 \dot{y}_2(t_0) + ... + c_n \dot{y}_n(t_0) = \dot{y}_0$$

$$c_1 \ddot{y}_1(t_0) + c_2 \ddot{y}_2(t_0) + ... + c_n \ddot{y}_n(t_0) = \ddot{y}_0 \tag{6.27}$$

$$...$$

$$c_1 y_1^{(n-1)}(t_0) + c_2 y_2^{(n-1)}(t_0) + ... + c_n y_n^{(n-1)}(t_0) = y_0^{(n-1)}$$

## 6.3 HIGHER-ORDER NONHOMOGENEOUS LINEAR EQUATIONS

As was the case for second-order linear time-invariant ODEs, three methods are available to solve *nonhomogeneous* higher-order linear time-invariant problems. The next two sections describe two of them. Chapter 7 presents the third.

### 6.3.1 VARIATION OF PARAMETERS (KERNEL) METHOD

Consider the general higher-order nonhomogeneous linear ODE ( $g(t) \neq 0$ ):

$$\frac{d^n y}{dt^n} + a_1(t)\frac{d^{n-1}y}{dt^{n-1}} + \ldots + a_{n-1}(t)\frac{dy}{dt} + a_n(t)y = g(t) \tag{6.28}$$

with initial conditions stated at $t = t_0$ as in Equation 6.27. The general solution to Equation 6.30 can be written as

$$y(t) = c_1 y_1 + c_2 y_2 + \ldots + c_{n-1}y_{n-1} + c_n y_n + y_P$$

where:

$y_i$ are fundamental solutions to the homogeneous ODE Equation 6.25
$c_i$ satisfy Equation 6.27
$y_P$ is a solution to Equation 6.28 with zero initial conditions
$y_P(t)$ can be written as

$$y_P(t) = \int_{t_0}^{t} K(t,\tau)g(\tau)d\tau \tag{6.29}$$

where the kernel $K(t,\tau)$ is given by

$$K(t,\tau) = \sum_{i=1}^{n} y_i(t)W_i(\tau)/W(\tau) \tag{6.30}$$

In Equation 6.30, $W(\tau)$ is the Wronskian (Equation 6.23) and $W_i(\tau)$ is the Wronskian with ith column replaced with

$$\begin{pmatrix} 0 \\ 0 \\ \ldots \\ 0 \\ 1 \end{pmatrix}$$

For linear time-invariant ODEs Equation 6.29 has the form

$$y_P(t) = \int_{t_0}^{t} K(t-\tau)g(\tau)d\tau$$

For the third-order time-invariant equation for satellite orbit decay (Equation 6.10) and its three fundamental solutions

$$y_1(t) = \exp(\gamma\omega_0 t)$$

$$y_2(t) = \exp\left(-\gamma\omega_0 t/2\right)\cos\omega_0 t$$

$$y_3(t) = \exp\left(-\gamma\omega_0 t/2\right)\sin\omega_0 t$$

the kernel, as determined by Equation 6.30, is

$$K(t-\tau) = \left(y_1(t-\tau) - y_2(t-\tau) - \left(3\gamma/2\right)y_3(t-\tau)\right)\Big/\omega_0^2 \tag{6.31}$$

where we have again approximated using the fact that $\gamma \ll 1$. The calculation using Equation 6.30 can be laborious. Chapter 7 will offer a quicker and easier method of determining the kernel for higher-order linear time-invariant ODEs.

### 6.3.2 UNDETERMINED COEFFICIENTS METHOD

Just as it was for second-order linear time-invariant ODEs, the undetermined coefficients method is a comparatively simple approach to try to solve higher-order equations; Table 5.1 applies for higher-order equations as well as for second order. The limitations of the method expressed in Section 5.3.3.6 also apply to higher-order ODEs.

> **Example: Find the steady-state solution to the general fourth-order linear time-invariant ODE with a sine-wave input:**
>
> $$\frac{d^4y}{dt^4} + a_1\frac{d^3y}{dt^3} + a_2\frac{d^2y}{dt^2} + a_3\frac{dy}{dt} + a_4y = a_5\sin\omega t \tag{6.32}$$

Following Table 5.1, we try $y(t) = C\sin\omega t + D\cos\omega t$. Substituting directly and rearranging, we find

$$\left[-\left(a_1\omega^3 - a_3\omega\right)C + \left(\omega^4 - a_2\omega^2 + a_4\right)D\right]\cos\omega t$$

$$+ \left[\left(\omega^4 - a_2\omega^2 + a_4\right)C + \left(a_1\omega^3 - a_3\omega\right)D\right]\sin\omega t = a_5\sin\omega t$$

We must have

$$-\left(a_1\omega^3 - a_3\omega\right)C + \left(\omega^4 - a_2\omega^2 + a_4\right)D = 0$$

$$\left(\omega^4 - a_2\omega^2 + a_4\right)C + \left(a_1\omega^3 - a_3\omega\right)D = a_5$$

This is two linear equations in two unknowns. The solutions are

$$C = \left(\frac{\omega^4 - a_2\omega^2 + a_4}{\Delta}\right)a_5$$

$$D = \left(\frac{a_1\omega^3 - a_3\omega}{\Delta}\right)a_5$$

$$\Delta = \left(\omega^4 - a_2\omega^2 + a_4\right)^2 + \left(a_1\omega^3 - a_3\omega\right)^2$$

As we did for solutions to first- and second-order equations with a sine-wave input, we seek a steady-state solution to Equation 6.32 in the form

$$y(t) = G\sin(\omega t - \theta) \tag{6.33}$$

Defining

$$\cos\theta = \frac{C}{\sqrt{C^2 + D^2}} \qquad \sin\theta = \frac{-D}{\sqrt{C^2 + D^2}}$$

we determine that

$$G = \frac{a_5}{\sqrt{\left(\omega^4 - a_2\omega^2 + a_4\right)^2 + \left(a_1\omega^3 - a_3\omega\right)^2}} \tag{6.34}$$

$$\theta = \arctan\left(-a_1\omega^3 + a_3\omega, \omega^4 - a_2\omega^2 + a_4\right) \tag{6.35}$$

## 6.4 EXISTENCE AND UNIQUENESS

**Theorem:**

Let the functions $a_1(t), a_2(t), \ldots, a_n(t), g(t)$ be continuous on the open interval $I$ containing $t_0$ and let initial conditions be defined there as in Equation 6.27. Then the ODE

$$\frac{d^n y}{dt^n} + a_1(t)\frac{d^{n-1}y}{dt^{n-1}} + \ldots + a_n(t)y = g(t) \tag{6.36}$$

has exactly one solution and that solution exists throughout $I$. (Stated without proof. See [3]).

**Corollary:**

If the functions $a_1(t), a_2(t), ..., a_n(t), g(t)$ are continuous everywhere then Equation 6.36 has a unique solution at every instant of time. In particular, linear time-invariant equations have unique solutions at every instant of time.

## 6.5 STABILITY OF HIGHER-ORDER SYSTEMS

We saw in Sections 2.3 and 5.1.5 that first- and second-order linear time-invariant systems are stable if all coefficients in the ODE are positive and unstable if any is negative. For higher-order linear time-invariant systems it is also true that if any coefficient is negative then the system is unstable. However, a higher-order system may have all positive coefficients in the ODE and yet be unstable. As a concrete example consider the characteristic polynomial

$$(r + a)(r^2 - r + b) = r^3 + (a-1)r^2 + (b-a)r + ab$$

All coefficients in the cubic are positive if $b > a > 1$. But the quadratic factor has a negative coefficient, hence the system is unstable.

A linear time-invariant system is stable and damped if all the roots of its characteristic polynomial have negative real parts.

*Routh's criteria* provide a means of determining whether a system is stable without solving for the roots, but with web-based "root calculators" available at our fingertips it is typically more practical to determine the roots.

## 6.6 SUMMARY

This chapter has considered linear ODEs of order greater than two. The general form is

$$\frac{d^n y}{dt^n} + a_1(t)\frac{d^{n-1} y}{dt^{n-1}} + ... + a_n(t)y = g(t)$$

where $n > 2$. Linear ODEs of order greater than two have many similarities to second-order linear ODEs and we will devote more attention to higher-order systems when we introduce a more powerful representation in Chapter 8. Hence this chapter itself, beyond the examples, has been a succinct summary. It may suffice, then, to limit our discussion here to highlighting the chief differences between linear ODEs of order two and those of higher order.

For linear time invariant ODEs of order higher than two, when a solution $r$ to the characteristic equation has multiplicity $m$, there are $m$ fundamental solutions $y_k(t)$ to the ODE with the form $t^k e^{rt}$, where $0 \le k \le m-1$. This includes the case where $r$ is complex.

The kernel for higher-order linear ODEs is more complicated to calculate than that for second-order ODEs. For a linear ODE of order $n$, the kernel is given by

$$K(t, \tau) = \sum_{i=1}^{n} y_i(t)W_i(\tau)/W(\tau)$$

where $y_i(t)$ are the ODE's $n$ fundamental solutions, the Wronskian

$$W(\tau) = \begin{vmatrix} y_1(\tau) & y_2(\tau) & .. & y_{n-1}(\tau) & y_n(\tau) \\ \dot{y}_1(\tau) & \dot{y}_2(\tau) & .. & \dot{y}_{n-1}(\tau) & \dot{y}_n(\tau) \\ .. & .. & .. & .. & .. \\ y_1^{(n-2)}(\tau) & y_2^{(n-2)}(\tau) & .. & y_{n-1}^{(n-2)}(\tau) & y_n^{(n-2)}(\tau) \\ y_1^{(n-1)}(\tau) & y_2^{(n-1)}(\tau) & .. & y_{n-1}^{(n-1)}(\tau) & y_n^{(n-1)}(\tau) \end{vmatrix},$$

$y_j^{(k)}(\tau)$ is the $k$th derivative of the $j$th fundamental solution, and $W_i(\tau)$ is the Wronskian with $i$th column replaced with

$$\begin{pmatrix} 0 \\ 0 \\ ... \\ 0 \\ 1 \end{pmatrix}$$

Chapter 7 will provide a quicker and easier way to calculate the kernel for time-invariant higher-order systems.

For linear time-invariant ODEs, it remains true for higher order ODEs, as it did for first and second order ones, that if one or more of the coefficients in the ODE is negative then the system is unstable. However, unlike in first- and second-order ODEs, all the coefficients in a higher-order ODE may be positive and yet the system be unstable. Solutions to a higher-order ODE are stable and damped if all the roots of its characteristic equation have negative real parts.

## Solved Problems

**SP6.1** Solve:

$$\frac{d^3y}{dt^3} + 8\frac{d^2y}{dt^2} + 45\frac{dy}{dt} + 116y = 0 \tag{6.37}$$

$$y(0) = 2 \quad \dot{y}(0) = -1 \quad \ddot{y}(0) = 3$$

*Solution*:
Since this is a linear time-invariant equation, we try $y(t) = e^{rt}$, resulting in the characteristic equation

$$r^3 + 8r^2 + 45r + 116 = 0 \tag{6.38}$$

There is an algebraic formula for factoring a third-order polynomial but it is cumbersome and engineers seldom use it. There are handy "root calculators" on the web, but it is useful for engineers to know how to solve higher-order polynomial equations by hand, and we will do so here. Equation 6.38 must have at least one real root. How do we know this? Because the roots of a polynomial with real coefficients must be either real or occur in complex conjugate pairs (Recall Section 1.1.2.2). The only possibilities for the solutions of Equation 6.38 are (a) three real roots or (b) one real root and one complex conjugate pair. We search for a real root by a logical trial and error process. Call $P_3(r) = r^3 + 8r^2 + 45r + 116$. We evaluate $P_3(r)$ for various values of $r$. The easiest value for which to evaluate $P_3(r)$ is $r = 0$. We find $P_3(0) = 116$. Values of $r$ greater than zero would only increase $P_3(r)$, so we must examine $r < 0$. Trying $r = -1$ yields $P_3(-1) = 78$, which is in the right direction (i.e., better than 116) but still far from zero. We try $r = -10$ and find $P_3(-10) = -534$, a change of sign. Because $P_3(r)$ is a continuous function, somewhere between $r = -1$ and $r = -10$ lies at least one root. (There could be three.) Trying $r = -5$ yields $P_3(-5) = -34$, so there is at least one real root between $r = -1$ and $r = -5$. Trying $r = -4$ yields $P_3(-4) = 0$ and we have found a root; that is, $r = -4$ is a solution of $P_3(r) = r^3 + 8r^2 + 45r + 116 = 0$. We can employ the *Newton–Raphson* method when, as is the usual case, the roots are decimal numbers, not integers, and the root must be accurate to at least several decimal places.

Once the first root $r_1$ is obtained, how do we proceed? We divide through, forming a polynomial

$$P_2(r) = \frac{P_3(r)}{r - r_1}$$

We do this by long division:

$$
\begin{array}{r}
r^2 + 4r + 29 \\
r + 4 \overline{)\, r^3 + 8r^2 + 45r + 116} \\
\underline{r^3 + 4r^2} \\
4r^2 + 45r \\
\underline{4r^2 + 16r} \\
29r + 116
\end{array}
$$

Hence $P_2(r) = r^2 + 4r + 29$. Completing the square, we see that

$$P_2(r) = r^2 + 4r + 29 = (r + 2)^2 + 5^2$$

and so

$$P_3(r) = r^3 + 8r^2 + 45r + 116 = (r + 4)((r + 2)^2 + 5^2)$$

and the roots of Equation 6.38 are $r_1 = -4$, $r_2 = -2 + i5$, and $r_2 = -2 - i5$. Hence the fundamental solutions to Equation 6.37 are $y_1(t) = e^{-4t}$, $y_2(t) = e^{-2t} \cos 5t$, and $y_3(t) = e^{-2t} \sin 5t$ and the general solution to Equation 6.37 is

$$y(t) = c_1 y_1(t) + c_2 y_2(t) + c_3 y_3(t) = c_1 e^{-4t} + c_2 e^{-2t} \cos 5t + c_3 e^{-2t} \sin 5t \qquad (6.39)$$

Now

$$\frac{dy}{dt} = -4c_1 e^{-4t} - 2c_2 e^{-2t} \cos 5t - 5c_2 e^{-2t} \sin 5t - 2c_3 e^{-2t} \sin 5t + 5c_3 e^{-2t} \cos 5t \qquad (6.40)$$

and

$$\frac{d^2 y}{dt^2} = 16c_1 e^{-4t} + 4c_2 e^{-2t} \cos 5t + 20c_2 e^{-2t} \sin 5t - 25c_2 e^{-2t} \cos 5t + 4c_3 e^{-2t} \sin 5t$$
$$- 20c_3 e^{-2t} \cos 5t - 25c_3 e^{-2t} \sin 5t \qquad (6.41)$$

From the initial conditions and Equations 6.39, 6.40, and 6.41,

$$y(0) = 2 = c_1 + c_2$$

$$\frac{dy}{dt}(0) = -1 = -4c_1 - 2c_2 + 5c_3 \qquad (6.42)$$

$$\frac{d^2 y}{dt^2}(0) = 3 = 16c_1 - 21c_2 - 20c_3$$

This is three linear equations in three unknowns. The solutions are

$$c_1 = 57/29$$

$$c_2 = 1/29$$

$$c_3 = 201/145$$

Hence the solution to Equation 6.37 is

$$y(t) = \left(\frac{57}{29}\right) e^{-4t} + \left(\frac{1}{29}\right) e^{-2t} \cos 5t + \left(\frac{201}{145}\right) e^{-2t} \sin 5t \qquad (6.43)$$

**SP6.2** Find the kernel $K(t - \tau)$ for the system governed by

$$\frac{d^3 y}{dt^3} + 16\frac{d^2 y}{dt^2} + 81\frac{dy}{dt} + 126y = g(t) \tag{6.44}$$

*Solution:*
Section 6.3.1 shows that the kernel derives from the system's fundamental solutions. Since Equation 6.44 is a linear time-invariant equation we find the fundamental solutions by trying the solution $y(t) = e^{rt}$ in the homogeneous equation

$$\frac{d^3 y}{dt^3} + 16\frac{d^2 y}{dt^2} + 81\frac{dy}{dt} + 126y = 0 \tag{6.45}$$

We quickly arrive at the characteristic equation

$$r^3 + 16r^2 + 81r + 126 = 0 \tag{6.46}$$

We could again employ the method used in Solved Problem SP6.1 to factor this polynomial. Alternatively, we could use a web-based "root calculator" and carefully check the result with hand calculation, and that is what we choose to do here. According to the web-based root-calculator, the solutions of Equation 6.46 are $r_1 = -3$, $r_2 = -6$, and $r_3 = -7$. How do we check this? The easiest way is to multiply the factors out:

$$(r + 3)(r + 6)(r + 7) = (r + 3)(r^2 + 13r + 42)$$

$$= r^3 + 13r^2 + 42r + 3r^2 + 39r + 126$$

$$= r^3 + 16r^2 + 81r + 126$$

Comparing this result to Equation 6.46 we see that the web-based root calculator has given us the correct result. Hence the fundamental solutions of Equation 6.45 are $y_1(t) = e^{-3t}$, $y_2(t) = e^{-6t}$ and $y_3(t) = e^{-7t}$
Now, from Equation 6.32,

$$K(t, \tau) = \sum_{i=1}^{3} y_i(t)W_i(\tau)/W(\tau) \tag{6.47}$$

and from Equation 6.26,

$$W(\tau) = \begin{vmatrix} y_1(\tau) & y_2(\tau) & y_3(\tau) \\ \dot{y}_1(\tau) & \dot{y}_2(\tau) & \dot{y}_3(\tau) \\ \ddot{y}_1(\tau) & \ddot{y}_2(\tau) & \ddot{y}_3(\tau) \end{vmatrix} = \begin{vmatrix} e^{-3\tau} & e^{-6\tau} & e^{-7\tau} \\ -3e^{-3\tau} & -6e^{-6\tau} & -7e^{-7\tau} \\ 9e^{-3\tau} & 36e^{-6\tau} & 49e^{-7\tau} \end{vmatrix}$$

$$= e^{-3\tau} \begin{vmatrix} -6e^{-6\tau} & -7e^{-7\tau} \\ 36e^{-6\tau} & 49e^{-7\tau} \end{vmatrix} - e^{-6\tau} \begin{vmatrix} -3e^{-3\tau} & -7e^{-7\tau} \\ 9e^{-3\tau} & 49e^{-7\tau} \end{vmatrix} + e^{-7\tau} \begin{vmatrix} -3e^{-3\tau} & -6e^{-6\tau} \\ 9e^{-3\tau} & 36e^{-6\tau} \end{vmatrix}$$

$$= e^{-3\tau} \Big[ (-6e^{-6\tau})(49e^{-7\tau}) - (36e^{-6\tau})(-7e^{-7\tau}) \Big] - e^{-6\tau} \Big[ (-3e^{-3\tau})(49e^{-7\tau}) - (9e^{-3\tau})(-7e^{-7\tau}) \Big]$$

$$+ e^{-7\tau} \Big[ (-3e^{-3\tau})(36e^{-6\tau}) - (9e^{-3\tau})(-6e^{-6\tau}) \Big]$$

$$= e^{-3\tau} e^{-6\tau} e^{-7\tau} \Big( [-294 + 252] - [-147 + 63] + [-108 + 54] \Big)$$

$$W(\tau) = -12 e^{-16\tau} \tag{6.48}$$

Similarly,

$$W_1(\tau) = \begin{vmatrix} 0 & y_2(\tau) & y_3(\tau) \\ 0 & \dot{y}_2(\tau) & \dot{y}_3(\tau) \\ 1 & \ddot{y}_2(\tau) & \ddot{y}_3(\tau) \end{vmatrix} = \begin{vmatrix} 0 & e^{-6\tau} & e^{-7\tau} \\ 0 & -6e^{-6\tau} & -7e^{-7\tau} \\ 1 & 36e^{-6\tau} & 49e^{-7\tau} \end{vmatrix} = e^{-6\tau}(-7e^{-7\tau}) - (-6e^{-6\tau})e^{-7\tau}$$

$$W_1(\tau) = -e^{-13\tau} \tag{6.49}$$

$$W_2(\tau) = \begin{vmatrix} y_1(\tau) & 0 & y_3(\tau) \\ \dot{y}_1(\tau) & 0 & \dot{y}_3(\tau) \\ \ddot{y}_1(\tau) & 1 & \ddot{y}_3(\tau) \end{vmatrix} = \begin{vmatrix} e^{-3\tau} & 0 & e^{-7\tau} \\ -3e^{-3\tau} & 0 & -7e^{-7\tau} \\ 9e^{-3\tau} & 1 & 49e^{-7\tau} \end{vmatrix} = -\Big( e^{-3\tau}(-7e^{-7\tau}) - (-3e^{-3\tau})e^{-7\tau} \Big)$$

$$W_2(\tau) = 4 e^{-10\tau} \tag{6.50}$$

$$W_3(\tau) = \begin{vmatrix} y_1(\tau) & y_2(\tau) & 0 \\ \dot{y}_1(\tau) & \dot{y}_2(\tau) & 0 \\ \ddot{y}_1(\tau) & \ddot{y}_2(\tau) & 1 \end{vmatrix} = \begin{vmatrix} e^{-3\tau} & e^{-6\tau} & 0 \\ -3e^{-3\tau} & -6e^{-6\tau} & 0 \\ 9e^{-3\tau} & 36e^{-6\tau} & 1 \end{vmatrix} = e^{-3\tau}(-6e^{-6\tau}) - (-3e^{-3\tau})e^{-6\tau}$$

$$W_3(\tau) = -3 e^{-9\tau} \tag{6.51}$$

Substituting Equations 6.48 through 6.51 into Equation 6.47 we arrive at

$$K(t,\tau) = \Big( e^{-3t}(-e^{-13\tau}) + e^{-6t}(4e^{-10\tau}) + e^{-7t}(-3e^{-9\tau}) \Big) \Big/ (-12e^{-16\tau})$$

$$K(t,\tau) = \left(e^{-3t}(-e^{3\tau}) + e^{-6t}(4e^{6\tau}) + e^{-7t}(-3e^{7\tau})\right)\Big/(-12)$$

$$K(t,\tau) = (1/12)\left(e^{-3(t-\tau)} - 4e^{-6(t-\tau)} + 3e^{-7(t-\tau)}\right) \tag{6.52}$$

As a rudimentary check on Equation 6.52, since Equation 6.44 is a linear time-invariant ODE, its kernel should be of the form $K(t,\tau) = K(t-\tau)$. Equation 6.52 passes this test.

**SP6.3** Find the steady-state solution to

$$\frac{d^3 y}{dt^3} + 2\frac{d^2 y}{dt^2} + 4\frac{dy}{dt} + 5y = 2e^{2t}\sin 3t \tag{6.53}$$

*Solution*:
From Table 5.1, we try

$$y(t) = e^{2t}(c\sin 3t + d\cos 3t) \tag{6.54}$$

Differentiating Equation 6.54 once, twice, and three times:

$$\frac{dy}{dt} = e^{2t}(2c\sin 3t + 2d\cos 3t) + e^{2t}(3c\cos 3t - 3d\sin 3t) \tag{6.55}$$

$$\frac{d^2 y}{dt^2} = e^{2t}(-5c\sin 3t - 5d\cos 3t) + e^{2t}(12c\cos 3t - 12d\sin 3t) \tag{6.56}$$

$$\frac{d^3 y}{dt^3} = e^{2t}(-46c\sin 3t - 46d\cos 3t) + e^{2t}(9c\cos 3t - 9d\sin 3t) \tag{6.57}$$

Substituting Equations 6.55 through 6.57 into Equation 6.54 and combining coefficients of $\sin 3t$ and $\cos 3t$ we have

$$e^{2t}(-43c - 45d)\sin 3t + e^{2t}(45c + 43d)\cos 3t = 2e^{2t}\sin 3t$$

We can divide both sides through by $e^{rt}$, since it is never zero, and then we are left with

$$(-43c - 45d)\sin 3t + (45c + 43d)\cos 3t = 2\sin 3t \tag{6.58}$$

For Equation 6.58 to be true for all $t$ implies that

$$-43c - 45d = 2$$
$$45c + 43d = 0 \tag{6.59}$$

Equation 6.59 are two linear equations in two unknowns. The solutions are

$$c = 43/88$$

$$d = -45/88$$

and so the steady-state solution to Equation 6.53 is

$$y(t) = e^{2t}\left((43/88)\sin 3t - (45/88)\cos 3t\right) \qquad (6.60)$$

**Problems to Solve**

**P6.1** Solve:

$$\frac{d^3 y}{dt^3} + 11\frac{d^2 y}{dt^2} + 44\frac{dy}{dt} + 60y = 0$$

$$y(0) = -2 \quad \dot{y}(0) = 3 \quad \ddot{y}(0) = 4$$

**P6.2** Solve:

$$\frac{d^3 y}{dt^3} + 16\frac{d^2 y}{dt^2} + 101\frac{dy}{dt} + 246y = 0$$

$$y(0) = 4 \quad \dot{y}(0) = -2 \quad \ddot{y}(0) = 1$$

**P6.3** Find the kernel for the system governed by

$$\frac{d^3 y}{dt^3} + 15\frac{d^2 y}{dt^2} + 66\frac{dy}{dt} + 80y = g(t)$$

**P6.4** Find the kernel for the system governed by

$$\frac{d^3 y}{dt^3} + 18\frac{d^2 y}{dt^2} + 105\frac{dy}{dt} + 196y = g(t)$$

**P6.5** Find the steady-state solution to

$$\frac{d^3 y}{dt^3} + 3\frac{d^2 y}{dt^2} + 4\frac{dy}{dt} + 6y = 8te^{2t}$$

**P6.6** Find the steady-state solution to

$$\frac{d^3 y}{dt^3} + 2\frac{d^2 y}{dt^2} + 5\frac{dy}{dt} + 7y = 6\cos 2t$$

**Word Problems**

**WP6.1** Figure 6.3 depicts a double pendulum. Suppose, as shown in the figure, the two masses are equal and the two pendulum lengths are equal. It can then be shown that, for small angles (i.e., the pendulum hanging nearly vertically), the angles $\theta_1$ and $\theta_2$ satisfy the equation

$$\frac{d^4 y}{dt^4} + 4\omega_0^2 \frac{d^2 y}{dt^2} + 2\omega_0^4 y = 0$$

where $\omega_0 = \sqrt{g/L}$. Find the two natural frequencies the system can exhibit.

**WP6.2** A lumped parameter model is an approximate mathematical representation of a system distributed in a continuum, using a finite number of discrete entities. Engineers often use lumped parameter models to simplify complicated situations in order to gain a preliminary, ball-park estimate of system behavior. Figure 6.4

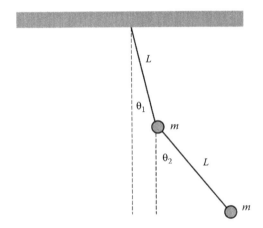

**FIGURE 6.3**   Double pendulum.

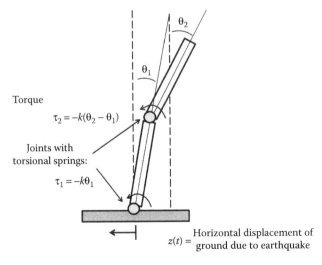

**FIGURE 6.4**   Lumped parameter model of a skyscraper in an earthquake.

pictures a skyscraper of mass $m$ and height $h$ in an earthquake. The ground is assumed to move horizontally in a sinusoidal manner with magnitude $z_0$ and frequency $\omega_E$. The building is represented by two segments joined with a torsional spring with spring constant $k$. The base of the building is assumed to be joined to the ground by a similar spring. Mass is assumed to be uniformly distributed in both segments. Under these circumstances it can be shown that the angles $\theta_1$ and $\theta_2$ satisfy an equation of the form

$$\frac{d^4\theta_n}{dt^4} + 9\omega_0^2 \frac{d^2\theta_n}{dt^2} + 4\omega_0^4\theta_n = c_n\left(\frac{z_0}{h}\right)\sin\omega_E t$$

where:

$$\omega_0 = \sqrt{\frac{12k}{mh^2}}$$

$c_n$ is a constant that depends on $\omega_0$ and $\omega_E$

Assuming that $c_n$ plays no part in the calculation, find the two values of the earthquake frequency $\omega_E$ that cause a resonance in the building's oscillation.

**WP6.3** Show that if $y_*(t)$ is the solution of the ODE

$$\frac{d^n y}{dt^n} + a_1\frac{d^{n-1}y}{dt^{n-1}} + \ldots + a_{n-1}\frac{dy}{dt} + a_n y = g(t)$$

($n \geq 1$) with zero initial conditions and $g(t)$ is differentiable then the solution to

$$\frac{d^n y}{dt^n} + a_1\frac{d^{n-1}y}{dt^{n-1}} + \ldots + a_{n-1}\frac{dy}{dt} + a_n y = \frac{dg}{dt}$$

with zero initial conditions is $dy_*/dt - g(0)K(t)$, where $K(t)$ is the convolution kernel for the ODE.

## Challenge Problems

**CP6.1** Given Equations 6.1, the satellite equations of motion in Cartesian $(x, y)$ coordinates, derive Equations 6.2, the equations of motion in polar coordinates.

*Guidance*:
Begin with the transformation

$$x = R\cos\theta$$
$$y = R\sin\theta \tag{6.61}$$

Differentiate Equations 6.61 and solve the resulting equations to obtain

$$\dot{R} = \dot{x}\cos\theta + \dot{y}\sin\theta$$
$$R\dot{\theta} = -\dot{x}\sin\theta + \dot{y}\cos\theta \tag{6.62}$$

Differentiate Equations 6.62 and substitute (a) Equations 6.1 for $\ddot{x}$ and $\ddot{y}$ and (b) equations you have derived for $\dot{x}$ and $\dot{y}$. Simplify and obtain Equation 6.2.

**CP6.2** Given Equation 6.2, assume small perturbations about the nominal trajectory and derive linearized equations of motion for the deviation in altitude $r$ from a circular orbit and the deviation $v$ from a constant orbital frequency.

*Guidance*: Let $R = R_0 + r$ and $\dot{\theta} = \omega = \omega_0 + v$, where $\dot{R}_0 - 0$ and $\omega_0^2 = GM/R_0^3$. To linearize Equation 6.2 we proceed term-by-term and employ whichever of the two methods seems most convenient.

One method is to write the term out in its full nonlinear form and then neglect higher-order terms such as $r^2, \dot{r}^2, v^2$, and $rv$. For example,

$$\sqrt{(R\dot{\theta})^2 + \dot{R}^2} = \sqrt{(R_0 + r)^2(\omega_0 + v)^2 + \dot{r}^2} \cong \sqrt{(R_0 + r)^2(\omega_0 + v)^2}$$

$$= (R_0 + r)(\omega_0 + v)$$

$$= R_0\omega_0 + \omega_0 r + R_0 v + rv \cong R_0\omega_0 + \omega_0 r + R_0 v$$

We are justified in this approximation as long as $\dot{r} \ll R_0\omega_0$ and $rv \ll \omega_0 r + R_0 v$, the latter of which will be true provided $r \ll R_0$ and $v \ll \omega_0$.

A second, more formal, way is sometimes more convenient. Consider a nonlinear function $F$ of the two variables $R$ and $\omega$ and small perturbations about $R = R_0$ and $\omega = \omega_0$. If $F$ is sufficiently smooth at that point, we can approximate

$$F(R_0 + r, \omega_0 + v) \cong F(R_0, \omega_0) + \left(\frac{\partial F}{\partial R}\right)_0 \cdot r + \left(\frac{\partial F}{\partial \omega}\right)_0 \cdot v$$

where the partial derivatives are evaluated at $R = R_0$ and $\omega = \omega_0$. For example,

$$R\dot{\theta}^2 = (R_0 + r)(\omega_0 + v)^2 \cong R_0\omega_0^2 + \frac{\partial}{\partial R}(R\omega^2)_0 \cdot r + \frac{\partial}{\partial \omega}(R\omega^2)_0 \cdot v = R_0\omega_0^2 + \omega_0^2 r + 2\omega_0 R_0 v$$

As another example in just one variable,

$$\frac{GM}{R^2} \cong \frac{GM}{R_0^2} + \frac{\partial}{\partial R}\left(\frac{GM}{R^2}\right)_0 \cdot r = \frac{GM}{R_0^2} + \frac{d}{dR}\left(\frac{GM}{R^2}\right)_0 \cdot r = \frac{GM}{R_0^2} - \left(\frac{2GM}{R_0^3}\right)r$$

For density $\rho$, using Equations 6.8 and 6.9 and the formal linearization method, we can write

$$\rho \cong \rho_0 + \left(\frac{\partial \rho}{\partial R}\right)_0 r = \rho_0 + \rho_0\frac{1}{\rho_0}\left(\frac{d\rho}{dR}\right)_0 r = \rho_0(1 - Zr)$$

Show that the linearized equations of motion are

$$\ddot{r} = 3\omega_0^2 r + 2\omega_0 R_0 v - kR_0\omega_0\dot{r}$$

$$R_0\dot{v} = -2\omega_0\dot{r} + kR_0(ZR_0 - 2)\omega_0^2 r - 2kR_0^2\omega_0 v - kR_0^2\omega_0^2$$

**CP6.3** Given the results of Challenge Problem CP6.2, find an equation for $r$ alone.

*Guidance*: This problem is solved much easier with the method introduced in Chapter 7, but we can do it with methods already in hand. Note that the first equation involves $v$ but no derivatives of $v$. Note also that we can rewrite the second equation so that the left-hand side is $R_0\dot{v} + 2kR_0^2\omega_0 v$ and the right-hand side involves only $r$ and its derivatives. If we differentiate the first equation and add to that result $2kR_0\omega_0$ times the first equation, we end up with an equation involving $R_0\dot{v} + 2kR_0^2\omega_0 v$, which, from the second equation, can be written as a function of $r$ and its derivatives. Show that the resulting equation for $r$ is

$$\dddot{r} + 3kR_0\omega_0\ddot{r} + (1 + 2(kR_0)^2)\omega_0^2\dot{r} - 2(kR_0)(ZR_0 + 2)\omega_0^3 r = -2kR_0^2\omega_0^3$$

**CP6.4** Show that the results of Challenge Problem CP6.3 lead to Equation 6.11 of the text.

*Guidance:* As outlined in the text, we can let $t_* = \omega_0 t$ to yield

$$\dddot{r} + (3kR_0)\ddot{r} + (1 + 2(kR_0)^2)\dot{r} - 2(kR_0)(ZR_0 + 2)r = -2kR_0^2 \tag{6.63}$$

where the derivatives are understood to be with respect to $t_*$. Next we do some order of magnitude calculations. Using [5] and [6] we determine that at 300 statute miles altitude (a) $Z \approx 10^{-5}$ 1/ft and (b) $\rho_0 \approx 10^{-15}$ slugs/cu ft. From [5] and [7] we learn that for a typical satellite at 300 statute miles altitude $C_D \approx 2$. From [6] and [7] we ascertain that for typical satellite dimensions and weights the ratio $S/m$ is usually in the range 0.5–2 sq ft/slug. With these facts in hand show that Equation 6.63 can be accurately approximated by

$$\dddot{r} + \dot{r} - (2kZR_0^2)r = -2kR_0^2 \tag{6.64}$$

Finally, show that the substitution $r_* = r - 1/Z$ leads to Equation 6.11, as was to be shown.

## COMPUTING PROJECTS: PHASE 2

We return here to solving nonlinear differential equations by numerical integration on digital computers. At the end of Chapter 3 we defined the first phase of several extensive computing projects intended to give the student first-hand experience with programming. We now continue those projects.

The objective of the first phase was to demonstrate an effective method for checking basic elements of our program. We considered our equations in a modified setting, simple enough to permit hand-calculation yet realistic enough to permit testing of a significant portion of our program.

In Phase 2 we program the full equations, which are too complicated for hand-calculation. Nevertheless we have an initial step to check our results. In both of the projects below, the first case we consider should produce output comparing closely to the output of Phase 1. We have built a multilevel safety net to give us confidence in our final results.

Phase 2 of both projects require greater accuracy in our models, including that of the atmosphere. The tables and equations to follow apply to both projects. Table 6.1 defines new terms beyond those defined in Phase 1.

A constant was a good approximation for the drag coefficient of the flight vehicles in Phase 1 but it is inadequate as a representation for the vehicles and trajectories in Phase 2. Table 6.2 and Figure 6.5 gives the drag coefficient (at zero lift) for all flight vehicles as it depends on the drag coefficient at subsonic speeds $C_{D0}$ and on Mach number, defined as

$$M = \frac{|V|}{a}$$

where:

   $V$ is the vehicle's speed

   $a$ is the speed of sound, which varies with temperature according to

$$a = a_0 \sqrt{\frac{TA(h)}{TA_0}}$$

---

**TABLE 6.1**

**Definition of New Terms for Phase 2 of Both Project A and Project B**

| Term | Definition |
|------|------------|
| $TA(h)$ | Air temperature at altitude $h$ |
| $TA_0$ | Air temperature at sea level = 519°R |
| $M$ | Mach number |
| $a$ | Speed of sound |
| $a_0$ | Speed of sound at sea level = 1117 ft/s |
| $R$ | Radius of Earth = $20.88 \cdot 10^6$ ft |
| $H(h)$ | Reference altitude for air density at altitude $h$ |

---

**TABLE 6.2**

**Equations for Drag Coefficient as Function of Mach Number**

| $M$ | $C_D$ (at zero lift) |
|-----|----------------------|
| $0 < M < 0.8$ | $C_{D0}$ |
| $0.8 < M < 1.0$ | $C_{D0}/(\sqrt{1-M^2}+0.4)$ |
| $1.0 < M < 1.5$ | $C_{D0}(2.5+8(M-1)(1.5-M))$ |
| $1.5 < M$ | $2.795 C_{D0}/\sqrt{M^2-1}$ |

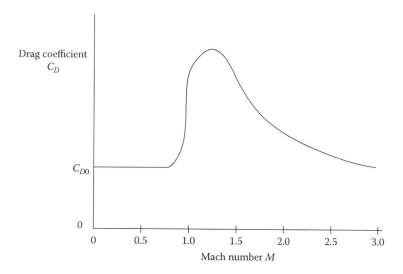

**FIGURE 6.5**   Drag coefficient (at zero lift) as a function of Mach number.

Figure 6.6 graphs air temperature as a function of altitude according to a commonly used standard model [1], which represents an average over geography and time.

Air density also varies with altitude and Figure 6.7 shows an efficient way to represent it. (Derived by the author from tables of the standard model in [1]. Word Problem WP2.6 in Chapter 2 points the way to derivation of equations for air density as a function of altitude. We have opted for the simpler, approximate representation in Figure 6.7 to reduce the burden on the student.). In Figure 6.7, $\rho_{a0}$ is the air density at sea level, quantified in Chapter 3.

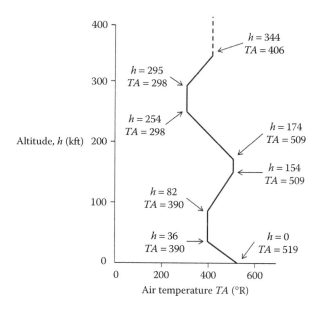

**FIGURE 6.6**   Air temperature as a function of altitude.

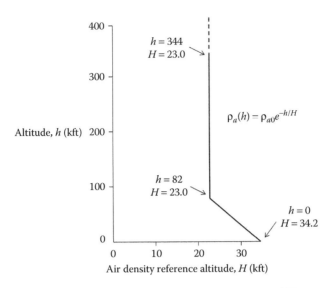

**FIGURE 6.7**  Air density reference altitude $H$ (defined by $\rho_a(h) = \rho_{a0}e^{-h/H}$) as it varies with altitude.

The numerical integration program you use probably has efficient ways to input data when, as in Figures 6.6 and 6.7, it can be represented with straight lines between a few coordinates. Use the program you wrote for Phase 1 as a start for the program you write here.

## Project A: Phase 2

As you will recall, Project A explores further the sounding rocket trajectory analyzed in Section 3.1.1. In Phase 1 we programmed the basic equations of motion and the basic relations for the forces (thrust, drag, and gravity). We programmed constant values for drag coefficient, air density and gravity, matching the simple, approximate model we analyzed in Section 3.1.1. We input the parameters of a small model rocket for which the simple equations should work well and compared the numerically integrated results with the closed form solutions we derived. That comparison tested that we programmed correctly the basic equations of motion and the basic equations for the forces.

In Phase 2, we will program equations for drag coefficient, air density and gravity that more closely resemble reality. Our first run will again input the parameters of the small model rocket. The output of this case should be very similar to the output of Phase 1's program. With success on this first run as an additional confidence builder, we then move on to input the parameters of two much larger rockets more similar to those used in NASA's science program.

### Desired Output

First, run your numerical integration program with the equations given below and the parameters for the small model rocket (Rocket 1). Compare the computed maximum speed during ascent, the maximum altitude, and the maximum speed during descent to the results you computed by numerical integration in Phase 1. You should expect Phase 2 results to be within 5% of those in Phase 1.

When your results with the small model rocket are within the expected bounds, run your numerical integration program with the equations below and the parameters for Rockets 2 and 3. Print out and hand in a copy of a graph of the computed rocket trajectories (altitude versus time) and your program, including all functions that you define. Hand in copies of the computer's instant-by-instant tabulated results at the appropriate times to record the maximum speed during ascent, the maximum height, and the maximum speed during descent.

## Equations

The basic equations of motion and basic equations for thrust and drag given at the end of Chapter 3 in the description of Phase 1 of Project A apply here as well. Program the equations given earlier in Tables 6.1 and 6.2 and Figures 6.5 through 6.7. We also need a more accurate equation for gravity because of the altitude reached by Rockets 2 and 3:

$$g = g_0 \left( \frac{R}{R+h} \right)^2 \qquad g_0 = 32.2 \text{ ft/s/s}$$

Table 6.3 lists parameters for each of the three rockets.

## Expected Results

Your results should be close to those shown in Table 6.4.

## Project B: Phase 2

As you will recall, Project B analyzes the flight performance of a high-performance fighter aircraft. In Phase 2 we will answer three fundamental questions: (a) How high can it fly (aircraft ceiling)? (b) How fast can it fly? And (c) How long does it take to go from the ground to the required speed and altitude (minimum time to climb)?

## TABLE 6.3
## Parameters for Each Rocket

| Parameter | Symbol | Rocket 1 | Rocket 2 | Rocket 3 |
|---|---|---|---|---|
| Initial weight (lbs) | $W_0$ | 8 | 120 | 1500 |
| Mass ratio | $MR = m_0/m_b$ | 1.15 | 1.5 | 5 |
| Cross-sectional area (sq ft) | $S$ | 0.05 | 0.25 | 1.0 |
| Gas exit speed (ft/s) | $c$ | 5000 | 6500 | 8000 |
| Subsonic drag coefficient | $C_{D0}$ | 0.03 | 0.025 | 0.020 |

## TABLE 6.4
## Computed Rocket Performance

| Performance Measure | Rocket 1 | Rocket 2 | Rocket 3 |
|---|---|---|---|
| Maximum velocity during ascent (ft/s) | 691 | 2553 | 12,267 |
| Maximum altitude (ft) | $7.14 \cdot 10^3$ | $8.78 \cdot 10^4$ | $2.69 \cdot 10^6$ |
| Maximum velocity during descent (ft/s) | −660 | −2135 | −12,180 |

Recall that in Phase 1 we had the more limited objective of checking out our program to ensure we programmed the basic equations of motion and basic thrust and drag equations correctly. To accomplish this we analyzed aircraft dynamics in a special setting: beginning just after the aircraft has rotated off the runway, when it has settled into a constant flight path angle and is gaining speed and altitude. For this special case we were able to solve the equations analytically, and then to compare our numerically integrated results with the analytical solutions.

The basic equations of motion and basic equations for the forces given at the end of Chapter 3 in the description of Phase 1 of Project B apply here as well. In Phase 2, we will program equations to allow the aircraft to perform maneuvers, as discussed shortly. We will also program the equations given above in Tables 6.1 and 6.2 and in Figures 6.5 through 6.7 to more realistically represent the drag force. In addition, a more accurate model for engine thrust is

$$T = T_0 \left( \rho_a / \rho_{a0} \right)^\lambda$$

where the exponent $\lambda$ is given by Table 6.5. Finally, the drag coefficient increases with lift according to

$$C_D = C_{D1} + 0.18 C_L^2$$

where $C_{D1}$ is the drag coefficient at zero lift, from Table 6.2. In these short flights we are not taking into account weight loss due to fuel expenditure.

## Run Objectives and Required Output

Run 1: Test the program. Program the same initial conditions as in Phase 1. You do not have to hand anything in. The values for altitude $h(t)$, speed $V(t)$, and flight path angle $\gamma(t)$ should be about the same (within 5%) as they were for Phase 1.

Run 2: Determine the ceiling. Hand in a graph of $h$ as a function of time. Hand in copies of the computer data page (its instant-by-instant tabulated results) at the end of the run to record the maximum altitude reached. Also, for this case only, hand in a copy of your program, the user-defined functions and the look-up tables.

Run 3: Determine the aircraft's maximum speed without afterburner. Hand in the graph of $V$ versus time and the data page showing $V$ at the end of the run.

Run 4: Determine maximum speed with afterburner (which increases thrust at the expense of fuel rate). Hand in the graph of $V$ versus time and the data page showing $V$ at the end of the run.

---

**TABLE 6.5**

**Thrust Variation with Mach Number**

| $M$ | $\lambda$ |
|---|---|
| $0 < M < 0.95$ | 0.7 |
| $0.95 < M < 1.2$ | $1.2M - 0.44$ |
| $1.2 < M$ | 1.0 |

---

**TABLE 6.6**

**Run Parameters for Project B, Phase 2**

| Run Number | $h(0)$ kft | $V(0)$ ft/s | $\gamma(0)$ Radians | $I_\gamma(0)$ | Flight Profile | Run Time (s) | $T_0$ klbs |
|---|---|---|---|---|---|---|---|
| 1 | 0 | 350 | 0.25 | 0 | $\gamma_D = 0.25$ | 25 | 70 |
| 2 | 0 | 350 | 0.25 | 0 | $\gamma_D = 0.25$ | 600 | 70 |
| 3 | 35 | 350 | 0 | 0 | $\gamma_D = 0$ | 600 | 70 |
| 4 | 35 | 350 | 0 | 0 | $\gamma_D = 0$ | 900 | 90 |
| 5 | 0 | 350 | 0 | 0 | Student design | 520 | 90 |

Run 5: Determine minimum time to climb. Design your own flight profile schedule (the commanded flight path angle $\gamma_D$ versus time) to see how quickly you can "pilot" the aircraft to the required altitude and speed (45 kft and $M = 1.8$). Hand in the graphs of $h$ and $V$ versus time, the user-defined function for the flight profile schedule, the data pages showing $h$ and $V$ at the end of the run, and data pages documenting your minimum time to climb. With reasonable design effort your minimum time to climb should be less than 520 s.

## Run Parameters

Table 6.6 lists initial conditions and key parameter values for Project B, Phase 2.

## Autopilot

Program the following equations to have the aircraft fly a desired trajectory, defined by $\gamma_D(t)$, desired flight path angle as a function of time. We are controlling the aircraft's angle of attack, and thereby the lift on the aircraft, to achieve the trajectory we desire.

$$\alpha = \alpha_0 - k_1(\gamma - \gamma_D) - k_2 I_\gamma$$

$$\frac{dI_\gamma}{dt} = \gamma - \gamma_D$$

$$\alpha_0 = W \cos \gamma_D / (L_\alpha + T)$$

$$k_1 = (2n\varsigma + \sin \gamma_D)\alpha_0 / \cos \gamma_D$$

$$k_2 = n^2 (g/V) \alpha_0 / \cos \gamma_D$$

$$n = 3$$

$$\varsigma = 0.7$$

If $\alpha > 0.14$ then $\alpha = 0.14$ (in radians).

If $\alpha < -0.14$ then $\alpha = -0.14$ (in radians).

A few comments on these equations. In the angle of attack equation the term $\alpha_0$ provides steady-state lift. The other two terms correct the error $\gamma - \gamma_D$. The "autopilot" is designed to enable smooth transitions from one flight path angle to another while limiting accelerations (g's) experienced by the pilot. We are assuming that angle of attack can be changed instantaneously. This is an approximate model that ignores "short-period" dynamics. A pilot or autopilot actually controls the elevator (a movable surface on the tail) to control the angle of attack and rapid, brief oscillations in aircraft orientation typically accompany any change. We will examine "short-period" dynamics in Chapter 8, Section 8.5.1.3.

### Remarks on the Trajectory to Achieve Minimum Time to Climb

When jet aircraft were first introduced military pilots flew trajectories similar to that shown in Figures 6.8 and 6.9 to minimize time to climb. Note that speed levels off when the aircraft is transiting through Mach 1 (the "sound barrier"). Recall the steep rise in drag coefficient at Mach 1 shown in Figure 6.5. Through advanced mathematical methods using a differential equation model very similar to the one we are employing here, including use of angle of attack as control, engineers found that a trajectory similar to that marked Trajectory II in Figures 6.10 and 6.11 was optimum and should greatly reduce time to climb. The dive midway in the trajectory assists the aircraft's transit through the high-drag-coefficient region. Pilots were understandably dubious, but flight tests proved remarkably successful.

The trajectories in Figures 6.8 and 6.10 were not generated by advanced mathematical methods but rather by the trial and error process the student is expected to employ.

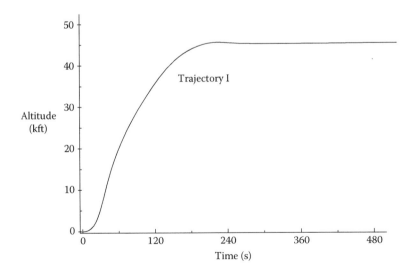

**FIGURE 6.8** Altitude profile in original trajectories to minimize time to climb.

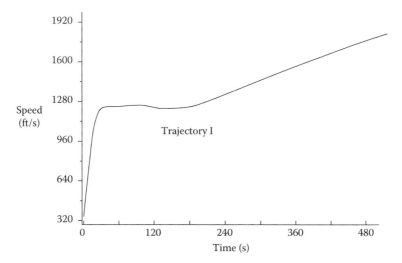

**FIGURE 6.9**    Speed profile in original trajectories to minimize time to climb.

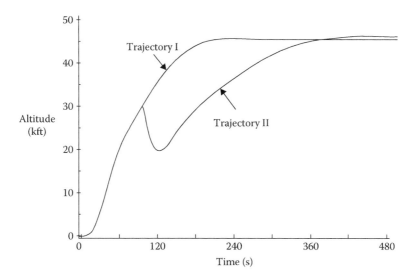

**FIGURE 6.10**    Comparison of time to climb trajectories.

Nevertheless the numerically computed time to climb is 470 s using Trajectory I and 396 s using Trajectory II.

## Expected Results

Your results should be close to those given in Table 6.7. We reiterate that the models and performance presented here are not intended to represent the characteristics or performance of any actual fighter aircraft.

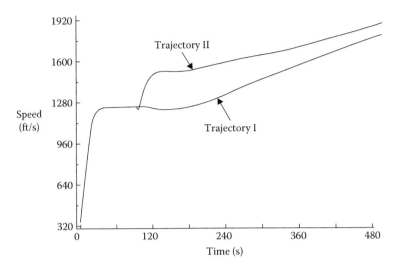

**FIGURE 6.11**  Comparison of speed profiles in time to climb trajectories.

**TABLE 6.7**
**Computed Fighter Aircraft Performance**

| Performance Measure | Computed Value |
| --- | --- |
| Maximum altitude (ft) | 62,000 |
| Maximum speed w/o afterburner (ft/s) | 1420 |
| Maximum speed with afterburner (ft/s) | 2410 |
| Minimum time to climb (s) | < 520 |

## Additional Challenge Problem

**CP6.5** Derive an approximate linear time-invariant differential equation for the flight path angle when the aircraft is under "autopilot" control.

*Guidance*: Substitute the autopilot equations given earlier into the equation for the derivative of the flight path angle:

$$\dot{\gamma} = (T \sin \alpha + L)/mV - (g/V)\cos \gamma$$

Assume that the angle of attack $\alpha$ and the error $\gamma - \gamma_D$ are small. Let $\gamma_E = \gamma - \gamma_D$ and assume $\gamma_D$ is constant. Show that, approximately, $\gamma_E$ satisfies the differential equation

$$\ddot{\gamma}_E + 2\varsigma\omega\dot{\gamma}_E + \omega^2\gamma_E = 0$$

where:

$$\omega = \frac{ng}{V}$$

Show that if $V$ is a constant, initial conditions are $\gamma_E(0) = \gamma_{E0}$, $\dot{\gamma}_E(0) = 0$, and $\varsigma = 1/\sqrt{2}$; then the maximum acceleration that the aircraft experiences normal to the flight path is approximately

$$a_N = V\dot{\gamma} = 0.456 \cdot ng\gamma_{E0}$$

Of course, in all of the trajectories we are considering the speed $V$ is varying, and in the time-to-climb trajectory $\gamma_D$ also varies considerably. Furthermore the angle difference $\gamma - \gamma_D$ may not always be small. Hence, the usefulness of the autopilot can only be determined by numerical experiment.

*Lessons to be Learned from the Computing Projects*

- Differential equations are at the heart of much of engineering and numerical integration routines are an essential tool.
- Real-life problems often require collection of complex data and finding the most efficient way to input it.
- Approximate closed-form solutions can provide useful test cases to check for programming errors.
- From Project B, advanced mathematical optimization methods based on differential equations can provide significant design insights.

# 7 Laplace Transforms

Laplace transforms simplify our work with differential equations. For *linear time-invariant* (*LTI*) ordinary differential equations (ODEs) of any order, Laplace transforms convert differential equations into algebraic ones, solve for initial conditions without extra steps, permit solution of nonhomogeneous equations without first solving homogeneous equations, and address impulsive forcing functions easily. As a result of these attributes, Laplace transforms are a handy means of solving differential equations. They are also useful for describing systems in constituent parts and for designing systems to meet specific requirements. They are a principal analytic tool for engineers.

## 7.1 PRELIMINARIES

Before we introduce Laplace transforms we must review a few definitions. The *improper integral* $\int_a^\infty f(t)dt$ is defined as

$$\int_a^\infty f(t)dt = \lim_{A\to\infty} \int_a^A f(t)dt$$

A function is *piecewise continuous* in the interval $\alpha \le t \le \beta$ if it is continuous there except for a finite number of jump discontinuities. A function is of *exponential order* if there exist $a, K$, and $M$ such that

$$|f(t)| < Ke^{at} \text{ if } t > M$$

## 7.2 INTRODUCING THE LAPLACE TRANSFORM

*Definition*:
  The Laplace transform $F(s)$ of a function $f(t)$ is

$$F(s) = L\{f(t)\} = \int_0^\infty e^{-st} f(t)dt \tag{7.1}$$

The following theorems are stated without proof.

  *Existence*: The transform $F(s)$ exists provided the function $f(t)$ is piecewise continuous and of exponential order.
  *Uniqueness*: If $f(t)$ is continuous and $F(s)$ is its transform there is no other continuous function with that transform.
  *Linearity*: By its definition the Laplace transform is a linear operation; that is, if the Laplace transform of $f_1(t)$ is $F_1(s)$ and the Laplace transform of $f_2(t)$ is $F_2(s)$ and $a$ and $b$ are constants then the Laplace transform of $af_1(t) + bf_2(t)$ is $aF_1(s) + bF_2(s)$.

## 7.3  LAPLACE TRANSFORMS OF SOME COMMON FUNCTIONS

Using Equation 7.1 we can compute the Laplace transform of many common functions. We will do that for several functions here as examples and then add to our *short table of Laplace transforms* as we proceed.

First, we compute the Laplace transform of $f(t) = e^{at}$:

$$L\{e^{at}\} = \int_0^\infty e^{-st}e^{at}dt = \lim_{T\to\infty}\int_0^T e^{-(s-a)t}dt = \lim_{T\to\infty}\left[\frac{1}{(s-a)}\left(1 - e^{-(s-a)T}\right)\right] = \frac{1}{s-a}$$

The integral is defined only for $s > a$, but the reader will discover that in our everyday use of Laplace transforms we never need to concern ourselves with this limitation.

Next, the Laplace transform of $f(t) = t$:

$$L\{t\} = \int_0^\infty e^{-st}tdt = \lim_{T\to\infty}\int_0^T e^{-st}tdt$$

Integrating by parts:

$$\lim_{T\to\infty}\int_0^T e^{-st}tdt = \lim_{T\to\infty}-\frac{Te^{-sT}}{s} + \frac{1}{s}\int_0^T e^{-st}dt = \lim_{T\to\infty}-\frac{Te^{-sT}}{s} + \frac{1}{s^2}\left(1 - e^{-sT}\right) = \frac{1}{s^2}$$

That is,

$$L\{t\} = \int_0^\infty e^{-st}tdt = \frac{1}{s^2}$$

Again, the result is valid only for $s > 0$ but henceforth this need not concern us.

Computing the Laplace transform of $f(t) = \sin\omega t$:

Recall that we encountered the integral $\int_0^T e^{-st}\sin\omega tdt$ in Chapter 2 (with $s = -\lambda$) but postponed its calculation, promising to pick it up again here in Chapter 7.

Integrating by parts twice, we find that

$$\int_0^T e^{-st}\sin\omega tdt = -\frac{1}{s}\left(e^{-sT}\sin\omega T\right) + \frac{\omega}{s}\int_0^T e^{-st}\cos\omega tdt =$$
$$-\frac{1}{s}\left(e^{-sT}\sin\omega T\right) + \frac{\omega}{s}\left[-\frac{1}{s}\left(e^{-sT}\cos\omega T - 1\right) - \frac{\omega}{s}\int_0^T e^{-st}\sin\omega tdt\right]$$

Note that the term $\int_0^T e^{-st}\sin\omega tdt$ is on both sides of this equation. Rearranging so that the term is only on the left-hand side:

$$\left(1 + \frac{\omega^2}{s^2}\right)\int_0^T e^{-st}\sin\omega tdt = -\frac{1}{s}(e^{-sT}\sin\omega T) - \frac{\omega}{s^2}(e^{-sT}\cos\omega T - 1)$$

Dividing through by $\left(1 + (\omega^2/s^2)\right)$:

$$\int_0^T e^{-st} \sin\omega t\, dt = \left(\frac{s^2}{s^2+\omega^2}\right)\left(-\frac{1}{s}(e^{-sT}\sin\omega T) - \frac{\omega}{s^2}(e^{-sT}\cos\omega T - 1)\right)$$

$$\int_0^T e^{-st} \sin\omega t\, dt = \left(\frac{1}{s^2+\omega^2}\right)\left(\omega - e^{-sT}(s\sin\omega T + \omega\cos\omega T)\right)$$

Therefore

$$L\{\sin\omega t\} = \int_0^\infty e^{-st} \sin\omega t\, dt = \lim_{T\to\infty} \int_0^T e^{-st} \sin\omega t\, dt = \frac{\omega}{s^2+\omega^2}$$

As the last of our examples here, we compute the Laplace transform of any function $f(t)$, provided its transform $F(s)$ exists, when the function is multiplied by $e^{-\lambda t}$:

$$L\{e^{-\lambda t}f(t)\} = \int_0^\infty e^{-st}\left[e^{-\lambda t}f(t)\right]dt = \int_0^\infty e^{-(s+\lambda)t}f(t)dt = F(s+\lambda)$$

## 7.4 LAPLACE TRANSFORMS OF DERIVATIVES

**Theorem:**

Let $f(t)$ be continuous and of exponential order and let $df/dt$ be piecewise continuous on any interval $0 \le t \le A$. Then the Laplace transform of $df/dt$ exists and

$$L\left\{\frac{df}{dt}\right\} = sL\{f(t)\} - f(0) \tag{7.2}$$

Challenge Problem CP7.2 guides the student through a proof of this theorem.

**Corollary:**

Let $f(t)$, $df/dt$, $d^2f/dt^2$, ..., and $d^{n-1}f/dt^{n-1}$ be continuous and of exponential order and let $d^n f/dt^n$ be piecewise continuous on any interval $0 \le t \le A$. Then the Laplace transform of $d^n f/dt^n$ exists and

$$L\left\{\frac{d^n f}{dt^n}\right\} = s^n L\{f(t)\} - s^{n-1}f(0) - s^{n-2}\left(\frac{df}{dt}(0)\right) - \ldots - s\left(\frac{d^{n-2}f}{dt^{n-2}}(0)\right) - \left(\frac{d^{n-1}f}{dt^{n-1}}(0)\right)$$

In particular,

$$L\left\{\frac{d^2 f}{dt^2}\right\} = s^2 L\{f(t)\} - sf(0) - \left(\frac{df}{dt}(0)\right) \tag{7.3}$$

$$L\left\{\frac{d^3 f}{dt^3}\right\} = s^3 L\{f(t)\} - s^2 f(0) - s\left(\frac{df}{dt}(0)\right) - \left(\frac{d^2 f}{dt^2}(0)\right) \tag{7.4}$$

**Proof:**

To prove the corollary we apply the theorem sequentially to $d^2f/dt^2$, $d^3f/dt^3$, ..., and $d^{n-1}f/dt^{n-1}$.

We can apply the theorem to compute the Laplace transforms of more common functions and add to our *short table of Laplace Transforms*.

Computing the Laplace transform of a constant $a$:

First note that $L\{at\} = aL\{t\} = a/s^2$. Then, applying the theorem,

$$L\{a\} = aL\{1\} = aL\left\{\frac{d}{dt}[t]\right\} = a\left(sL\{t\} - 0\right) = a\frac{s}{s^2} = \frac{a}{s}$$

Computing the Laplace transform of $\cos\omega t$:

$$L\{\cos\omega t\} = L\left\{\frac{1}{\omega}\frac{d}{dt}(\sin\omega t)\right\} = \frac{1}{\omega}\left[sL\{\sin\omega t\} - \sin\omega(0)\right] =$$

$$\frac{1}{\omega}\left[\frac{s\omega}{s^2+\omega^2} - 0\right] = \frac{s}{s^2+\omega^2}$$

The results of Sections 7.3 and 7.4 are sufficient to give us a start on our *short table of Laplace transforms*. See Table 7.1.

**TABLE 7.1**

**Short Table of Laplace Transforms**

**(Initial Version)**

| $f(t)$ | $F(s)$ |
|---|---|
| 1 | $\dfrac{1}{s}$ |
| $t$ | $\dfrac{1}{s^2}$ |
| $e^{-\lambda t}$ | $\dfrac{1}{s+\lambda}$ |
| $\sin\omega t$ | $\dfrac{\omega}{s^2+\omega^2}$ |
| $\cos\omega t$ | $\dfrac{s}{s^2+\omega^2}$ |
| $te^{-\lambda t}$ | $\dfrac{1}{(s+\lambda)^2}$ |
| $e^{-\lambda t}\sin\omega t$ | $\dfrac{\omega}{(s+\lambda)^2+\omega^2}$ |
| $e^{-\lambda t}\cos\omega t$ | $\dfrac{s+\lambda}{(s+\lambda)^2+\omega^2}$ |

## 7.5   LAPLACE TRANSFORMS IN HOMOGENEOUS ORDINARY DIFFERENTIAL EQUATIONS

The general approach to solving ODEs with Laplace transforms is a four step process:

1. Transform each element of the given ODE from a function of time into an algebraic function of $s$.
2. Solve the resulting algebraic equation.
3. Manipulate the algebraic solution into a linear combination of forms that appear in our short table of Laplace transforms.
4. Use the table to determine the function of time corresponding to the algebraic solution.

The clearest way to demonstrate these steps is by example.

**Example 1:**

*Problem*: Using Laplace transforms, solve the first-order ODE

$$\frac{dy}{dt} + 2y = 0, \quad y(0) = 5 \tag{7.5}$$

*Solution*: First, we take the Laplace transform of both sides of the ODE:

$$L\left\{\frac{dy}{dt} + 2y\right\} = L\{0\} \tag{7.6}$$

Since the Laplace transform is a linear operation,

$$L\left\{\frac{dy}{dt} + 2y\right\} = L\left\{\frac{dy}{dt}\right\} + 2L\{y\} \tag{7.7}$$

Defining

$$L\{y(t)\} = Y(s) \tag{7.8}$$

we have, from Equations 7.2, 7.5, and 7.8:

$$L\left\{\frac{dy}{dt}\right\} = sL\{y\} - y(0) = sY - 5 \tag{7.9}$$

From the definition of the Laplace transform (Equation 7.1),

$$L\{0\} = \int_0^\infty e^{-st}(0)dt = 0 \tag{7.10}$$

Substituting Equations 7.7 through 7.10 into Equation 7.6 we arrive at

$$(sY - 5) + 2Y = 0$$

Rearranging and solving:

$$Y = \frac{5}{s+2} = 5 \cdot \left( \frac{1}{s+2} \right) \tag{7.11}$$

Equation 7.11 is the solution to the ODE in the *Laplace transform domain*, which engineers sometimes refer to as the *frequency domain*. We need to convert Equation 7.11 into the solution in the *time domain*. The mathematical procedure for this, called the *inverse Laplace transform*, would take us into the realm of functions of a complex variable. Fortunately, it is not necessary for us to perform the procedure. The statement of uniqueness in Section 7.2 ensures that Table 7.1 suffices. We can enter the table in the column on the right and move left to find the corresponding function of time. In the case of Equation 7.11 we find that the function of time corresponding to $1/(s + 2)$ is $e^{-2t}$. (Here $\lambda = 2$.) The inverse transform of

$$Y(s) = \frac{5}{s+2} = 5 \cdot \left( \frac{1}{s+2} \right)$$

is

$$y(t) = L^{-1}\{Y(s)\} = 5e^{-2t}$$

and this is the solution to Equation 7.5.

Note that we satisfied initial conditions in the process of solving the algebraic equation. Recall that initial conditions required an extra step in time-domain methods.

The process of solving ODEs by Laplace transforms outlined at the beginning of this section rests on two propositions we have not proved. As just discussed, we ask the student to accept without proof the statement of uniqueness in Section 7.2 that enables us to convert the algebraic solution back into the time domain. An additional question, however, is whether the algebraic operations we performed necessarily lead to the Laplace transform of the time-domain solution. They do, which we will address in Section 7.6 and Word Problem WP7.4.

**Example 2:**

*Problem*: Using Laplace transforms, solve the second-order ODE

$$\frac{d^2y}{dt^2} + 4\frac{dy}{dt} + 13y = 0, \quad y(0) = 4, \quad \dot{y}(0) = -1 \tag{7.12}$$

*Solution*: First, we transform both sides of the ODE:

$$L\left\{ \frac{d^2y}{dt^2} + 4\frac{dy}{dt} + 13y \right\} = L\{0\} \tag{7.13}$$

Because the Laplace transform is a linear operation:

$$L\left\{\frac{d^2y}{dt^2} + 4\frac{dy}{dt} + 13y\right\} = L\left\{\frac{d^2y}{dt^2}\right\} + 4L\left\{\frac{dy}{dt}\right\} + 13L\{y\} \tag{7.14}$$

Defining

$$L\{y(t)\} = Y(s) \tag{7.15}$$

we have, from Equations 7.2, 7.12, and 7.15:

$$L\left\{\frac{dy}{dt}\right\} = sL\{y\} - y(0) = sY - 4 \tag{7.16}$$

From Equation 7.2, 7.12, and 7.16,

$$L\left\{\frac{d^2y}{dt^2}\right\} = sL\left\{\frac{dy}{dt}\right\} - \frac{dy}{dt}(0) = s(sY - 4) - (-1) = s^2Y - 4s + 1 \tag{7.17}$$

Substituting Equations 7.15, 7.16, and 7.17 into Equation 7.14, we arrive at

$$(s^2Y - 4s + 1) + 4(sY - 4) + 13Y = 0 \tag{7.18}$$

Rearranging Equation 7.18 and solving,

$$Y(s) = \frac{4s + 15}{s^2 + 4s + 13} \tag{7.19}$$

Equation 7.19 is the solution to the ODE in the Laplace transform domain. We must convert Equation 7.19 into the time domain. To proceed we begin factoring the denominator in Equation 7.19. Completing the square, we see that

$$s^2 + 4s + 13 = (s + 2)^2 + 3^2 \tag{7.20}$$

which is sufficient in this case. From Table 7.1 we see that the solution must be of the form

$$y(t) = c_1 e^{-2t} \cos 3t + c_2 e^{-2t} \sin 3t \tag{7.21}$$

(Here $\lambda = 2$ and $\omega = 3$). From Table 7.1

$$L\{e^{-2t} \cos 3t\} = \frac{s + 2}{(s + 2)^2 + 3^2} \tag{7.22}$$

and

$$L\{e^{-2t}\sin 3t\} = \frac{3}{(s+2)^2+3^2} \tag{7.23}$$

We must manipulate Equation 7.19 into a combination of terms matching Equations 7.22 and 7.23:

$$Y(s) = \frac{4s+15}{s^2+4s+13} = \frac{4(s+2)+7}{(s+2)^2+3^2} = 4\cdot\frac{(s+2)}{(s+2)^2+3^2} + \left(\frac{7}{3}\right)\frac{3}{(s+2)^2+3^2} \tag{7.24}$$

Hence the inverse Laplace transform of $Y(s)$, and the solution to Equation 7.12, is

$$y(t) = 4e^{-2t}\cos 3t + \left(\frac{7}{3}\right)e^{-2t}\sin 3t \tag{7.25}$$

## Example 3:

*Problem*: Using Laplace transforms, solve the third-order ODE

$$\frac{d^3y}{dt^3} + 3\frac{d^2y}{dt^2} + 4\frac{dy}{dt} + 2y = 0 \tag{7.26}$$

$$y(0)=1, \dot{y}(0)=2, \ddot{y}(0)=3$$

*Solution*: Transforming both sides of Equation 7.26 leads immediately (because the Laplace transform is a linear operation) to transforming each term in the equation. Defining

$$L\{y(t)\} = Y(s) \tag{7.27}$$

we have, from Equations 7.2 and 7.26,

$$L\left\{\frac{dy}{dt}\right\} = sY-1$$

$$L\left\{\frac{d^2y}{dt^2}\right\} = sL\left\{\frac{dy}{dt}\right\} - \dot{y}(0) = s(sY-1)-2 = s^2Y-s-2 \tag{7.28}$$

$$L\left\{\frac{d^3y}{dt^3}\right\} = sL\left\{\frac{d^2y}{dt^2}\right\} - \ddot{y}(0) = s(s^2Y-s-2)-3 = s^3Y-s^2-2s-3$$

Using Equations 7.27 and 7.28, the transformed equation is

$$(s^3Y-s^2-2s-3) + 3(s^2Y-s-2) + 4(sY-1) + 2Y = 0$$

Solving:

$$(s^3+3s^2+4s+2)Y = s^2+5s+13$$

$$Y = \frac{s^2 + 5s + 13}{s^3 + 3s^2 + 4s + 2} \tag{7.29}$$

The next step in solving the problem is to factor the denominator. Solved Problem 6.1 in Chapter 6 factored a third-order polynomial by hand, which is useful for engineers to know how to do, but here we opt for the easier solution of going out on the Internet to a "root calculator", which tells us the roots are $s_1 = -1$, $s_2 = -1+i$, and $s_3 = -1-i$. We can check this by multiplying the factors:

$$P(s) = (s+1)(s+1-i)(s+1-i) = (s+1)((s+1)^2 + 1) = (s+1)^3 + (s+1)$$

$$= s^3 + 3s^2 + 3s + 1 + s + 1 = s^3 + 3s^2 + 4s + 2$$

$P(s)$ matches the denominator of Equation 7.29, so the "root calculator" has given us the correct answer. Looking at Table 7.1, our short table of Laplace transforms, we see on the right side no entries with imaginary numbers, but the calculation of $P(s)$ has shown us we can write

$$P(s) = (s+1)((s+1)^2 + 1)$$

so

$$Y = \frac{s^2 + 5s + 13}{s^3 + 3s^2 + 4s + 2} = \frac{s^2 + 5s + 13}{(s+1)((s+1)^2 + 1)} \tag{7.30}$$

Equation 7.30 is not in Table 7.1 either, but we can rearrange Equation 7.30 by partial fraction expansion

$$Y = \frac{c_1}{s+1} + \frac{c_2 s + c_3}{(s+1)^2 + 1} \tag{7.31}$$

To determine the unknown coefficients we recombine Equation 7.31:

$$Y = \frac{c_1\left[(s+1)^2 + 1\right] + (s+1)(c_2 s + c_3)}{(s+1)((s+1)^2 + 1)} = \frac{c_1\left[s^2 + 2s + 2\right] + (c_2 s^2 + c_3 s + c_2 s + c_3)}{(s+1)((s+1)^2 + 1)} \tag{7.32}$$

Combining like terms in powers of $s$ in the numerator of Equation 7.32:

$$Y = \frac{(c_1 + c_2)s^2 + (2c_1 + c_2 + c_3)s + (2c_1 + c_3)}{(s+1)((s+1)^2 + 1)} \tag{7.33}$$

Equating Equations 7.30 and 7.33, we must have

$$s^2 + 5s + 13 = (c_1 + c_2)s^2 + (2c_1 + c_2 + c_3)s + (2c_1 + c_3)$$

Recognizing that polynomials of like degree are equal everywhere only if their coefficients are equal, we must have

$$c_1 + c_2 = 1$$

$$2c_1 + c_2 + c_3 = 5$$

$$2c_1 + c_3 = 13$$

This is three linear equations in three unknowns. The solutions are

$$c_1 = 9$$
$$c_2 = -8$$
$$c_3 = -5$$

Hence Equation 7.31 becomes

$$Y(s) = \frac{9}{s+1} - \frac{8s+5}{(s+1)^2 + 1} = \frac{9}{s+1} - \frac{8(s+1)}{(s+1)^2 + 1} + \frac{3}{(s+1)^2 + 1}$$

and from our short table of Laplace transforms, Table 7.1:

$$y(t) = L^{-1}\{Y(s)\} = 9e^{-t} - 8e^{-t}\cos t + 3e^{-t}\sin t \tag{7.34}$$

Equation 7.34 is the solution to Equation 7.26.

## 7.6 LAPLACE TRANSFORMS IN NONHOMOGENEOUS ORDINARY DIFFERENTIAL EQUATIONS

We consider now ODEs of the form

$$\frac{d^n y}{dt^n} + a_1 \frac{d^{n-1} y}{dt^{n-1}} + a_2 \frac{d^{n-2} y}{dt^{n-2}} + \ldots + a_n y = g(t)$$

$$y(0) = y_0, \dot{y}(0) = \dot{y}_0, \ddot{y}(0) = \ddot{y}_0, \ldots, y^{(n-1)}(0) = y_0^{(n-1)} \tag{7.35}$$

### 7.6.1 GENERAL SOLUTION USING LAPLACE TRANSFORMS

**Theorem:**

The solution to Equation 7.35 is

$$y(t) = L^{-1}\{Y(s)\}$$

where

$$Y(s) = \frac{p_1 s^{n-1} + p_2 s^{n-2} + \ldots + p_{n-1} s + p_n}{s^n + a_1 s^{n-1} + \ldots + a_{n-1} s + a_n} + \frac{G(s)}{s^n + a_1 s^{n-1} + \ldots + a_{n-1} s + a_n} \tag{7.36}$$

$$p_1 = y_0$$

$$p_2 = \dot{y}_0 + a_1 y_0$$

$$p_3 = \ddot{y}_0 + a_1 \dot{y}_0 + a_2 y_0 \qquad (7.37)$$

$$\ldots$$

$$p_n = y_0^{(n-1)} + a_1 y_0^{(n-2)} + \ldots + a_{n-2} \dot{y}_0 + a_{n-1} y_0$$

and

$$G(s) = L\{g(t)\} \qquad (7.38)$$

**Proof:**

Step 1: Transform each element of the ODE:

$$L\{y(t)\} = Y(s)$$

$$L\left\{\frac{dy}{dt}\right\} = sY(s) - y(0) = sY(s) - y_0$$

$$\ldots$$

$$L\left\{\frac{d^n y}{dt^n}\right\} = s^n Y(s) - s^{n-1} y_0 - \ldots - s y_0^{(n-2)} - y_0^{(n-1)}$$

$$L\{g(t)\} = G(s)$$

Step 2: Substitute the transformed elements into the ODE:

$$(s^n Y(s) - s^{n-1} y_0 - \ldots - y_0^{(n-1)})$$

$$+ a_1 (s^{n-1} Y(s) - s^{n-2} y_0 - \ldots - y_0^{(n-2)}) + \ldots$$

$$+ a_{n-2}(s^2 Y(s) - s y_0 - \dot{y}_0) + a_{n-1}(sY(s) - y_0) + a_n Y(s) = G(s)$$

Step 3: Collect all terms involving $Y(s)$ on the left-hand side of the equation, everything else on the other:

$$(s^n + a_1 s^{n-1} + \ldots + a_{n-1} s + a_n )Y(s) = s^{n-1} y_0 + \ldots + y_0^{(n-1)}$$

$$+ a_1 (s^{n-2} y_0 + \ldots + y_0^{(n-2)}) + \ldots + a_{n-2}(s y_0 + \dot{y}_0) + a_{n-1} y_0 + G(s)$$

Step 4: On the right-hand side, collect coefficients of each power of $s$:

$$(s^n + a_1 s^{n-1} + \ldots + a_{n-1} s + a_n )Y(s) = y_0 s^{n-1} + (\dot{y}_0 + a_1 y_0)s^{n-2} + (\ddot{y}_0 + a_1 \dot{y}_0 + a_2 y_0)s^{n-3} + \ldots$$

$$+ (y_0^{(n-2)} + a_1 y_0^{(n-3)} + a_2 y_0^{(n-4)} + \ldots + a_{n-2} y_0)s$$

$$+ (y_0^{(n-1)} + a_1 y_0^{(n-2)} + \ldots + a_{n-1} y_0) + G(s) = p_1 s^{n-1} + p_2 s^{n-2} + \ldots + p_{n-1} s + p_n + G(s)$$

where the $p_i$ are given by Equations 7.37.

Step 5: Divide through by $s^n + a_1 s^{n-1} + ... + a_{n-1}s + a_n$ and the result is Equation 7.36.

We have determined that $Y(s)$ is the algebraic solution to the transformed ODE. We must also prove that $y(t) = L^{-1}\{Y(s)\}$ is the solution to the ODE. Word Problem WP7.4 guides the student through this proof, which ultimately depends on the uniqueness statement in Section 7.2.

### Example

*Problem*: Using Laplace transforms, solve the ODE

$$\frac{d^2y}{dt^2} + 8\frac{dy}{dt} + 15y = 2t \tag{7.39}$$

$$y(0) = 0, \quad \dot{y}(0) = 0$$

*Solution:*

We begin by transforming each element of the equation:

$$L\{y(t)\} = Y(s)$$

$$L\left\{\frac{dy}{dt}\right\} = sY(s) - y(0) = sY(s)$$

$$L\left\{\frac{d^2y}{dt^2}\right\} = s^2Y(s) - sy_0 - \dot{y}_0 = s^2Y(s)$$

$$L\{2t\} = \frac{2}{s^2}$$

Substituting the transformed elements into the ODE:

$$(s^2 + 8s + 15)Y = \frac{2}{s^2}$$

Solving:

$$Y = \frac{2}{s^2(s^2 + 8s + 15)} \tag{7.40}$$

We must factor the quadratic in the denominator of Equation 7.40. We seek two integers whose sum is 8 and whose product is 15 and find the desired integers in 3 and 5. (Had this search been unsuccessful we would next have tried completing the square.) Hence

$$Y = \frac{2}{s^2(s+3)(s+5)} \tag{7.41}$$

Equation 7.41 is not in our short list of Laplace transforms (Table 7.1). We must expand it in partial fractions:

$$\frac{2}{s^2(s+3)(s+5)} = \frac{c_1s+c_2}{s^2} + \frac{c_3}{s+3} + \frac{c_4}{s+5} \tag{7.42}$$

Note that we must include an $s$ term in the first numerator. To determine the unknown coefficients $c_i$ we recombine the three terms on the right-hand side of Equation 7.42:

$$\frac{c_1s+c_2}{s^2} + \frac{c_3}{s+3} + \frac{c_4}{s+5} = \frac{(c_1s+c_2)(s+3)(s+5)+c_3s^2(s+5)+c_4s^2(s+3)}{s^2(s+3)(s+5)} \tag{7.43}$$

Multiplying out the numerator and collecting coefficients of powers of $s$ results in

$$\frac{(c_1s+c_2)(s+3)(s+5)+c_3s^2(s+5)+c_4s^2(s+3)}{s^2(s+3)(s+5)} =$$
$$\frac{(c_1+c_3+c_4)s^3+(8c_1+c_2+5c_3+3c_4)s^2+(15c_1+8c_2)s+15c_2}{s^2(s+3)(s+5)} \tag{7.44}$$

From Equations 7.42, 7.43, and 7.44,

$$(c_1+c_3+c_4)s^3+(8c_1+c_2+5c_3+3c_4)s^2+(15c_1+8c_2)s+15c_2 = 2 \tag{7.45}$$

Equation 7.45 is true for all values of $s$ if, equating coefficients in like powers of $s$:

$$c_1+c_3+c_4 = 0$$
$$8c_1+c_2+5c_3+3c_4 = 0$$
$$15c_1+8c_2 = 0 \tag{7.46}$$
$$15c_2 = 2$$

This is four linear equations in four unknowns but, due to its form, easily solved. The result is

$$c_1 = -\frac{16}{225}$$
$$c_2 = \frac{2}{15}$$
$$c_3 = \frac{1}{9} \tag{7.47}$$
$$c_4 = -\frac{1}{25}$$

Combining Equations 7.41, 7.42, and 7.47,

$$Y(s) = -\left(\frac{16}{225}\right)\frac{1}{s} + \left(\frac{2}{15}\right)\frac{1}{s^2} + \left(\frac{1}{9}\right)\frac{1}{s+3} - \left(\frac{1}{25}\right)\frac{1}{s+5} \tag{7.48}$$

From Table 7.1:

$$y(t) = L^{-1}\{Y(s)\} = -\left(\frac{16}{225}\right) + \left(\frac{2}{15}\right)t + \left(\frac{1}{9}\right)e^{-3t} - \left(\frac{1}{25}\right)e^{-5t} \tag{7.49}$$

Equation 7.49 is the solution to Equation 7.39.

### 7.6.2 LAPLACE TRANSFORM OF A CONVOLUTION

**Theorem:**

Let $H(s) = L\{h(t)\}$ and $G(s) = L\{g(t)\}$. Then

$$H(s)G(s) = L\left\{\int_0^t h(t-\tau)g(\tau)d\tau\right\} \tag{7.50}$$

In words, the Laplace transform of the convolution of two functions is the product of their Laplace transforms. This is another very important attribute of Laplace transforms.

**Proof:**

$$L\left\{\int_0^t h(t-\tau)g(\tau)d\tau\right\} = \int_0^\infty e^{-st}\int_0^t h(t-\tau)g(\tau)d\tau dt = \int_0^\infty \int_0^t e^{-st}h(t-\tau)g(\tau)d\tau dt$$

First we change the order of integration, ensuring as we do that $t-\tau$ is always positive:

$$L\left\{\int_0^t h(t-\tau)g(\tau)d\tau\right\} = \int_0^\infty \int_\tau^\infty e^{-st}h(t-\tau)g(\tau)dt d\tau$$

Next we change variables

$$u = t - \tau$$

$$du = dt$$

for fixed $\tau$. Then we have

$$L\left\{\int_0^t h(t-\tau)g(\tau)d\tau\right\} = \int_0^\infty \int_0^\infty e^{-s(u+\tau)}h(u)g(\tau)du d\tau = \int_0^\infty e^{-su}h(u)du\int_0^\infty e^{-s\tau}g(\tau)d\tau = H(s)G(s)$$

as was to be shown.

**Corollary:**

The kernel for the ODE

$$\frac{d^n y}{dt^n} + a_1 \frac{d^{n-1} y}{dt^{n-1}} + a_2 \frac{d^{n-2} y}{dt^{n-2}} + \ldots + a_n y = g(t) \tag{7.51}$$

is

$$h(t) = L^{-1}\left\{H(s)\right\} = L^{-1}\left\{\frac{1}{s^n + a_1 s^{n-1} + a_2 s^{n-2} + \ldots + a_n}\right\} \tag{7.52}$$

**Proof:**

Recall from Section 6.3.1 that if $y(t)$ is the solution to Equation 7.51 with zero initial conditions then $y(t)$ can be written as

$$y(t) = \int_0^t h(t - \tau) g(\tau) d\tau \tag{7.53}$$

(We are letting $K(t - \tau) = h(t - \tau)$ here.) Taking the Laplace transform of Equation 7.53 and using the theorem (Equation 7.50):

$$L\left\{y(t)\right\} = Y(s) = H(s)G(s) \tag{7.54}$$

From Equation 7.36, assuming zero initial conditions,

$$Y(s) = \frac{G(s)}{s^n + a_1 s^{n-1} + a_2 s^{n-2} + \ldots + a_n} \tag{7.55}$$

From Equations 7.54 and 7.55,

$$H(s) = \frac{1}{s^n + a_1 s^{n-1} + a_2 s^{n-2} + \ldots + a_n} \tag{7.56}$$

Hence

$$h(t) = L^{-1}\left\{H(s)\right\} = L^{-1}\left\{\frac{1}{s^n + a_1 s^{n-1} + a_2 s^{n-2} + \ldots + a_n}\right\} \tag{7.57}$$

as was to be shown.

Equation 7.57 is typically a much easier way of determining the kernel of a higher-order LTI ODE than is Equation 6.30 (Section 6.3). Engineers call $H(s)$ the *transfer function* of the system that Equation 7.51 represents.

### 7.6.3  Differential Equations with Impulsive and Discontinuous Inputs

In this section, we explore the utility of the Laplace transform when the input (forcing function) in LTI ODEs is *impulsive* or discontinuous. We must begin by defining what we mean by *impulsive*.

#### 7.6.3.1  Defining the Dirac Delta Function

Consider the sequence of functions defined for $t_0 \geq 0$

$$d_n(t - t_0) = n/\tau \quad 0 < t - t_0 < \tau/n$$

$$d_n(t - t_0) = 0 \quad \text{otherwise} \tag{7.58}$$

See Figure 7.1. Note that the integral

$$I_n = \int_0^\infty d_n(t - t_0)dt = 1$$

independent of $n$.

Consider now the sequence of integrals

$$J_n = \int_0^\infty d_n(t - t_0)f(t)dt$$

From Equations 7.58,

$$J_n = \int_0^\infty d_n(t - t_0)f(t)dt = \frac{n}{\tau}\int_{t_0}^{t_0 + \tau/n} f(t)dt$$

If $f(t)$ is a continuous function, by the mean value theorem of integral calculus,

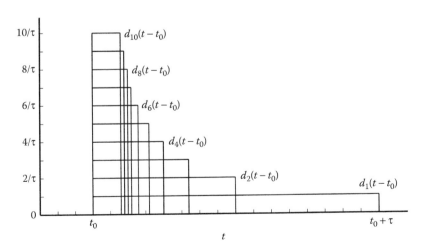

**FIGURE 7.1**  Dirac delta function as $\lim_{n \to \infty} d_n(t - t_0)$.

$$J_n = \int_0^\infty d_n(t-t_0)f(t)dt = \frac{n}{\tau}\int_{t_0}^{t_0+\tau/n} f(t)dt = \frac{n}{\tau}\cdot\frac{\tau}{n}\cdot f(t^*) = f(t^*)$$

for some $t^*$ such that

$$t_0 < t^* < t_0 + \tau/n$$

Hence,

$$\lim_{n\to\infty} J_n = \lim_{n\to\infty}\int_0^\infty d_n(t-t_0)f(t)dt = f(t_0) \tag{7.59}$$

Since their duration is short and their integral finite the functions $d_n(t-t_0)$ are often called "impulses".

The limit $$\lim_{n\to\infty} d_n(t-t_0) = \delta(t-t_0) \tag{7.60}$$

is also referred to as an impulse and carries the name *Dirac delta function* at $t_0$. It is not a true function (mapping, e.g., the real line into the real line) but rather what mathematicians call a "generalized function" or "functional", defined by

$$\int_0^\infty \delta(t-t_0)f(t)dt = f(t_0) \tag{7.61}$$

Although it is not a true function, the Dirac delta function has a well defined Laplace transform. Letting $f(t) = e^{-st}$ in Equation 7.61, we find that

$$L\{\delta(t-t_0)\} = \int_0^\infty e^{-st}\delta(t-t_0)dt = e^{-st_0} \tag{7.62}$$

In particular, a Dirac delta function at the origin $t_0 = 0$ has the Laplace transform

$$L\{\delta(t)\} = 1 \tag{7.63}$$

We will add Equation 7.63 to our short table of Laplace transforms.

### 7.6.3.2 Ordinary Differential Equations with Impulsive Inputs

Consider now the nth order LTI ODE with no initial conditions and an input consisting of just the Dirac delta function at the origin:

$$\frac{d^n y}{dt^n} + a_1\frac{d^{n-1}y}{dt^{n-1}} + a_2\frac{d^{n-2}y}{dt^{n-2}} + \dots + a_n y = \delta(t) \tag{7.64}$$

By the time-domain methods of Section 6.3.2 we know that the solution $y(t)$ can be written as

$$y(t) = \int_0^t h(t-\tau)\delta(\tau)d\tau \tag{7.65}$$

From Equation 7.61,

$$y(t) = h(t)$$

We can show the same thing by Laplace transform methods. From Equation 7.36,

$$Y(s) = \frac{G(s)}{s^n + a_1 s^{n-1} + \ldots + a_n}$$

From Equation 7.63,

$$G(s) = 1$$

Therefore

$$y(t) = L^{-1} \left\{ \frac{1}{s^n + a_1 s^{n-1} + \ldots + a_n} \right\}$$

and using Equation 7.57 we again find that $y(t) = h(t)$. Because of this, in addition to being called the kernel of the ODE, $h(t)$ is often referred to by engineers as the "impulse response" of the system.

### 7.6.3.3  Ordinary Differential Equations with Discontinuous Inputs

We define a *unit step function* beginning at time $T$ as $u_T(t)$:

$$u_T(t) = 0 \quad t \leq T$$

$$u_T(t) = 1 \quad t > T$$

Figure 7.2 depicts $u_T(t)$.

**Theorem:**
If $F(s)$ is the Laplace transform of a function $f(t)$, then the Laplace transform of the function delayed by time $T$ is

$$L\{u_T(t)f(t-T)\} = \int_0^\infty e^{-st} u_T(t) f(t-T) dt = e^{-sT} F(s) \qquad (7.66)$$

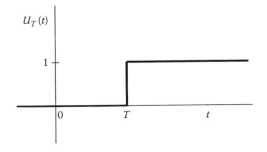

**FIGURE 7.2**   Unit step function $u_T(t)$.

**Proof:**

$$L\{u_T(t)f(t-T)\} = \int_0^\infty e^{-st}u_T(t)f(t-T)dt = \int_T^\infty e^{-st}f(t-T)dt \qquad (7.67)$$

Let $t' = t - T$. Then Equation 7.67 becomes

$$L\{u_T(t)f(t-T)\} = \int_T^\infty e^{-st}f(t-T)dt = \int_0^\infty e^{-s(t'+T)}f(t')dt' = e^{-sT}\int_0^\infty e^{-st'}f(t')dt'$$

$$L\{u_T(t)f(t-T)\} = e^{-sT}F(s)$$

as was to be shown.

The theorem requires some explanation. By convention, the starting point for our ODEs and therefore for Laplace transform integration is $t = 0$. But commonly used functions are usually defined for $t < 0$ as well as $t > 0$, hence we multiply $f(t-T)$ by $u_T(t)$ so that the delayed function is zero prior to $t = T$.

**Corollary:**

Using Equation 7.66 and Table 7.1, the Laplace transform of $u_T(t)$ is the Laplace transform of 1 delayed by $T$:

$$L\{u_T(t)\} = L\{u_T(t) \cdot 1\} = e^{-sT}L\{1\} = \frac{e^{-sT}}{s} \qquad (7.68)$$

---

**TABLE 7.2**

**Short Table of Laplace Transforms (Final Version)**

| $f(t)$ | $F(s)$ |
|---|---|
| $\delta(t)$ | $1$ |
| $1$ | $\dfrac{1}{s}$ |
| $t$ | $\dfrac{1}{s^2}$ |
| $e^{-\lambda t}$ | $\dfrac{1}{s+\lambda}$ |
| $\sin \omega t$ | $\dfrac{\omega}{s^2+\omega^2}$ |
| $\cos \omega t$ | $\dfrac{s}{s^2+\omega^2}$ |
| $te^{-\lambda t}$ | $\dfrac{1}{(s+\lambda)^2}$ |
| $e^{-\lambda t}\sin \omega t$ | $\dfrac{\omega}{(s+\lambda)^2+\omega^2}$ |
| $e^{-\lambda t}\cos \omega t$ | $\dfrac{s+\lambda}{(s+\lambda)^2+\omega^2}$ |
| $u_T(t)f(t-T)$ | $e^{-sT}F(s)$ |

Adding Equation 7.66 completes our short list of Laplace transforms (Table 7.2). Table 7.2 together with solution techniques this chapter demonstrates suffice for the examples and problems in this text. One can find much more extensive tables of Laplace transforms in the literature or on the web. See, for example, [8].

## Example

*Problem*: Determine the output $y(t)$ of an underdamped second-order system with the non-repeating square wave input $w(t)$ shown in Figure 7.3.

*Solution*:

How do we represent $w(t)$ in a differential equation? We can describe $w(t)$ as a combination of step functions occurring at different times:

$$w(t) = u_0(t) - 2u_T(t) + u_{2T}(t) \tag{7.69}$$

Using Equations 5.42 and 7.69, we can describe a general underdamped second-order system with input $w(t)$ by the ODE

$$\frac{d^2y}{dt^2} + 2\lambda\frac{dy}{dt} + (\lambda^2 + \omega^2)y = u_0(t) - 2u_T(t) + u_{2T}(t) \tag{7.70}$$

We want to solve this problem using Laplace transforms. What is the Laplace transform of $w(t)$? Using Equation 7.68:

$$L\{w(t)\} = L\{u_0(t)\} - L\{2u_T(t)\} + L\{u_{2T}(t)\} = \frac{1}{s}\left(1 - 2e^{-sT} + e^{-2sT}\right) \tag{7.71}$$

Assuming zero initial conditions, we have

$$\left(s^2 + 2\lambda s + (\lambda^2 + \omega^2)\right)Y = \frac{1}{s}\left(1 - 2e^{-sT} + e^{-2sT}\right) \tag{7.72}$$

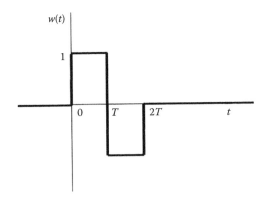

**FIGURE 7.3**   Nonrepeating square wave.

Solving,

$$Y(s) = \frac{1 - 2e^{-sT} + e^{-2sT}}{s\left(s^2 + 2\lambda s + (\lambda^2 + \omega^2)\right)} \tag{7.73}$$

Equation 7.73 is not in our short list of Laplace transforms (Table 7.2). In this case it suffices to expand $Y_0(s)$ in a partial fraction expansion, where

$$Y_0(s) = \frac{1}{s\left(s^2 + 2\lambda s + (\lambda^2 + \omega^2)\right)}$$

We have

$$Y_0(s) = \frac{1}{s\left(s^2 + 2\lambda s + (\lambda^2 + \omega^2)\right)} = \frac{C_1}{s} + \frac{C_2 s + C_3}{s^2 + 2\lambda s + (\lambda^2 + \omega^2)}$$

$$= \frac{C_1(s^2 + 2\lambda s + (\lambda^2 + \omega^2)) + C_2 s^2 + C_3 s}{s(s^2 + 2\lambda s + (\lambda^2 + \omega^2))} =$$

$$\frac{s^2(C_1 + C_2) + s(2\lambda C_1 + C_3) + C_1(\lambda^2 + \omega^2)}{s(s^2 + 2\lambda s + (\lambda^2 + \omega^2))}$$

Equating coefficients of like powers of $s$ :

$$C_1 + C_2 = 0, \qquad 2\lambda C_1 + C_3 = 0, \qquad C_1(\lambda^2 + \omega^2) = 1$$

This is three linear equations in three unknowns. The solutions are

$$C_1 = \frac{1}{\lambda^2 + \omega^2}, \qquad C_2 = \frac{-1}{\lambda^2 + \omega^2}, \qquad C_3 = \frac{-2\lambda}{\lambda^2 + \omega^2}$$

Therefore

$$Y_0(s) = \frac{1}{\lambda^2 + \omega^2}\left[\frac{1}{s} - \frac{(s + 2\lambda)}{s^2 + 2\lambda s + (\lambda^2 + \omega^2)}\right] =$$

$$\frac{1}{\lambda^2 + \omega^2}\left[\frac{1}{s} - \frac{(s + \lambda)}{(s + \lambda)^2 + \omega^2} - \frac{\lambda}{(s + \lambda)^2 + \omega^2}\right]$$

Referring to Table 7.2 we see that

$$y_0(t) = L^{-1}\{Y_0(s)\} = \frac{1}{\lambda^2 + \omega^2}\left[1 - e^{-\lambda t}\cos\omega t - \frac{\lambda}{\omega}e^{-\lambda t}\sin\omega t\right] \tag{7.74}$$

and

$$y(t) = L^{-1}\{Y(s)\} = L^{-1}\left\{\frac{1 - 2e^{-sT} + e^{-2sT}}{s\left(s^2 + 2\lambda s + (\lambda^2 + \omega^2)\right)}\right\} = y_0(t) - 2u_T y_0(t - T) + u_{2T}y_0(t - 2T) \tag{7.75}$$

Equations 7.74 and 7.75 are the solution to Equation 7.70.

### 7.6.4　Initial and Final Value Theorems

This section states and proves two theorems frequently used by engineers.

**Initial Value Theorem**

Let $F(s)$ be the Laplace transform of a continuous function $f(t)$ and let $|df/dt| \leq M$ for all $t$. Then

$$f(0) = \lim_{s \to \infty} sF(s) \qquad (7.76)$$

**Proof:**

We begin with Equation 7.2

$$L\left\{\frac{df}{dt}\right\} = sL\{f(t)\} - f(0)$$

or

$$L\left\{\frac{df}{dt}\right\} = sF(s) - f(0)$$

Now

$$L\left\{\frac{df}{dt}\right\} = \int_0^\infty \left(\frac{df}{dt}\right) e^{-st} dt$$

so

$$sF(s) = f(0) + \int_0^\infty \left(\frac{df}{dt}\right) e^{-st} dt \qquad (7.77)$$

Since

$$\left|\int_0^\infty \left(\frac{df}{dt}\right) e^{-st} dt\right| \leq \int_0^\infty \left|\frac{df}{dt}\right| e^{-st} dt \leq M \int_0^\infty e^{-st} dt = \frac{M}{s}$$

we have

$$\lim_{s \to \infty} \left|\int_0^\infty \left(\frac{df}{dt}\right) e^{-st} dt\right| \leq \lim_{s \to \infty} \frac{M}{s} = 0 \qquad (7.78)$$

Hence, from Equations 7.77 and 7.78,

$$\lim_{s \to \infty} sF(s) = f(0)$$

as was to be shown.

Note that the theorem proof began with Equation 7.2, which stipulated that the function $f(t)$ is continuous. It can be shown that the theorem also holds for functions continuous everywhere except at the origin, provided that the theorem is amended to

$$\lim_{s \to \infty} sF(s) = f(0^+)$$

where $f(0^+)$ is understood to be the value of the function just *after* $t = 0$. This is useful for systems with an impulsive input at $t = 0$. Solved Problem SP7.4 demonstrates an example of the utility of the initial value theorem in such a case.

**Final Value Theorem**

Let $F(s)$ be the Laplace transform of a continuous function $f(t)$ that has a limit as $t \to \infty$. Then

$$\lim_{t \to \infty} f(t) = \lim_{s \to 0} sF(s) \tag{7.79}$$

**Proof:**

We again begin with

$$L\left\{\frac{df}{dt}\right\} = sF(s) - f(0)$$

and as before arrive at

$$sF(s) = f(0) + \int_0^\infty \left(\frac{df}{dt}\right) e^{-st} dt \tag{7.80}$$

Now

$$\int_0^\infty \left(\frac{df}{dt}\right) e^{-st} dt = \lim_{T \to \infty} \int_0^T \left(\frac{df}{dt}\right) e^{-st} dt$$

so

$$\lim_{s \to 0} \int_0^\infty \left(\frac{df}{dt}\right) e^{-st} dt = \lim_{s \to 0} \lim_{T \to \infty} \int_0^T \left(\frac{df}{dt}\right) e^{-st} dt = \lim_{T \to \infty} \lim_{s \to 0} \int_0^T \left(\frac{df}{dt}\right) e^{-st} dt$$

$$= \lim_{T \to \infty} \int_0^T \left(\frac{df}{dt}\right) dt = \lim_{T \to \infty} f(T) - f(0)$$

or

$$\lim_{s \to 0} \int_0^\infty \left(\frac{df}{dt}\right) e^{-st} dt = \lim_{T \to \infty} f(T) - f(0) \tag{7.81}$$

From Equations 7.80 and 7.81,

$$\lim_{s \to 0} sF(s) = f(0) + \lim_{T \to \infty} f(T) - f(0) = \lim_{T \to \infty} f(T)$$

as was to be shown.

A word of caution about the final value theorem: Some engineering texts state and prove the theorem without noting the condition that the function limit as time goes to infinity must exist.

For example, applying the theorem to $F(s) = \omega/(s^2 + \omega^2)$ yields $\lim_{s \to 0} sF(s) = 0$, but we know that the inverse Laplace transform of $F(s)$ is $\sin \omega t$, which does not have a limit.

### 7.6.5 ORDINARY DIFFERENTIAL EQUATIONS WITH PERIODIC INPUTS

This section examines the response of nth order LTI ODEs to a periodic input; that is, one that satisfies $g(t + T) = g(t)$ for all $t > 0$. The Laplace transform of a periodic input $g(t)$ is

$$
\begin{aligned}
G(s) = L\{g(t)\} &= \int_0^\infty e^{-st} g(t) dt \\
&= \int_0^T e^{-st} g(t) dt + \int_T^{2T} e^{-st} g(t) dt + \int_{2T}^{3T} e^{-st} g(t) dt + \ldots + \int_{nT}^{(n+1)T} e^{-st} g(t) dt + \ldots \quad (7.82) \\
&= \sum_{n=0}^\infty G_n(s)
\end{aligned}
$$

Consider the nth term

$$
G_n(s) = \int_{nT}^{(n+1)T} e^{-st} g(t) dt
$$

Let $t' = t - nT$. Then, using the periodicity of $g(t)$:

$$
G_n(s) = \int_0^T e^{-s(t'+nT)} g(t' + nT) dt' = \int_0^T e^{-s(t'+nT)} g(t') dt'
$$

$$
= e^{-nsT} \int_0^T e^{-st'} g(t') dt' = (e^{-sT})^n G_0(s) \quad (7.83)
$$

where

$$
G_0(s) = \int_0^T e^{-st} g(t) dt \quad (7.84)
$$

From Equations 7.82, 7.83, and 7.84,

$$
G(s) = G_0(s) \sum_{n=0}^\infty (e^{-sT})^n \quad (7.85)
$$

Now

$$
\sum_{n=0}^\infty x^n = \frac{1}{1-x} \quad \text{for } |x| < 1
$$

so

$$
G(s) = \frac{G_0(s)}{1 - e^{-sT}} \quad (7.86)
$$

### 7.6.5.1 Response to a Periodic Input

Using Equations 7.36 and 7.86, assuming zero initial conditions, the Laplace transform $Y(s)$ of the response of an nth order LTI ODE to a periodic input is

$$Y(s) = \frac{G_0(s)}{(1 - e^{-sT})(s^n + a_1 s^{n-1} + \ldots + a_n)} \tag{7.87}$$

We can use Equation 7.87 to derive an important result.

**Theorem:**

The response of a stable, damped nth order LTI ODE to a bounded periodic input is eventually periodic.

**Proof:**

From Equation 7.87,

$$Y(s)(1 - e^{-sT}) = \frac{G_0(s)}{s^n + a_1 s^{n-1} + \ldots + a_n} \tag{7.88}$$

Now

$$L^{-1}\left\{Y(s)(1 - e^{-sT})\right\} = L^{-1}\left\{Y(s)\right\} - L^{-1}\left\{e^{-sT}Y(s)\right\} \tag{7.89}$$

and from Table 7.2,

$$L^{-1}\left\{e^{-sT}Y(s)\right\} = u_T y(t - T) \tag{7.90}$$

hence the inverse Laplace transform of Equation 7.88 is

$$y(t) - u_T y(t - T) = L^{-1}\left\{\frac{G_0(s)}{s^n + a_1 s^{n-1} + \ldots + a_n}\right\}$$

For $t > T$ this is

$$y(t) - y(t - T) = L^{-1}\left\{\frac{G_0(s)}{s^n + a_1 s^{n-1} + \ldots + a_n}\right\} \tag{7.91}$$

To show that the response to a periodic input eventually becomes periodic, we must show that

$$\lim_{t \to \infty} \left(y(t) - y(t - T)\right) = 0$$

which is to say, from Equation 7.91, we must show that

$$\lim_{t \to \infty} L^{-1}\left\{\frac{G_0(s)}{s^n + a_1 s^{n-1} + \ldots + a_n}\right\} = 0$$

We could use the final value theorem (Equation 7.79) if we were certain that the function

$$f(t) = y(t) - y(t-T) = L^{-1}\left\{\frac{G_0(s)}{s^n + a_1 s^{n-1} + \ldots + a_n}\right\}$$

has a limit. In examining whether it does we see that $f(t)$ can be thought of as the output of a stable and damped system when the input is a function $g_0(t)$ that is zero after time $T$. At time $T$ the system has a set of bounded "initial" conditions and from our definition of a stable and damped LTI system (Chapter 2, Section 2.3), the output approaches zero as time goes to infinity. Hence we have determined that the function $f(t)$ has a limit and in the process established that its value is zero. That is,

$$\lim_{t\to\infty}(y(t) - y(t-T)) = 0$$

which was to be shown.

As an exercise, we can apply the final value theorem (Equation 7.79) to see if we arrive at the same answer:

$$\lim_{t\to\infty} L^{-1}\left\{\frac{G_0(s)}{s^n + a_1 s^{n-1} + \ldots + a_n}\right\} = \lim_{s\to 0} \frac{s G_0(s)}{s^n + a_1 s^{n-1} + \ldots + a_n} \tag{7.92}$$

Now

$$\lim_{s\to 0} \frac{s G_0(s)}{s^n + a_1 s^{n-1} + \ldots + a_n} = \lim_{s\to 0} \frac{s G_0(s)}{a_n} \tag{7.93}$$

From Equation 7.84, and using the fact that $g(t)$ is bounded,

$$|G_0(s)| \le \int_0^T |g(t)| e^{-st} dt \le M \int_0^T e^{-st} dt = \frac{M}{s}(1 - e^{-sT}) \tag{7.94}$$

From Equations 7.93 and 7.94,

$$\lim_{s\to 0} \left|\frac{s G_0(s)}{a_n}\right| \le \lim_{s\to 0} \frac{M}{a_n}(1 - e^{-sT}) = \frac{M}{a_n}\lim_{s\to 0}(1 - e^{-sT}) = 0 \tag{7.95}$$

From Equations 7.91 through 7.95,

$$\lim_{t\to\infty}\left(y(t) - y(t-T)\right) = 0$$

as indeed we had determined earlier.

### 7.6.5.2 Resonance with a Nonsinusoidal Periodic Input

This section explores the magnitude of an LTI system's response to a nonsinusoidal periodic input. It shows by example that a periodic input can cause resonance if the periodicity of the input is commensurate with the system's natural frequency, no matter the waveform.

Consider a second-order underdamped system excited by a periodic square wave $\chi_T(t)$:

$$\chi_T(t) = +1 \quad 2nT < t < (2n+1)T$$

$$\chi_T(t) = -1 \quad (2n+1)T < t < (2n+2)T \quad n = 0,1,2,.... \tag{7.96}$$

Figure 7.4 illustrates. Note that $\chi_T(t)$ has period $2T$.

From Equation 5.42 the ODE for a second-order underdamped system with input $\chi_T(t)$ is

$$\frac{d^2 y}{dt^2} + 2\lambda \frac{dy}{dt} + (\lambda^2 + \omega^2)y = \chi_T(t) \tag{7.97}$$

From Equations 7.84 and 7.86

$$L\{\chi_T(t)\} = \frac{\int_0^{2T} e^{-st}\left(u_0(t) - 2u_T(t) + u_{2T}(t)\right)dt}{1 - e^{-2sT}} \tag{7.98}$$

From Equation 7.71,

$$\int_0^{2T} e^{-st}\left(u_0(t) - 2u_T(t) + u_{2T}(t)\right)dt = \frac{1 - 2e^{-sT} + e^{-2sT}}{s} = \frac{(1 - e^{-sT})^2}{s} \tag{7.99}$$

From Equations 7.98 and 7.99,

$$L\{\chi_T(t)\} = \frac{(1 - e^{-sT})^2}{s(1 - e^{-2sT})} = \frac{(1 - e^{-sT})^2}{s(1 - e^{-sT})(1 + e^{-sT})} = \frac{1 - e^{-sT}}{s(1 + e^{-sT})} \tag{7.100}$$

From Equations 7.36 and 7.100, assuming zero initial conditions, the Laplace transform of the solution to Equation 7.97 is

$$Y(s) = L\{y(t)\} = \frac{1 - e^{-sT}}{s(1 + e^{-sT})(s^2 + 2\lambda s + (\lambda^2 + \omega^2))} \tag{7.101}$$

Equation 7.101 is not in our short list of Laplace tranforms (Table 7.2) and not amenable to the techniques we have been using to make it so, such as a partial fraction expansion. The problem is the transcendental function $1 + e^{-sT}$ in the denominator. Challenge Problem

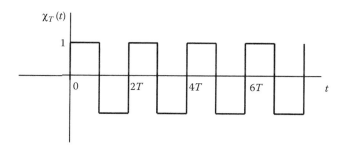

**FIGURE 7.4** Periodic square wave.

CP7.1 takes results we have already obtained via Laplace transforms and combines them with time-domain methods to derive the *steady-state* solution to Equation 7.97. The result can be written as

$$y_{SS}(t) = \frac{(-1)^N}{c^2}\left[1 - \frac{(2c/\omega)}{D}q(\delta t)\right] \tag{7.102}$$

where $N$ begins at zero and advances by one with each switch of the square wave, $\delta t$ is the time elapsed since the last switch,

$$c = \sqrt{\lambda^2 + \omega^2}$$

$$q(\delta t) = e^{-\lambda \delta t}\left(\cos(\omega \delta t - \theta) + e^{-\lambda T}\cos(\omega(\delta t - T) + \theta)\right) \tag{7.103}$$

$$\theta = \tan^{-1}(\lambda/\omega)$$

and, most importantly for the purpose at hand,

$$D = 1 + 2e^{-\lambda T}\cos\omega T + e^{-2\lambda T} \tag{7.104}$$

It is the value of $D$ that determines the peak magnitude of the steady-state response: the smaller is $D$, the larger is the response. We are interested in how $D$, and hence the peak response, varies with the period of the square wave, $2T$. From Equation 7.104 we readily see that $D$ reaches a local minimum when

$$\omega T = (2n+1)\pi \tag{7.105}$$

There are an infinite number of solutions to this. Call them $T_n$. At these points,

$$D = 1 - 2e^{-\lambda T_n} + e^{-2\lambda T_n} = (1 - e^{-\lambda T_n})^2 \tag{7.106}$$

From Equation 7.106 we see that $T_0 = \pi/\omega$ provides the global minimum of $D$. Then

$$D = (1 - \exp(-\pi\lambda/\omega))^2 \tag{7.107}$$

In those cases where the system has very low damping; that is, $\lambda/\omega \ll 1$, $D$ can be very small and hence the peak steady-state response quite large. When $T = T_0$ the *radian frequency* of the square-wave input is

$$\omega_{input} = \frac{2\pi}{period} = \frac{2\pi}{2T_0} = \frac{\pi}{(\pi/\omega)} = \omega$$

Hence *resonance* can occur when the frequency of the periodic input equals or is nearly equal to the natural frequency of the system, independent of the input's waveform. Figure 7.5 graphs the result of numerically integrating Equation 7.97 for three different square-wave input frequencies for the same underdamped second-order system represented in Figure 5.8.

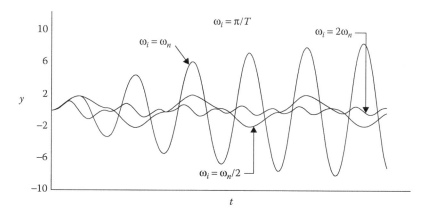

**FIGURE 7.5**  Output of an example underdamped second-order system of natural frequency $\omega_n$ when excited by square-wave inputs of unit amplitude but different frequencies $\omega_i$.

## 7.7   SUMMARY

Laplace transforms are a principal analytic tool for engineers. For LTI ODEs of any order, Laplace transforms convert differential equations into algebraic ones, solve for initial conditions without extra steps, permit solution of nonhomogeneous equations without first solving homogeneous equations, and address impulsive and discontinuous forcing functions easily.

The Laplace transform $F(s)$ of a function $f(t)$ is

$$F(s) = L\{f(t)\} = \int_0^\infty e^{-st} f(t)dt$$

The transform $F(s)$ exists provided the function $f(t)$ is piecewise continuous and of exponential order (i.e., for $t > M$, $|f(t)| < Ke^{at}$ for some $K$, $a$, and $M$). By its definition the Laplace transform is a linear operation, meaning that, for constants $a$ and $b$,

$$L\{af(t) + bg(t)\} = aL\{f(t)\} + bL\{g(t)\}$$

This chapter computed the transform of a number of common functions, including functions multiplied by $e^{-\lambda t}$, functions delayed by time $T$, derivatives of a function, the convolution of two functions, and periodic functions.

If $f(t)$ is continuous and $F(s)$ is its transform then there is no other continuous function with that transform. This property enables us to construct a table using the transform's defining equation in which the left-hand column is the function of time and the right-hand column its computed Laplace transform. Our process for solving ODEs is then

1. Transform each element of the given ODE from a function of time into an algebraic function of $s$.
2. Solve the resulting algebraic equation.
3. Manipulate the algebraic solution into a linear combination of forms that appear in the right-hand column of the table of Laplace transforms.
4. Move from right- to left-hand column to determine the function of time corresponding to the algebraic solution.

Assuming zero initial conditions and that $L\{g(t)\} = G(s)$, the Laplace transform $Y(s)$ of the solution $y(t)$ to

$$\frac{d^n y}{dt^n} + a_1 \frac{d^{n-1} y}{dt^{n-1}} + a_2 \frac{d^{n-2} y}{dt^{n-2}} + \ldots + a_n y = g(t)$$

is

$$Y(s) = H(s)G(s)$$

where

$$H(s) = \frac{1}{s^n + a_1 s^{n-1} + a_2 s^{n-2} + \ldots + a_n}$$

is called the transfer function of the system. The transfer function is the Laplace transform of the kernel.

The generalized function known as the Dirac delta function is defined by

$$\int_0^\infty \delta(t - t_0) f(t) dt = f(t_0)$$

It can be thought of as the limit of a sequence of impulses having unit integral but higher magnitude and shorter duration as the sequence progresses. The number one is the Laplace transform of the Dirac delta function. The solution of the ODE

$$\frac{d^n y}{dt^n} + a_1 \frac{d^{n-1} y}{dt^{n-1}} + a_2 \frac{d^{n-2} y}{dt^{n-2}} + \ldots + a_n y = \delta(t)$$

is

$$h(t) = L^{-1}\{H(s)\} = L^{-1}\left\{ \frac{1}{s^n + a_1 s^{n-1} + a_2 s^{n-2} + \ldots + a_n} \right\}$$

so, in addition to being called the kernel of the ODE, $h(t)$ is also called its impulse response.

Two properties of Laplace transforms frequently useful to engineers are the initial and final value theorems:

Let $F(s)$ be the Laplace transform of a function $f(t)$ continuous everywhere except possibly at the origin and let $|df/dt| \leq M$ for all $t$. Then

$$\lim_{t \to 0} f(t) = f(0^+) = \lim_{s \to \infty} sF(s)$$

Let $F(s)$ be the Laplace transform of a continuous function $f(t)$ that has a limit as $t \to \infty$. Then

$$\lim_{t \to \infty} f(t) = \lim_{s \to 0} sF(s)$$

This chapter showed that the response of a stable and damped LTI system to a periodic input eventually becomes periodic. It also showed by example that a periodic input can cause resonance if the periodicity of the input is commensurate with the system's natural frequency, no matter the waveform.

## Solved Problems

**SP7.1** Use Laplace transforms to solve

$$\frac{d^2 y}{dt^2} + 10\frac{dy}{dt} + 25y = 0 \tag{7.108}$$

$$y(0) = -3 \quad \dot{y}(0) = 1$$

*Solution*:
We begin by transforming both sides of the ODE. Because the Laplace transform is a linear operation we have

$$L\left\{\frac{d^2 y}{dt^2} + 10\frac{dy}{dt} + 25y\right\} = L\left\{\frac{d^2 y}{dt^2}\right\} + 10L\left\{\frac{dy}{dt}\right\} + 25L\{y\} = L\{0\} \tag{7.109}$$

Now define

$$L\{y(t)\} = Y(s) \tag{7.110}$$

Then

$$L\left\{\frac{dy}{dt}\right\} = sL\{y\} - y(0) = sY + 3 \tag{7.111}$$

$$L\left\{\frac{d^2 y}{dt^2}\right\} = sL\left\{\frac{dy}{dt}\right\} - \dot{y}(0) = s(sY + 3) - 1 = s^2 Y + 3s - 1 \tag{7.112}$$

From the definition of the Laplace transform

$$L\{0\} = \int_0^\infty e^{-st}(0)dt = 0 \tag{7.113}$$

Substituting Equations 7.110 through 7.113 into Equation 7.109, we have

$$(s^2 Y + 3s - 1) + 10(sY + 3) + 25Y = 0$$

Rearranging:

$$(s^2 + 10s + 25)Y = -3s - 29$$

Solving:

$$Y = \frac{-3s - 29}{s^2 + 10s + 25} \tag{7.114}$$

Equation 7.114 is the Laplace transform of the (time-domain) solution to Equations 7.108. It is not in a form we can find on the right-hand side of Table 7.2. We must manipulate it into a linear combination of those forms.

First, we factor the denominator:

$$s^2 + 10s + 25 = (s+5)^2 \tag{7.115}$$

Then, looking at the form of the denominator, we write the numerator as

$$-3s - 29 = -3(s+5) - 14 \tag{7.116}$$

Substituting Equations 7.115 and 7.116 into Equation 7.114 we arrive at

$$Y = \frac{-3(s+5) - 14}{(s+5)^2} = -\frac{3}{s+5} - \frac{14}{(s+5)^2} = -3\left(\frac{1}{s+5}\right) - 14\left(\frac{1}{(s+5)^2}\right) \tag{7.117}$$

This *is* a linear combination of forms we can find on the right-hand side of Table 7.2 and the inverse Laplace transform of Equation 7.117 is

$$y(t) = L^{-1}\{Y(s)\} = -3L^{-1}\left\{\frac{1}{s+5}\right\} - 14L^{-1}\left\{\frac{1}{(s+5)^2}\right\} = -3e^{-5t} - 14te^{-5t} \tag{7.118}$$

Equation 7.118 is the solution to Equations 7.108. We can check our result by examining initial conditions. Using Equation 7.118 we find that $y(0) = -3$ and $\dot{y}(0) = 1$, as desired.

**SP7.2** Use Laplace transforms to find the kernel $K(t - \tau)$ for the system governed by

$$\frac{d^3 y}{dt^3} + 16\frac{d^2 y}{dt^2} + 81\frac{dy}{dt} + 126y = g(t) \tag{7.119}$$

Compare the difficulty of this method with that of the method used in Chapter 6, Equation 6.30.

*Solution*:
From Section 7.6.3.2, the kernel of an LTI system is its impulse response; that is, the solution to Equation 7.119 when $g(t) = \delta(t)$, the Dirac delta function. Taking the Laplace transform of each term in Equation 7.119, and assuming zero initial conditions, we have

$$L\{y(t)\} = Y(s), \quad L\left\{\frac{dy}{dt}\right\} = sY, \quad L\left\{\frac{d^2 y}{dt^2}\right\} = s^2 Y, \quad L\left\{\frac{d^3 y}{dt^3}\right\} = s^3 Y, \quad L\{\delta(t)\} = 1$$

Then the transformed Equation 7.119 becomes

$$(s^3 + 16s^2 + 81s + 126)Y(s) = 1$$

We solve for $Y(s)$:

$$Y(s) = \frac{1}{s^3 + 16s^2 + 81s + 126} \tag{7.120}$$

We must factor the denominator. Using a web-based root calculator,

$$s^3 + 16s^2 + 81s + 126 = (s+3)(s+6)(s+7)$$

Expanding Equation 7.120 in partial fractions:

$$Y(s) = \frac{1}{s^3 + 16s^2 + 81s + 126} = \frac{1}{(s+3)(s+6)(s+7)} = \frac{c_1}{s+3} + \frac{c_2}{s+6} + \frac{c_3}{s+7}$$

$$= \frac{c_1(s+6)(s+7) + c_2(s+3)(s+7) + c_3(s+3)(s+6)}{(s+3)(s+6)(s+7)} =$$

$$\frac{c_1(s^2 + 13s + 42) + c_2(s^2 + 10s + 21) + c_3(s^2 + 9s + 18)}{(s+3)(s+6)(s+7)} =$$

$$\frac{(c_1 + c_2 + c_3)s^2 + (13c_1 + 10c_2 + 9c_3)s + (42c_1 + 21c_2 + 18c_3)}{(s+3)(s+6)(s+7)}$$

Equating coefficients in like powers of $s$:

$$c_1 + c_2 + c_3 = 0$$

$$13c_1 + 10c_2 + 9c_3 = 0$$

$$42c_1 + 21c_2 + 18c_3 = 1$$

This is three linear equations in three unknowns. The solutions are $c_1 = 1/12$, $c_2 = -4/12$, and $c_3 = 3/12$. Hence

$$Y(s) = \frac{1}{12}\left( \frac{1}{s+3} - \frac{4}{s+6} + \frac{3}{s+7} \right)$$

and using Table 7.2, the kernel is given by

$$K(t) = L^{-1}\{Y(s)\} = \frac{1}{12}(e^{-3t} - 4e^{-6t} + 3e^{-7t})$$

Even taking into account that we did not write out the steps in solving the three linear equations in three unknowns, this procedure is simpler than that used in Problem SP6.2. The difference is more marked when the fundamental solutions are not just simple exponentials.

**SP7.3** Challenge Problems CP6.1 through CP6.4 analyzed the stability of an Earth's satellite launched into a low-altitude circular orbit. It derived coupled differential equations for the radial deviation $r(t)$ and angular velocity deviation $v(t)$ and then used time domain techniques that were admittedly not easy to apply to find a differential equation in $r(t)$ alone. Our objective here is to show that Laplace transforms make this process very easy. We will solve equations in the same form but for clarity we will use simple coefficients. The equations for $r(t)$ and $v(t)$ were in the form

$$\ddot{r} = -a\dot{r} + br + cv$$
$$\dot{v} = -d\dot{r} + er - fv - g$$

(7.121)

where $a, b, c, d, e, f,$ and $g$ are constants. Taking the Laplace transform of Equations 7.121, assuming zero initial conditions and letting

$$L\{r(t)\} = \hat{r}(s) \quad L\{v(t)\} = \hat{v}(s)$$

we have

$$s^2\hat{r} = -as\hat{r} + b\hat{r} + c\hat{v}$$
$$s\hat{v} = -ds\hat{r} + e\hat{r} - f\hat{v} - g/s$$

or

$$(s^2 + as - b)\hat{r} - c\hat{v} = 0$$
$$(ds - e)\hat{r} + (s + f)\hat{v} = -g/s$$

(7.122)

Equations 7.122 are two equations in two unknowns. (Here, as always, we are treating $s$ as a known algebraic quantity.)

Using the first equation to express $\hat{v}$ in terms of $\hat{r}$,

$$\hat{v} = \left(\frac{s^2 + as - b}{c}\right)\hat{r}$$

Substituting this into the second equation:

$$(ds - e)\hat{r} + (s + f)\left(\frac{s^2 + as - b}{c}\right)\hat{r} = -g/s$$

Multiplying through by $c$ and multiplying out the product:

$$(cds - ce)\hat{r} + (s + f)(s^2 + as - b)\hat{r} = -cg/s$$
$$(cds - ce)\hat{r} + (s^3 + as^2 - bs + fs^2 + fas - bf)\hat{r} = -cg/s$$

Combining terms in powers of $s$:

$$\left(s^3 + (a + f)s^2 + (cd + af - b)s - (ce + bf)\right)\hat{r} = -cg/s$$

(7.123)

Taking the inverse Laplace transform of Equation 7.123

$$\ddot{r} + (a+f)\ddot{r} + (cd+af-b)\dot{r} - (ce+bf)r = -cg \qquad (7.124)$$

Equation 7.124 is the desired differential equation for $r(t)$ alone.

**SP7.4** Using the initial value theorem, find the value of $y(t)$ just after $t = 0$ given that

$$\frac{dy}{dt} + ay = b\delta(t) \quad y(0) = y_0 \qquad (7.125)$$

where $a$ and $b$ are constants and $\delta(t)$ is the Dirac delta function.

*Solution*:
Applying the Laplace transform to Equations 7.125 and letting $Y(s) = L\{y(t)\}$:

$$sY - y_0 + aY = b$$

$$Y = \frac{y_0 + b}{s + a}$$

$$y(0^+) = \lim_{s \to \infty} sY(s) = \lim_{s \to \infty} s\left(\frac{y_0 + b}{s + a}\right) = y_0 + b$$

Note that this result is reached much faster using the initial value theorem than it would with a time domain solution using an integrating factor.

## Problems to Solve

In Problems 7.1 through 7.13, use the Laplace transform method to find $y(t)$. Where needed, use web-based programs for finding roots of higher-order polynomials or solutions to higher-order simultaneous linear equations.

**P7.1** $\dfrac{dy}{dt} + 4y = 2e^{-6t}, \quad y(0) = 3$

**P7.2** $\dfrac{dy}{dt} + 5y = 7t, \quad y(0) = 8$

**P7.3** $\dfrac{d^2 y}{dt^2} + 8\dfrac{dy}{dt} + 16y = 0, \quad y(0) = 3, \quad \dot{y}(0) = -1$

**P7.4** $\dfrac{d^2 y}{dt^2} + 8\dfrac{dy}{dt} + 80y = 0, \quad y(0) = -2, \quad \dot{y}(0) = 1$

**P7.5** $\dfrac{d^2 y}{dt^2} + 15\dfrac{dy}{dt} + 54y = 0, \quad y(0) = 6, \quad \dot{y}(0) = 4$

**P7.6** $\dfrac{d^2 y}{dt^2} + 10\dfrac{dy}{dt} + 21y = 0, \quad y(0) = 2, \quad \dot{y}(0) = -4$

**P7.7** $\dfrac{d^2 y}{dt^2} + 18 \dfrac{dy}{dt} + 97 y = 0, \quad y(0) = -3, \quad \dot{y}(0) = 3$

**P7.8** $\dfrac{d^2 y}{dt^2} + 22 \dfrac{dy}{dt} + 121 y = 0, \quad y(0) = 4, \quad \dot{y}(0) = 6$

**P7.9** $\dfrac{d^2 y}{dt^2} + 9 \dfrac{dy}{dt} + 20 y = 3 e^{-2t}, \quad y(0) = 0, \quad \dot{y}(0) = 0$

**P7.10** $\dfrac{d^2 y}{dt^2} + 12 \dfrac{dy}{dt} + 100 y = 4, \quad y(0) = 0, \quad \dot{y}(0) = 0$

**P7.11** $\dfrac{d^2 y}{dt^2} + 20 \dfrac{dy}{dt} + 125 y = 5 \cos 2t, \quad y(0) = 0, \quad \dot{y}(0) = 0$

**P7.12** $\dfrac{d^3 y}{dt^3} + 15 \dfrac{d^2 y}{dt^2} + 66 \dfrac{dy}{dt} + 80 y = 0, \quad y(0) = -1, \quad \dot{y}(0) = 2, \quad \ddot{y}(0) = 3$

**P7.13** $\dfrac{d^3 y}{dt^3} + 14 \dfrac{d^2 y}{dt^2} + 49 \dfrac{dy}{dt} + 36 y = 0, \quad y(0) = 3, \quad \dot{y}(0) = -1, \quad \ddot{y}(0) = -2$

In Problems 7.14 and 7.15, find the value of $dy/dt$ at $t = 0^+$, where $a, b, c$, and $e$ are constants and $\delta(t)$ is the Dirac delta function.

**P7.14** $\dfrac{d^2 y}{dt^2} + b \dfrac{dy}{dt} + cy = e\delta(t), \quad y(0) = y_0, \quad \dot{y}(0) = \dot{y}_0$

**P7.15** $\dfrac{d^3 y}{dt^3} + a \dfrac{d^2 y}{dt^2} + b \dfrac{dy}{dt} + cy = e\delta(t), \quad y(0) = y_0, \quad \dot{y}(0) = \dot{y}_0, \quad \ddot{y}(0) = \ddot{y}_0$

In Problems 7.16 and 7.17, find the limiting value of $y(t)$ as $t \to \infty$, where $a, b, c$, and $e$ are constants and $u_0(t)$ is the unit step function, and given that the limits exist.

**P7.16** $\dfrac{d^2 y}{dt^2} + b \dfrac{dy}{dt} + cy = e u_0(t), \quad y(0) = y_0, \quad \dot{y}(0) = \dot{y}_0$

**P7.17** $\dfrac{d^3 y}{dt^3} + a \dfrac{d^2 y}{dt^2} + b \dfrac{dy}{dt} + cy = e u_0(t), \quad y(0) = y_0, \quad \dot{y}(0) = \dot{y}_0, \quad \ddot{y}(0) = \ddot{y}_0$

In Problems 7.18 through 7.20, determine the differential equation that $u(t)$ satisfies when $v(t)$ has been eliminated from the equations. Assume zero initial conditions.

**P7.18** $\begin{aligned} \dot{u} + 2\dot{v} + 8u &= 1 \\ \dot{v} + 6v - 3u &= 2 \end{aligned}$

**P7.19** $\ddot{u} + 4\dot{u} + 5u - 3v = 4$
$\dot{v} + 9v + 7\dot{u} = 2$

**P7.20** $\ddot{u} + 2\dot{u} + 4u + 6\dot{v} + 8v = t$
$\ddot{v} + 3\dot{v} + 5\dot{u} + 7u = 0$

In Problems P7.21 through P7.23, find the kernel for the given ODE. Use web-based root-finders and linear equation solvers.

**P7.21** $\dfrac{d^3 y}{dt^3} + 21\dfrac{d^2 y}{dt^2} + 140\dfrac{dy}{dt} + 288y = g(t)$

**P7.22** $\dfrac{d^3 y}{dt^3} + 10\dfrac{d^2 y}{dt^2} + 68\dfrac{dy}{dt} + 104y = g(t)$

**P7.23** $\dfrac{d^4 y}{dt^4} + 16\dfrac{d^3 y}{dt^3} + 110\dfrac{d^2 y}{dt^2} + 400\dfrac{dy}{dt} + 625y = g(t)$

**P7.24** Show that if $L\{f(t)\} = F(s)$ then $L\left\{\displaystyle\int_0^t f(\tau)d\tau\right\} = \dfrac{F(s)}{s}$.

## Word Problems

**WP7.1** A spacecraft is traveling with its velocity and its centerline parallel to a reference direction as it nears the asteroid it will land on. It begins a rocket burn to slow its descent. With its onboard sensors, the spacecraft determines that, to enable a safe landing, it needs to add a velocity perpendicular to the reference direction. It must perform this correction while ensuring that, at the end of the maneuver, its centerline is again parallel to the reference direction. The spacecraft will accomplish both objectives simultaneously through thrust vector control of its gimbaled rocket nozzle. See Figure 7.6.

Applied to this system, Newton's laws of translational and rotational motion are

$$m\dot{v} = F$$
$$I\ddot{\theta} = \tau$$

where:
$m$ is the mass of the spacecraft
$v$ is the velocity perpendicular to the reference direction
$I$ is the moment of inertia
$\theta$ is the angle of the spacecraft's centerline with respect to the reference direction
$F$ and $\tau$ are the force and moment (torque) it experiences

The force on the spacecraft perpendicular to the reference direction is given (for small angles) by

$$F = T(\theta + \phi)$$

where:
$T$ is the (constant) rocket thrust
$\phi$ is the angle of the rocket nozzle with respect to the spacecraft's centerline

When the rocket nozzle is not aligned with the spacecraft's centerline the spacecraft experiences a moment given by

$$\tau = -Tl\phi$$

where $l$ is the distance from the rocket gimbal to the spacecraft's center of gravity.

To effect the trajectory change and simultaneously control its orientation, the spacecraft uses its onboard sensors and control system to move the rocket nozzle angle according to the law

$$\phi = K_1(v - v_C) + K_2\dot{\theta} + K_3\theta$$

where $v_C$ is the commanded velocity change. (The student should be aware that this is only the start of a practical control law. Other influences, including disturbance forces and moments, and bending modes in a light and flexible vehicle, require additional "compensation," in the language of control system engineers. Section 5.3.1 shows how use of an integral term can deal with a disturbance force.)

Assume that the gimbal's response to a commanded change is essentially instantaneous. This is a system with input $v_C$ and outputs $v$ and $\theta$. Let the Laplace transforms of $v_C, v(t)$ and $\theta(t)$ be $v_C(s), v(s)$, and $\theta(s)$.

Part (a): Show that the transfer functions are

$$v(s)/v_C(s) = \frac{-aK_1 s^2 + abK_1}{s^3 + (bK_2 - aK_1)s^2 + bK_3 s + abK_1}$$

$$\theta(s)/v_C(s) = \frac{bK_1 s}{s^3 + (bK_2 - aK_1)s^2 + bK_3 s + abK_1}$$

where $a = T/m$ and $b = Tl/I$.

Part (b): Show that the system is certainly unstable if $aK_1 > bK_2$. Using the initial and final value theorems, and assuming that the gains $K_1, K_2, K_3$ are set so that the system is stable, show that $v(t)$ starts out in the wrong direction (i.e., its derivative is initially negative), but that $v(t)$ and $\theta(t)$ end up at their commanded values.

Part (c): Let $a = 3, b = 1, K_1 = 0.1, K_2 = 2$, and $K_3 = 1$. Show that the roots of the denominator are $-1$ and $-0.350 \pm i0.421$.

Part (d): Using the parameter values in Part (c), and setting $v_C = 5$, program the equations of motion and show that the motions are as shown in Figures 7.7 and 7.8.

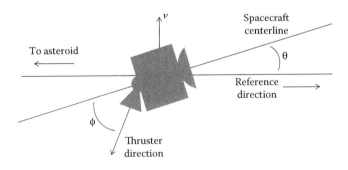

**FIGURE 7.6**  Spacecraft with thrust vector control.

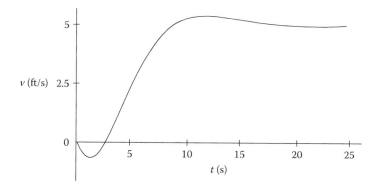

**FIGURE 7.7** Spacecraft velocity in response to commanded velocity change.

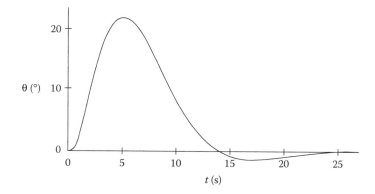

**FIGURE 7.8** Spacecraft orientation in response to commanded velocity change.

**WP7.2** Show that the kernel $K(t)$ of an $n$th order LTI system, for $n \geq 3$, satisfies

$$K(0^+) = 0, \qquad \frac{d^m K}{dt^m}(0^+) = 0 \quad \text{for } 1 \leq m \leq n-2, \qquad \frac{d^{n-1} K}{dt^{n-1}}(0^+) = 1$$

Hint: Recall that the kernel is the impulse response and use the initial value theorem. Use an induction argument. This is a useful check on the calculation of a kernel for an LTI system.

**WP7.3** An electrical low-pass filter is a circuit that responds to low frequency inputs with an output that closely matches the input, and responds to high-frequency inputs with an output that is nearly zero. That is, it allows low-frequency signals to pass through while blocking high-frequency signals. As an example application, if a signal consists of a low-frequency measurement corrupted with high-frequency noise, the measurement can be recovered by putting the signal through a low-pass filter. An nth order Butterworth filter is a circuit that performs the low-pass function as well as possible in a circuit with an nth order transfer function. A third-order Butterworth filter with cut-off frequency of one radian per second has transfer function

$$H(s) = \frac{v_{\text{out}}(s)}{v_{\text{in}}(s)} = \frac{1}{s^3 + 2s^2 + 2s + 1} \tag{7.126}$$

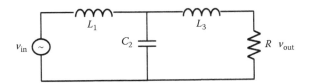

**FIGURE 7.9**   Electrical circuit to implement a third-order Butterworth filter

Determine the circuit element values to make the circuit shown in Figure 7.9 a
  Butterworth filter.

*Guidance*: Review the equations relating voltage to current for resistor, capacitor,
  and inductor, and also Kirchhoff's voltage law, as given, for example, in Chapter 5,
  Section 5.1. Write differential equations for the voltage drops around each of the
  two loops in the circuit shown in Figure 7.9. Apply the Laplace transform to the
  differential equations. Solve for $v_{out}(s)$ as a function of $v_{in}(s)$. Show that

$$\frac{v_{out}(s)}{v_{in}(s)} = \frac{R}{(L_1 C_2 L_3)s^3 + (L_1 C_2 R)s^2 + (L_1 + L_3)s + R} \tag{7.127}$$

Determine circuit element values to make the transfer functions in Equations 7.126
  and 7.127 equal.

**WP7.4** Following on to Section 7.6.1, show that the solution to

$$\frac{d^n y}{dt^n} + a_1 \frac{d^{n-1} y}{dt^{n-1}} + a_2 \frac{d^{n-2} y}{dt^{n-2}} + \ldots + a_n y = g(t)$$

$$y(0) = y_0,\, \dot{y}(0) = \dot{y}_0,\, \ddot{y}(0) = \ddot{y}_0, \ldots, y^{(n-1)}(0) = y_0^{(n-1)}$$

is

$$y(t) = L^{-1}\{Y(s)\}$$

where

$$Y(s) = \frac{p_1 s^{n-1} + p_2 s^{n-2} + \ldots + p_{n-1}s + p_n}{s^n + a_1 s^{n-1} + \ldots + a_{n-1}s + a_n} + \frac{G(s)}{s^n + a_1 s^{n-1} + \ldots + a_{n-1}s + a_n}$$

$$p_1 = y_0$$

$$p_2 = \dot{y}_0 + a_1 y_0$$

$$p_3 = \ddot{y}_0 + a_1 \dot{y}_0 + a_2 y_0$$

$$\ldots$$

$$p_n = y_0^{(n-1)} + a_1 y_0^{(n-2)} + \ldots + a_{n-2} \dot{y}_0 + a_{n-1} y_0$$

and

$$G(s) = L\{g(t)\}$$

*Guidance*: Section 7.6.1 showed that $Y(s)$ as given earlier is the algebraic solution to the transformed ODE. It remains to show that the inverse transform $y(t) = L^{-1}\{Y(s)\}$ satisfies the ODE. Since the ODE and the Laplace transform are linear operations, we may divide the problem into two parts and then sum the results: part (1) – nonzero input $g(t)$ and zero initial conditions; part (2) – zero $g(t)$ and nonzero initial conditions.

Part 1: We must show that

$$y(t) = L^{-1}\left\{\frac{G(s)}{s^n + a_1 s^{n-1} + \ldots + a_{n-1}}\right\} \tag{7.128}$$

satisfies

$$\frac{d^n y}{dt^n} + a_1 \frac{d^{n-1} y}{dt^{n-1}} + \ldots + a_{n-1} y = g(t) \tag{7.129}$$

$$y(0) = 0, \dot{y}(0) = 0, \ldots, y^{(n-1)}(0) = 0$$

Consider $d^m y/dt^m$. If $L\{f(t)\} = F(s)$ we know from the Theorem in Section 7.4 that $L\{d^m f/dt^m\} = s^m F(s)$ (when initial conditions are zero). Then $d^m f/dt^m = L^{-1}\{s^m F(s)\}$. Hence

$$\frac{d^m}{dt^m}\left(L^{-1}\left\{\frac{G(s)}{s^n + a_1 s^{n-1} + \ldots + a_{n-1}}\right\}\right) = L^{-1}\left\{\frac{s^m G(s)}{s^n + a_1 s^{n-1} + \ldots + a_{n-1}}\right\} \tag{7.130}$$

Then from Equations 7.128 and 7.130, using the linearity of the inverse Laplace transform:

$$\frac{d^n y}{dt^n} + a_1 \frac{d^{n-1} y}{dt^{n-1}} + \ldots + a_{n-1} y = L^{-1}\left\{\frac{(s^n + a_1 s^{n-1} + \ldots + a_{n-1})G(s)}{s^n + a_1 s^{n-1} + \ldots + a_{n-1}}\right\} = L^{-1}\{G(s)\} = g(t)$$

Part 2: Develop an argument paralleling that in Part 1 to prove that

$$y(t) = L^{-1}\left\{\frac{p_1 s^{n-1} + p_2 s^{n-2} + \ldots + p_n}{s^n + a_1 s^{n-1} + \ldots + a_{n-1}}\right\}$$

solves

$$\frac{d^n y}{dt^n} + a_1 \frac{d^{n-1} y}{dt^{n-1}} + \ldots + a_{n-1} y = 0$$

$$y(0) = y_0, \dot{y}(0) = \dot{y}_0, \ldots, y^{(n-1)}(0) = y_0^{(n-1)}$$

where

$$p_1 = y_0$$

$$p_2 = \dot{y}_0 + a_1 y_0$$

$$p_3 = \ddot{y}_0 + a_1 \dot{y}_0 + a_2 y_0$$

...

$$p_n = y_0^{(n-1)} + a_1 y_0^{(n-2)} + ... + a_{n-2} \dot{y}_0 + a_{n-1} y_0$$

## Challenge Problems

**CP7.1** Following on to Section 7.6.5.2, show that the steady-state solution to

$$\frac{dy}{dt} + 2\lambda \frac{dy}{dt} + (\lambda^2 + \omega^2) y = \chi_T(t) \tag{7.131}$$

is

$$y_{SS}(t) = \frac{(-1)^N}{c^2} \left( 1 - \frac{(2c/\omega)}{D} q(\delta t) \right) \tag{7.132}$$

where $\chi_T(t)$ is the periodic square wave described by Figure 7.4 and $c$, $D$, and $q(\delta t)$ are given by Equations 7.103 and 7.104.

*Guidance*:

(a) Extending the results in Section 7.6.3.3, show that the solution to Equation 7.131 can be written as

$$y(t) = \frac{1}{\lambda^2 + \omega^2} \left( 1 - e^{-\lambda t} \cos \omega t - \frac{\lambda}{\omega} e^{-\lambda t} \sin \omega t \right) +$$

$$\frac{2}{\lambda^2 + \omega^2} \sum_{n=1}^{N} (-1)^n u_{nT}(t) \left( 1 - e^{-\lambda(t-nT)} \cos \omega(t - nT) - \frac{\lambda}{\omega} e^{-\lambda(t-nT)} \sin \omega(t - nT) \right)$$

For large $t$ the terms

$$e^{-\lambda t} \cos \omega t - \frac{\lambda}{\omega} e^{-\lambda t} \sin \omega t$$

will have died away, and we can write the remainder, using $t = NT + \delta t$ and counting backward from the last switch of the square wave, as

$$y(t) = \frac{(-1)^N}{\lambda^2 + \omega^2} \left[ 1 - 2 \sum_{m=0}^{N-1} (-1)^m e^{-\lambda(\delta t + mT)} (\cos \omega(\delta t + mT) + \frac{\lambda}{\omega} \sin \omega(\delta t + mT)) \right] \tag{7.133}$$

(b) Show that, for large $N$, Equation 7.131 can be approximated as

$$y(t) = \frac{(-1)^N}{\lambda^2 + \omega^2} \left[ 1 - 2e^{-\lambda \delta t} (K_1 \cos \omega \delta t + K_2 \sin \omega \delta t) \right] \tag{7.134}$$

where

$$K_1 = \sum_{m=0}^{\infty} (-1)^m e^{-\lambda mT} \left( \cos m\omega T + \frac{\lambda}{\omega} \sin m\omega T \right) \tag{7.135}$$

$$K_2 = \sum_{m=0}^{\infty} (-1)^m e^{-\lambda mT} \left( -\sin m\omega T + \frac{\lambda}{\omega} \cos m\omega T \right) \tag{7.136}$$

(c) Show that, in order for $y(t)$ and $dy/dt$ to be continuous (required in a second-order ODE with bounded input), we must have

$$K_1 = \frac{1 + e^{-\lambda T} \cos \omega T - (\lambda/\omega)e^{-\lambda T} \sin \omega T}{1 + 2e^{-\lambda T} \cos \omega T + e^{-2\lambda T}} \tag{7.137}$$

$$K_2 = \frac{e^{-\lambda T} \sin \omega T + (\lambda/\omega)(1 + e^{-\lambda T} \cos \omega T)}{1 + 2e^{-\lambda T} \cos \omega T + e^{-2\lambda T}} \tag{7.138}$$

(d) As a check: show that, in the special cases below, the series in Equations 7.135 and 7.136 can be summed to obtain the results in Table 7.3.
Show that Equations 7.137 and 7.138 confirm these results.
(e) Use Equations 7.134, 7.137, and 7.138 to derive Equation 7.132.

**CP7.2** Prove that if $f(t)$ is continuous and of exponential order and $df/dt$ is piecewise continuous on any interval $0 \le t \le A$, then the Laplace transform of $df/dt$ exists and

$$L\left\{ \frac{df}{dt} \right\} = sL\{f(t)\} - f(0) \tag{7.139}$$

*Guidance*: Begin with the fundamental definition

$$L\left\{ \frac{df}{dt} \right\} = \lim_{A \to \infty} \int_0^A e^{-st} \frac{df}{dt} dt \tag{7.140}$$

---

**TABLE 7.3**
**Tabulation of Equations 7.135 and 7.136**
**for Particular Values of $\omega T$**

| $\omega T$ | $K_1$ | $K_2$ |
|---|---|---|
| $2\pi$ | $\dfrac{1}{1 + e^{-\lambda T}}$ | $\dfrac{\lambda/\omega}{1 + e^{-\lambda T}}$ |
| $\pi/2$ | $\dfrac{1 - (\lambda/\omega)e^{-\lambda T}}{1 + e^{-2\lambda T}}$ | $\dfrac{(\lambda/\omega) + e^{-\lambda T}}{1 + e^{-2\lambda T}}$ |
| $\pi$ | $\dfrac{1}{1 - e^{-\lambda T}}$ | $\dfrac{\lambda/\omega}{1 - e^{-\lambda T}}$ |

and for simplicity assume at first that $df/dt$ is continuous everywhere. Integrating by parts:

$$\int_0^A e^{-st}\frac{df}{dt}\,dt = f(A)e^{-sA} - f(0)e^{-s(0)} - \int_0^A \frac{d}{dt}\left(e^{-st}\right)f(t)\,dt =$$
$$f(A)e^{-sA} - f(0)e^{-s(0)} + s\int_0^A e^{-st}f(t)\,dt \tag{7.141}$$

If $f(t)$ is of exponential order than there is an $s$ large enough that

$$\lim_{A\to\infty} f(A)e^{-sA} = 0 \tag{7.142}$$

Then by Equations 7.141 and 7.142, Equation 7.140 becomes

$$L\left\{\frac{df}{dt}\right\} = \lim_{A\to\infty}\int_0^A e^{-st}\frac{df}{dt}\,dt = s\lim_{A\to\infty}\left(\int_0^A e^{-st}f(t)\,dt\right) - f(0)$$

$$L\left\{\frac{df}{dt}\right\} = sL\{f(t)\} - f(0)$$

Now consider the case where $df/dt$ is piecewise continuous. Break the integral into segments over which $df/dt$ is continuous. Denote the endpoints of the nth segment as $A_n$ and $A_{n+1}$. Integrate by parts as in Equation 7.141 and note cancellations.

**CP7.3** Consider an experimental robotic arm, as in Figure 7.10, lying flat on a frictionless table, pinned at the "shoulder" (the base) and able to reach with its "hand" (the end of the second segment) any point within a radius of $2l$ of the base.

In the two joints there are computer-controlled electric motors. When it is desired to move the hand from one position to a very different one the major part of the move is done "open loop", that is, based on a calculation with no observation of the result. When the hand is close to its objective it is placed under feedback control, that is, the difference between the hand's true position and the commanded position is observed and used to control the torque exerted by the two electric motors. The challenge is to design and assess a feedback control law to bring the hand from the nearby position to its commanded final destination.

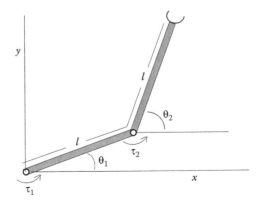

**FIGURE 7.10** Experimental robotic arm.

*Guidance*:

The Cartesian coordinates of the hand are related to the angles of the first and second
  segments by

$$x = l(\cos\theta_1 + \cos\theta_2)$$

$$y = l(\sin\theta_1 + \sin\theta_2)$$

At the nearby position $(x_0, y_0)$, that is, when feedback control starts,

$$x_0 = l(\cos\theta_{10} + \cos\theta_{20})$$

$$y_0 = l(\sin\theta_{10} + \sin\theta_{20})$$

and the incremental progress from that starting position is given by

$$\delta x = x - x_0 = \left(\frac{\partial x}{\partial\theta_1}\right)_0 \delta\theta_1 + \left(\frac{\partial x}{\partial\theta_2}\right)_0 \delta\theta_2 = -l\sin\theta_{10}\delta\theta_1 - l\sin\theta_{20}\delta\theta_2$$

$$(7.143)$$

$$\delta y = y - y_0 = \left(\frac{\partial y}{\partial\theta_1}\right)_0 \delta\theta_1 + \left(\frac{\partial y}{\partial\theta_2}\right)_0 \delta\theta_2 = l\cos\theta_{10}\delta\theta_1 + l\cos\theta_{20}\delta\theta_2$$

We desire

$$\lim_{t\to\infty} \delta x(t) = x_C - x_0 = \delta x_C$$

$$(7.144)$$

$$\lim_{t\to\infty} \delta y(t) = y_C - y_0 = \delta y_C$$

where $x_C, y_C$ is the hand's commanded final position. The equations of motion can be
  shown to be

$$\left(\frac{ml^2}{3}\right)\ddot{\theta}_1 = \tau_1 - \tau_2$$

$$\left(\frac{ml^2}{2}\cos(\theta_2 - \theta_1)\right)\ddot{\theta}_1 + \left(\frac{ml^2}{3}\right)\ddot{\theta}_2 = \tau_2 - ml^2\dot{\theta}_1^2\sin(\theta_2 - \theta_1)$$

where $\tau_1$ and $\tau_2$ are the computer-controlled torques. Assuming small changes from the
  starting position we can write $\delta\theta_i = \theta_i - \theta_{i0}$ and linearize the equations of motion:

$$\left(\frac{ml^2}{3}\right)\delta\ddot{\theta}_1 = \tau_1 - \tau_2$$

$$(7.145)$$

$$\left(\frac{ml^2}{2}\cos(\theta_{20} - \theta_{10})\right)\delta\ddot{\theta}_1 + \left(\frac{ml^2}{3}\right)\delta\ddot{\theta}_2 = \tau_2$$

To begin design of a control law, we try

$$\tau_1 = -\left(\frac{ml^2}{3}\right)\left(K_1\delta\dot{\theta}_1 + K_3(\delta x_C - \delta x) + K_4(\delta y_C - \delta y)\right)$$

(7.146)

$$\tau_2 = -\left(\frac{ml^2}{3}\right)\left(K_2\delta\dot{\theta}_2 + K_5(\delta x_C - \delta x) + K_6(\delta y_C - \delta y)\right)$$

The terms involving $K_i\delta\dot{\theta}_i$ are for damping the response and the other terms are to satisfy Equations 7.144.

Part (a): Apply the Laplace transform to Equations 7.143, 7.145, and 7.146 and solve the resulting equations to obtain equations in the form

$$\delta x(s) = \left(\frac{N_{XX}(s)}{D(s)}\right)\frac{\delta x_C}{s} + \left(\frac{N_{XY}(s)}{D(s)}\right)\frac{\delta y_C}{s}$$

$$\delta y(s) = \left(\frac{N_{YX}(s)}{D(s)}\right)\frac{\delta x_C}{s} + \left(\frac{N_{YY}(s)}{D(s)}\right)\frac{\delta y_C}{s}$$

Use the final value theorem to find the conditions under which Equation 7.144 is satisfied.

Part (b): To take the next step toward a defined control law try

$$K_3 = K_0\sin\theta_{10} \quad K_4 = -K_0\cos\theta_{10}$$
$$K_5 = K_0\sin\theta_{20} \quad K_6 = -K_0\cos\theta_{20} \qquad K_0 > 0$$

$$K_1 = K_2 = K_D > 0$$

and show that the expression for $D(s)$ is simplified and that all the coefficients of $s^n$, $n = 0$ to $n = 4$, in $D(s)$ are greater than or equal to zero.

Part (c): To investigate the stability of the system, scale (nondimensionalize) $D(s)$ to reduce the examination to a two-parameter study: Let $s = K_D r$, $K_0 = pK_D^2/l$, and $\cos(\theta_{20} - \theta_{10}) = q$. Show that the equation $D(s) = 0$ reduces to

$$r^4 + (2 + 3q/2)r^3 + \left(1 + p(2 + q/2 - 3q^2/2)\right)r^2 + 2pr + p^2(1 - q^2) = 0 \qquad (7.147)$$

Part (d): Do a numerical study of the roots of this equation. Use a web-based root finder to create a graph in the $p, q$ plane showing the regions where the system is stable and damped and those where the system is unstable.

Part (e): What do the results of Part(d) mean for the utility of the trial control law? Can you say conclusively that it works for all geometries? Are there starting locations $x_0, y_0$ that give you concern?

# 8 Systems of First-Order Ordinary Differential Equations

This chapter discusses systems of first-order ordinary differential equations (ODEs), known as the *state space* representation by engineers and the *normal form* by mathematicians. This representation and form enables mathematical description of more complex systems than do methods previously presented. It is also better suited to computer-based numerical integration. It provides a unifying framework for analysis of linear systems, and it has been the basis for many of the advances in differential equations and engineering (e.g., in control system theory) over the past half century. The general form is

$$\frac{dx_1}{dt} = f_1(t, x_1, x_2, \ldots, x_n)$$

$$\frac{dx_2}{dt} = f_2(t, x_1, x_2, \ldots, x_n)$$

$$\ldots$$

$$\frac{dx_n}{dt} = f_n(t, x_1, x_2, \ldots, x_n)$$

(8.1)

The chapter begins with a discussion of nomenclature and a demonstration that nth order ODEs as addressed in Chapter 6 can be represented via Equation 8.1. Section 8.2 investigates existence and uniqueness of solutions to Equation 8.1 and demonstrates strong commonality with Chapter 4's results for first-order equations in one dependent variable. Computer programs for numerical integration of ordinary ODEs typically require the format of Equation 8.1 and Section 8.3 presents the essentials of these programs and discusses several commonly used algorithms. Section 8.4 reviews matrix algebra as necessary to provide a foundation for Section 8.5, which examines linear systems. We will devote most of our attention in this chapter to linear systems. At the end of the chapter, Section 8.6 takes a brief look at nonlinear systems.

## 8.1 PRELIMINARIES

For notational simplicity we will represent Equation 8.1 in vector form:

$$\frac{d\underline{x}}{dt} = \underline{f}(t, \underline{x})$$

(8.2)

where

$$\underline{x}(t) = \begin{pmatrix} x_1(t) \\ x_2(t) \\ \cdots \\ x_{n-1}(t) \\ x_n(t) \end{pmatrix}, \quad \frac{d\underline{x}}{dt} = \begin{pmatrix} \dfrac{dx_1}{dt} \\ \cdots \\ \cdots \\ \cdots \\ \dfrac{dx_n}{dt} \end{pmatrix}, \quad \underline{f}(t,\underline{x}) = \begin{pmatrix} f_1(t,x_1,x_2,\ldots,x_{n-1},x_n) \\ f_2(t,x_1,x_2,\ldots,x_{n-1},x_n) \\ \cdots \\ f_{n-1}(t,x_1,x_2,\ldots,x_{n-1},x_n) \\ f_n(t,x_1,x_2,\ldots,x_{n-1},x_n) \end{pmatrix}$$

In a system context, the vector $\underline{x}(t)$ is called the *state vector* because it defines fully the state of the system; knowing the value of $\underline{x}(t)$ at any time suffices to define its value at any future time.

We can easily represent an nth order linear ODE in state space form. Consider

$$\frac{d^n y}{dt^n} + a_1(t)\frac{d^{n-1}y}{dt^{n-1}} + \ldots + a_n(t)y = g(t) \tag{8.3}$$

Let $x_1 = y$, $x_2 = dy/dt$, $x_3 = d^2y/dt^2$, …, $x_n = d^{n-1}y/dt^{n-1}$.
   Then

$$\frac{dx_1}{dt} = x_2$$

$$\frac{dx_2}{dt} = x_3$$

$$\cdots \tag{8.4}$$

$$\frac{dx_{n-1}}{dt} = x_n$$

$$\frac{dx_n}{dt} = -a_1(t)x_n - a_2(t)x_{n-1} - \ldots - a_n(t)x_1 + g(t)$$

Equation 8.4 is in the desired format.

## 8.2   EXISTENCE AND UNIQUENESS

Statement of sufficient conditions for a solution to Equation 8.2 to exist and to be unique follows closely the theorem given in Chapter 4 for first-order equations in one dependent variable, with the $n$-dimensional column vector $\underline{x}$ replacing the scalar $y$.

   We must first extend a few definitions to $n$-dimensional space. The *magnitude* of an $n$-dimensional vector $\underline{x}$ is defined as

$$|\underline{x}| = \sqrt{x_1^2 + x_2^2 + \ldots + x_n^2}$$

The *distance* between two vectors $\underline{x}$ and $\underline{z}$ is the magnitude of their vector difference:

$$\left|\underline{x}-\underline{z}\right| = \sqrt{(x_1-z_1)^2+(x_2-z_2)^2+...+(x_n-z_n)^2}$$

Consider a vector-valued function $\underline{f}(t,\underline{x})$ defined on a set R such that

$$\left|t-t_0\right| \le a, \qquad \left|\underline{x}-\underline{x}_0\right| \le b$$

The function $\underline{f}(t,\underline{x})$ is said to satisfy a *Lipschitz condition* with respect to $\underline{x}$ on R if there exists a constant $K>0$ such that

$$\left|\underline{f}(t,\underline{x})-\underline{f}(t,\underline{z})\right| \le K\left|\underline{x}-\underline{z}\right|$$

for all points $(t,\underline{x})$, $(t,\underline{z})$ in R. A sufficient condition for $\underline{f}(t,\underline{x})$ to satisfy a Lipschitz condition is as follows:

If $\partial \underline{f}(t,\underline{x})/\partial x_i$ is continuous in R for $i=1,...,n$ and

$$\left|\frac{\partial \underline{f}(t,\underline{x})}{\partial x_i}\right| \le K \quad i=1,...,n$$

in R, then $\underline{f}(t,\underline{x})$ satisfies a Lipschitz condition in R with constant $K$.

With this background, we are prepared for a statement of the theorem.

**Theorem**

Let the vector-valued function $\underline{f}(t,\underline{x})$ be defined in a region

$$R: \left|t-t_0\right| \le a, \qquad \left|\underline{x}-\underline{x}_0\right| \le b$$

and in R let $\underline{f}(t,\underline{x})$:

1. Be continuous
2. Be bounded by M
3. Satisfy a Lipschitz condition with constant $K$ with respect to $\underline{x}$

Then a unique solution exists to

$$\frac{d\underline{x}}{dt} = \underline{f}(t,\underline{x})$$
(8.5)
$$\underline{x}(t_0) = \underline{x}_0$$

in the interval

$$I : \left|t-t_0\right| \le \min(a,b/M)$$

We will not provide a proof of this theorem, given its similarity with that for first-order equations in one dependent variable, as outlined in Chapter 4 and its challenge problems. However, we remark that the essence of the proof again lies in successive approximations:

The sequence of approximations given by

$$\underline{x}_0(t) = \underline{x}_0$$

$$\underline{x}_1(t) = \underline{x}_0 + \int_{t_0}^{t} \underline{f}(\tau, \underline{x}_0(\tau)) d\tau$$

$$\underline{x}_2(t) = \underline{x}_0 + \int_{t_0}^{t} \underline{f}(\tau, \underline{x}_1(\tau)) d\tau \qquad (8.6)$$

$$\cdots$$

$$\underline{x}_m(t) = \underline{x}_0 + \int_{t_0}^{t} \underline{f}(\tau, \underline{x}_{m-1}(\tau)) d\tau$$

converges to a unique solution of Equation 8.5.

## 8.3   NUMERICAL INTEGRATION METHODS

The great majority of the times, engineers solve differential equations by numerical integration on digital computers. Chapters 3 and 6 defined several extensive computing projects intended to give the student an opportunity to program a computer to achieve a practical result. The objective of this section is to provide the student a basic understanding of methods employed in numerical integration programs widely available to compute the solution of a system of first-order ODEs.

To solve

$$\frac{d\underline{x}}{dt} = \underline{f}(t, \underline{x}), \qquad \underline{x}(t_0) = \underline{x}_0$$

we initialize $t$ and $\underline{x}$ via

$$t = t_0, \qquad \underline{x} = \underline{x}_0$$

and then break the integration into small, sequential time steps. Let $h = \Delta t$ and let $\underline{x}(t_n)$ represent the solution at time step $n$. Then

$$t_{n+1} = t_n + h$$

$$\underline{x}(t_{n+1}) = \underline{x}(t_n) + \int_{t_n}^{t_{n+1}} \underline{f}(t, \underline{x}) dt \qquad (8.7)$$

The available numerical integration methods differ in how they represent integration over the time step.

### 8.3.1 EULER METHOD

The simplest method is due to Euler:

$$x(t_{n+1}) = x(t_n) + f(t_n, x(t_n))h$$

Euler's method represents $f(t, x)$ over the time step by its value at the beginning of the time step.

**Example**

Solve

$$\frac{dx}{dt} = -\lambda x, \qquad x(0)=x_0 \tag{8.8}$$

For this simple one-dimensional linear case we can both carry through the Euler method analytically and also solve the problem exactly, enabling an accuracy evaluation. Let us call the solution to Equation 8.8 by the Euler method $x_E(t_n)$. Note that

$$f(t_n, x_E(t_n)) = -\lambda x_E(t_n)$$

We have

$$x_E(t_1) = x_0 - \lambda x_0 h = (1 - \lambda h)x_0$$

$$x_E(t_2) = x_E(t_1) - \lambda x_E(t_1)h = (1 - \lambda h)x_E(t_1) = (1 - \lambda h)^2 x_0$$

$$x_E(t_3) = x_E(t_2) - \lambda x_E(t_2)h = (1 - \lambda h)x_E(t_2) = (1 - \lambda h)^3 x_0 \tag{8.9}$$

$$\dots$$

$$x_E(t_N) = x_E(t_{N-1}) - \lambda x_E(t_{N-1})h = (1 - \lambda h)x_E(t_{N-1}) = (1 - \lambda h)^N x_0$$

Now

$$h = \frac{T}{N} \tag{8.10}$$

where $T$ is the endpoint of the integration. From Equations 8.9 and 8.10

$$x_E(t_N) = x_0(1 - \lambda h)^N = x_0\left(1 - \frac{\lambda T}{N}\right)^N \tag{8.11}$$

From Chapter 1, Equation 1.9, we recognize

$$\lim_{N \to \infty} \left(1 - \frac{\lambda T}{N}\right)^N = e^{-\lambda T} \tag{8.12}$$

Hence

$$\lim_{N \to \infty} x_E(t_N) = x_E(T) = x_0 e^{-\lambda T} \tag{8.13}$$

From Chapter 2, we also know how to solve Equation 8.8 analytically and that result confirms Equation 8.13. That is, for this example Euler's method indeed converges to the correct answer. But how accurate is it for finite $N$? We can compare Equations 8.11 and 8.13. But before we do, we introduce another commonly used method.

### 8.3.2 RUNGE–KUTTA METHOD

In the widely used Runge–Kutta method, the general time step equation

$$\underline{x}(t_{n+1}) = \underline{x}(t_n) + \int_{t_n}^{t_{n+1}} \underline{f}(t, \underline{x}) dt$$

is approximated by

$$\underline{x}(t_{n+1}) = \underline{x}(t_n) + h\left(\frac{\underline{k}_{n1} + 2\underline{k}_{n2} + 2\underline{k}_{n3} + \underline{k}_{n4}}{6}\right) \tag{8.14}$$

where

$$\underline{k}_{n1} = \underline{f}(t_n, \underline{x}(t_n))$$

$$\underline{k}_{n2} = \underline{f}\left(t_n + h/2, \underline{x}(t_n) + (h/2)\underline{k}_{n1}\right)$$

$$\underline{k}_{n3} = \underline{f}\left(t_n + h/2, \underline{x}(t_n) + (h/2)\underline{k}_{n2}\right) \tag{8.15}$$

$$\underline{k}_{n4} = \underline{f}\left(t_n + h, \underline{x}(t_n) + h\underline{k}_{n3}\right)$$

Note that $\underline{k}_{n1}$ is the same as the single term in the Euler method, but then $k_{n2}$ uses $k_{n1}$ to estimate the derivative halfway through the time step, $k_{n3}$ uses $k_{n2}$ to derive an even better estimate of the derivative in the middle of the time step, and finally $k_{n4}$ uses $k_{n3}$ to estimate the derivative at the end of the time step.

We note in passing that if $\underline{f}(t, \underline{x})$ does not depend on $\underline{x}$ then

$$\underline{x}(t_{n+1}) = \underline{x}(t_n) + h\left(\frac{\underline{f}(t_n) + 4\underline{f}(t_n + (h/2)) + \underline{f}(t_n + h))}{6}\right) \tag{8.16}$$

which you may recall from integral calculus is Simpson's rule for approximate evaluation of an integral.

### Example

We apply the Runge–Kutta method on Equation 8.8, the first-order linear problem we considered for the Euler method. Let us call the solution to Equation 8.8 by the Runge–Kutta method $x_{RK}(t_n)$. We find

$$k_{n1} = -\lambda x_{RK}(t_n)$$

$$k_{n2} = -\lambda(x_{RK}(t_n) + (h/2)(-\lambda x_{RK}(t_n)))$$

$$= (1-(\lambda h/2))(-\lambda x_{RK}(t_n))$$

$$k_{n3} = -\lambda(x_{RK}(t_n) + (h/2)(1-(\lambda h/2))(-\lambda x_{RK}(t_n))) \qquad (8.17)$$

$$= (1-(\lambda h/2) + (\lambda h)^2/4)(-\lambda x_{RK}(t_n))$$

$$k_{n4} = -\lambda(x_{RK}(t_n) + h(1-(\lambda h/2) + (\lambda h)^2/4)(-\lambda x_{RK}(t_n))$$

$$= (1-(\lambda h) + (\lambda h)^2/2 - (\lambda h)^3/4)(-\lambda x_{RK}(t_n))$$

Using Equations 8.17 in Equation 8.14, we arrive at

$$x_{RK}(t_{n+1}) = x_{RK}(t_n) - \alpha x_{RK}(t_n) = (1-\alpha)x_{RK}(t_n) \qquad (8.18)$$

where

$$\alpha = \lambda h - \frac{1}{2}(\lambda h)^2 + \frac{1}{6}(\lambda h)^3 - \frac{1}{24}(\lambda h)^4 \qquad (8.19)$$

The reader will recognize that $1-\alpha$ is just the first four terms in the power series expansion of $e^{-\lambda h}$. From Equation 8.18, for our example problem, the Runge–Kutta method results in

$$x_{RK}(t_1) = (1-\alpha)x_0$$

$$x_{RK}(t_2) = (1-\alpha)x_{RK}(t_1) = (1-\alpha)^2 x_0$$

$$\qquad (8.20)$$

$$\dots$$

$$x_{RK}(t_N) = (1-\alpha)x_{RK}(t_{N-1}) = (1-\alpha)^N x_0$$

### 8.3.3 Comparing Accuracy of Euler and Runge–Kutta Methods

For our example first-order linear time-invariant (LTI) ODE (Equation 8.8), the Euler method produced

$$x_E(t_N) = x_0(1-\lambda h)^N$$

whereas the Runge–Kutta method produced

$$x_{RK}(t_N) = x_0\left(1-\alpha\right)^N$$

where $\alpha$ is given by Equation 8.19. Table 8.1 compares fractional errors of these two results using $\lambda T = 1$ and $h = T/N$. The fractional errors are defined as

---

**TABLE 8.1**

**Comparing Accuracy of Euler and Runge–Kutta Methods for a Simple Example**

| Number of Time Steps | Fractional Error for Euler Method | Fractional Error for Runge–Kutta Method |
|---|---|---|
| 2 | $3.20 \cdot 10^{-1}$ | $7.92 \cdot 10^{-4}$ |
| 3 | $1.95 \cdot 10^{-1}$ | $1.36 \cdot 10^{-4}$ |
| 4 | $1.40 \cdot 10^{-1}$ | $4.01 \cdot 10^{-5}$ |
| 5 | $1.09 \cdot 10^{-1}$ | $1.58 \cdot 10^{-5}$ |
| 6 | $8.97 \cdot 10^{-2}$ | $7.39 \cdot 10^{-6}$ |
| 8 | $6.60 \cdot 10^{-2}$ | $2.26 \cdot 10^{-6}$ |
| 10 | $5.22 \cdot 10^{-2}$ | $9.06 \cdot 10^{-7}$ |
| 20 | $2.25 \cdot 10^{-2}$ | $5.43 \cdot 10^{-8}$ |

---

$$\delta_E = \left| \frac{x_E(t_N) - x_0 e^{-\lambda T}}{x_0 e^{-\lambda T}} \right|$$

and

$$\delta_{RK} = \left| \frac{x_{RK}(t_N) - x_0 e^{-\lambda T}}{x_0 e^{-\lambda T}} \right|$$

These are called the *truncation errors* associated with the algorithms. We will meet another type of error shortly.

Table 8.1 shows that the Euler method is not competitive with the Runge–Kutta method. For $N = 10$, for example, the truncation error in the Euler method is more than one in 20, compared to less than one in a million for the Runge–Kutta method. The Euler method is widely used, but only in the classroom. The Runge–Kutta method, on the other hand, is widely available in numerical integration programs used by engineers for professional purposes.

The Euler method is useful as a learning tool, illustrating the fundamental concepts involved in numerical integration programs while not overburdening the student with algebra. The problems at the end of this chapter dealing with numerical integration ask the student to reach for his or her calculator and to apply the Euler method.

### 8.3.4 Variable Step Size

All widely available numerical integration programs feature automated variation of step size, responding to the needs of the ODE as it is being executed.

A computer program for the trajectory of a sounding rocket is a good example where a variable step size is needed. (Recall Section 3.1.1 and Computing Project A.) At the beginning, when the rocket is thrusting, the derivatives of mass, velocity, and altitude are high and the variables change rapidly. The time step must be small to keep up with the action. When the rocket has ceased thrusting all derivatives are comparatively small. Continued use of a small step size would increase computer run-time and *round off error* due to finite word length on digital computers. The computer represents a number using a certain number of bits. Numerical integration programs typically employ double precision (64 bits) to minimize round off error. But a smaller step size causes the computer to employ more integration steps and a greater round off error accumulates.

### 8.3.5 OTHER NUMERICAL INTEGRATION ALGORITHMS

As we have said, the Runge–Kutta algorithm is widely used for professional purposes, but it is by no means the only one available and not always the best for a given system of ODEs.

Some problems are inherently difficult to solve accurately. One such type is a *stiff* problem, where several motions occur simultaneously at markedly different speeds. Fortunately, numerical integration programs used by engineers are typically capable of informing the user when the truncation error is large.

The Adams–Moulton method is another algorithm widely used by engineers for professional purposes. It is called a multistep program because it uses information from past time steps to approximate

$$\underline{x}(t_{n+1}) = \underline{x}(t_n) + \int_{t_n}^{t_{n+1}} \underline{f}(t,\underline{x})dt$$

It is also called an implicit method in that it includes the term $\underline{f}(t_{n+1}, \underline{x}(t_{n+1}))$ on the right-hand side, meaning the computer must solve an implicit equation. The basis of the Adams–Moulton method is the fitting of a polynomial to $\underline{f}(t,\underline{x})$ over several time steps.

Some programs combine algorithms into *predictor–corrector* methods, utilizing one algorithm to predict a value of $\underline{x}(t_{n+1})$ and another to use the value in $\int_{t_n}^{t_{n+1}} \underline{f}(t,\underline{x})dt$ to correct the result.

A different approach intended to achieve greater accuracy is to observe the trend of the calculation of $\underline{x}(t_{n+1})$ as step size is reduced and then extrapolate to the limit of zero step size. The Bulirsch–Stoer method is one that employs this approach.

These remarks barely scratch the surface of the body of available numerical integration techniques. Professional engineers sometimes have to devote considerable study and experiment to determine the right method for a given problem.

## 8.4 REVIEW OF MATRIX ALGEBRA

This section contains a cursory review of the elements of matrix algebra, prior to their extensive use in the next section. It covers only those aspects that Section 8.5 requires.

Let $\underline{x}$ and $\underline{y}$ be n-dimensional column vectors and $\underline{A}$ be an n- by n-dimensional matrix. The *ith* component of

$$\underline{y} = \underline{A}\underline{x} \tag{8.21}$$

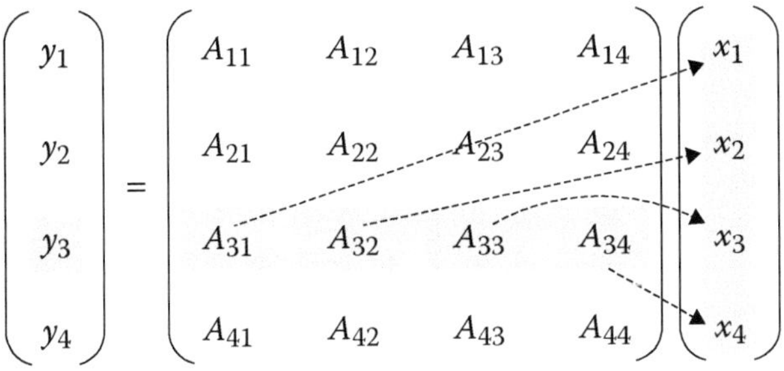

$$
\begin{pmatrix} y_1 \\ y_2 \\ y_3 \\ y_4 \end{pmatrix} = \begin{pmatrix} A_{11} & A_{12} & A_{13} & A_{14} \\ A_{21} & A_{22} & A_{23} & A_{24} \\ A_{31} & A_{32} & A_{33} & A_{34} \\ A_{41} & A_{42} & A_{43} & A_{44} \end{pmatrix} \begin{pmatrix} x_1 \\ x_2 \\ x_3 \\ x_4 \end{pmatrix}
$$

**FIGURE 8.1**   Illustrating multiplication of a vector by a matrix.

is given by

$$
y_i = \sum_{j=1}^{n} A_{ij} x_j \tag{8.22}
$$

As an example, Figure 8.1 illustrates the pattern of multiplication and addition for the third component of a four-dimensional vector.

Now let $\underline{A}$, $\underline{B}$, and $\underline{C}$ be n- by n-dimensional matrices. The $i, j$th component of

$$
\underline{B} = \underline{A}\underline{C} \tag{8.23}
$$

is given by

$$
B_{ij} = \sum_{k=1}^{n} A_{ik} C_{kj} \tag{8.24}
$$

As an example, Figure 8.2 illustrates the pattern of multiplication and addition for the 3, 2 component of a four- by four-dimensional matrix. Think of the $\underline{B}$ and $\underline{C}$ matrices as rows of column vectors.

As reviewed in Section 1.1.2.3, given a system of $n$ linear equations

$$
\underline{y} = \underline{A}\underline{x} \tag{8.25}
$$

where $\underline{y}$ and $\underline{A}$ are known, a unique solution

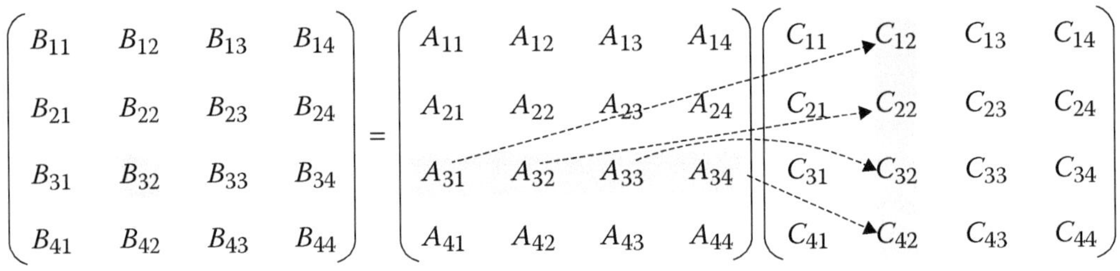

$$
\begin{pmatrix} B_{11} & B_{12} & B_{13} & B_{14} \\ B_{21} & B_{22} & B_{23} & B_{24} \\ B_{31} & B_{32} & B_{33} & B_{34} \\ B_{41} & B_{42} & B_{43} & B_{44} \end{pmatrix} = \begin{pmatrix} A_{11} & A_{12} & A_{13} & A_{14} \\ A_{21} & A_{22} & A_{23} & A_{24} \\ A_{31} & A_{32} & A_{33} & A_{34} \\ A_{41} & A_{42} & A_{43} & A_{44} \end{pmatrix} \begin{pmatrix} C_{11} & C_{12} & C_{13} & C_{14} \\ C_{21} & C_{22} & C_{23} & C_{24} \\ C_{31} & C_{32} & C_{33} & C_{34} \\ C_{41} & C_{42} & C_{43} & C_{44} \end{pmatrix}
$$

**FIGURE 8.2**   Illustrating multiplication of a matrix by a matrix.

$$x = \underline{A}^{-1}\underline{y} \tag{8.26}$$

exists provided the *determinant* of $\underline{A}$ is nonzero.

Section 1.1.2.3 gave equations for determinants of two-by-two and three-by-three matrices. The determinant of the *n*-by-*n* matrix

$$\underline{A} = \begin{pmatrix} A_{11} & A_{12} & \dots & A_{1n} \\ A_{21} & A_{22} & \dots & A_{2n} \\ \dots & \dots & \dots & \dots \\ A_{n1} & A_{n2} & \dots & A_{nn} \end{pmatrix}$$

can be written as

$$\det \underline{A} = \sum_{j=1}^{n}(-1)^{i+j}A_{ij}M_{ij} \tag{8.27}$$

where the *minor* $M_{ij}$ is the determinant of the $(n-1)$ by $(n-1)$ matrix formed by eliminating the *i*th row and *j*th column from $\underline{A}$. The determinant can also be written as

$$\det \underline{A} = \sum_{i=1}^{n}(-1)^{i+j}A_{ij}M_{ij} \tag{8.28}$$

Together Equations 8.27 and 8.28 tell us we can choose to go across any row or down any column to compute the determinant, which can simplify the calculation for a higher-order matrix when the matrix has elements that are zero.

The matrix $\underline{A}^{-1}$ is called the *inverse* of the matrix $\underline{A}$ and it satisfies the equation

$$\underline{A} \cdot \underline{A}^{-1} = \underline{I} \tag{8.29}$$

where $\underline{I}$ is the *n* by *n identity matrix*: $I_{ij} = 1$ if $i = j$, $I_{ij} = 0$ otherwise:

$$\underline{I} = \begin{pmatrix} 1 & 0 & 0 & 0 \\ 0 & 1 & 0 & 0 \\ \dots & \dots & \dots & \dots \\ 0 & 0 & 0 & 1 \end{pmatrix} \tag{8.30}$$

**Example**

*Problem*: Find the inverse of

$$\underline{A} = \begin{pmatrix} 2 & 5 \\ 3 & 4 \end{pmatrix} \tag{8.31}$$

*Solution*:
Let

$$A^{-1} = \begin{pmatrix} a & b \\ c & d \end{pmatrix} \tag{8.32}$$

We use the defining equation, Equation 8.29, and the technique of *row reduction* to find $a, b, c, d$. From Equations 8.29 through 8.32,

$$\begin{pmatrix} 2 & 5 \\ 3 & 4 \end{pmatrix} \begin{pmatrix} a & b \\ c & d \end{pmatrix} = \begin{pmatrix} 1 & 0 \\ 0 & 1 \end{pmatrix} \tag{8.33}$$

Writing out Equation 8.33 component-by-component:

$$\begin{aligned} 2a + 5c = 1 \quad & 2b + 5d = 0 \\ 3a + 4c = 0 \quad & 3b + 4d = 1 \end{aligned} \tag{8.34}$$

Note that Equations 8.34 break down into two sets of two equations in two unknowns, which is less onerous to solve than four linked equations in four unknowns.

Solving Equations 8.34 by row reduction:

$$\begin{aligned} c &= -3a/4 & d &= -2b/5 \\ 2a + 5(-3a/4) &= 1 & 3b + 4(-2b/5) &= 1 \\ a &= -4/7 & b &= 5/7 \\ c &= 3/7 & d &= -2/7 \end{aligned}$$

Hence

$$A^{-1} = \frac{1}{7} \begin{pmatrix} -4 & 5 \\ 3 & -2 \end{pmatrix}$$

This suffices to prepare us for the next section.

## 8.5   LINEAR SYSTEMS IN STATE SPACE FORMAT

This section considers linear ODEs of the general form

$$\dot{x} = \underline{A}(t)\underline{x} + \underline{B}(t)\underline{u}$$

$$\underline{x}(t_0) = \underline{x}_0 \tag{8.35}$$

In Equation 8.35, the unknown $\underline{x}(t)$ is an $n$-dimensional column vector (the *output* of the system), $\underline{x}_0$ is a known initial condition, $\underline{A}(t)$ is a known, possibly time-varying $n$ by $n$

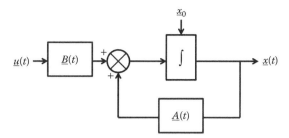

**FIGURE 8.3**  Block diagram for linear system in state space format.

matrix, $\underline{B}(t)$ is a known, possibly time-varying $n$ by $m$ matrix, and $\underline{u}(t)$ is a known, possibly time-varying $m$-dimensional *input* vector. Figure 8.3 is a system block diagram.

Many systems of engineering interest are naturally linear (within limits on the magnitude of the state vector) in the way defined by Equations 8.35. Many more are nearly so, the ODEs having been arrived at by means of linearization.

The state space representation enables us to define linearization more clearly than we have been able to do heretofore. Let $\underline{z}(t)$ be the *known* solution of a nonlinear nth order system described by

$$\underline{\dot{z}} = \underline{f}(t,\underline{z}), \qquad \underline{z}(t_0) = \underline{z}_0$$

and let $\underline{w}(t)$ be the *unknown* solution of a slightly different nth order ODE with slightly different initial condition:

$$\underline{\dot{w}} = \underline{g}(t,\underline{w}), \qquad \underline{w}(t_0) = \underline{w}_0$$

Consider the difference $\underline{x} = \underline{w} - \underline{z}$:

$$\underline{\dot{x}} = \underline{\dot{w}} - \underline{\dot{z}} = \underline{g}(t,\underline{w}) - \underline{f}(t,\underline{z}), \qquad \underline{x}(t_0) = \underline{w}_0 - \underline{z}_0 \tag{8.36}$$

At each point in time $t$ we can expand each component of $\underline{g}(t,\underline{w})$ about the known state vector $\underline{z}(t)$:

$$g_i(t,\underline{w}) \cong g_i(t,\underline{z}(t)) + \sum_{j=1}^{n} \left( \frac{\partial g_i}{\partial w_j}(t,\underline{z}(t)) \right)(w_j - z_j(t))$$

The $n$ by $n$ matrix

$$\frac{\partial g_i}{\partial w_j}(t,\underline{z}(t))$$

is a known function of time. Call it

$$\frac{\partial g_i}{\partial w_j}(t,\underline{z}(t)) = A_{ij}(t) \tag{8.37}$$

Then in matrix notation, using Equations 8.36 and 8.37, we can write

$$\dot{\underline{x}} = \dot{\underline{w}} - \dot{\underline{z}} = \underline{g}(t,\underline{w}) - \underline{f}(t,\underline{z}) \cong \underline{A}(t)(\underline{w} - \underline{z}(t)) + \underline{g}(t,\underline{z}(t)) - \underline{f}(t,\underline{z}(t))$$

Identifying $\underline{x} = \underline{w} - \underline{z}(t)$, calling $\underline{x}(t_0) = \underline{w}_0 - \underline{z}_0 = \underline{x}_0$, and calling the known function of time

$$\underline{g}(t,\underline{z}(t)) - \underline{f}(t,\underline{z}(t)) = \underline{b}(t)$$

we have

$$\dot{\underline{x}} = \underline{A}(t)\underline{x} + \underline{b}(t), \qquad \underline{x}(t_0) = \underline{x}_0 \qquad (8.38)$$

It is a small step from Equation 8.38 to Equation 8.35. Note that we can recover Equation 8.38 from Equation 8.35 if we let $\underline{B}(t)$ equal the identity matrix $\underline{I}$ and $\underline{u} = \underline{b}(t)$.

Suppose the known trajectory $\underline{z}(t) = \underline{z}_0$, a constant, because $\underline{z}_0$ is an equilibrium point of the system $\dot{\underline{z}} = \underline{f}(t,\underline{z})$. Also suppose that $(\partial g_i/\partial w_j)(t,\underline{z}_0)$ is not a function of time. Under these two conditions, $\underline{A}(t)$ is not a function of time and the linear system Equation 8.38 is time-invariant.

Many engineering applications meet these two conditions.

### 8.5.1 Homogeneous Linear Time-Invariant Systems

We begin with *homogeneous* LTI systems, that is, where $\underline{u}(t) = 0$ and $\underline{A}(t) = \underline{A}$, a matrix of constants, in Equations 8.35.

#### 8.5.1.1 Real, Distinct Eigenvalues: Heat Transfer Example

Figure 8.4 depicts two thin wafers of identical size and material separately heated or cooled to initial temperatures $T_{10}$ and $T_{20}$ and then brought together with a thin wall between them into an airstream initially at temperature $T_\infty$. Assume that the wafers are very good conductors of heat and the separating wall a good conductor with very low density and specific

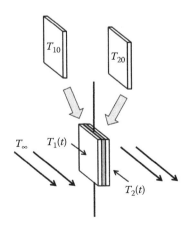

**FIGURE 8.4** Heat transfer example.

heat. With some help from the background provided by Section 2.2, the equations for the temperatures of the two wafers are

$$mc\frac{dT_1}{dt} = -hA(T_1 - T_\infty) - \frac{kA}{d}(T_1 - T_2)$$

$$mc\frac{dT_2}{dt} = -hA(T_2 - T_\infty) + \frac{kA}{d}(T_1 - T_2)$$

(8.39)

where:

   $m$ is the mass of each wafer

   $c$ is the specific heat of each wafer

   $h$ is the convection coefficient

   $A$ is the area of each wafer exposed to the air and also the area of the surface connecting
        the separating wall and the wafers

   $k$ is the thermal conductivity of the separating wall

   $d$ is the width of separating wall

The first term on the right side of each equation is the heat transferred by convection to or from the airstream. The second term is the heat transferred by conduction across the separating wall.

As in Section 2.2's example, we let $T_{*i} = T_i - T_\infty$ and divide through by $mc$, leading to

$$\frac{dT_{*1}}{dt} = -aT_{*1} - b(T_{*1} - T_{*2})$$

$$\frac{dT_{*2}}{dt} = -aT_{*2} + b(T_{*1} - T_{*2})$$

(8.40)

where

$$a = \frac{hA}{mc} \quad b = \frac{kA}{mcd}$$

Equations 8.40 are of the form

$$\frac{d\underline{x}}{dt} = \underline{A}\underline{x}$$

(8.41)

where, here,

$$\underline{x} = \begin{pmatrix} T_{*1} \\ T_{*2} \end{pmatrix}$$

$$\underline{A} = \begin{pmatrix} -(a+b) & b \\ b & -(a+b) \end{pmatrix}$$

(8.42)

How do we solve such an equation? We can solve it in the time domain or, as in Chapter 7, the Laplace transform domain. We postpone consideration of matrix Laplace transforms until later in this chapter and begin here with solution in the time domain.

Based on our experience with second- and higher-order LTI equations, we try

$$\underline{x}(t) = \underline{\xi} e^{rt} \tag{8.43}$$

where $\underline{\xi}$ is a vector of constants. Substituting Equation 8.43 into Equation 8.41 we have

$$r\underline{\xi} e^{rt} = \underline{A}\underline{\xi} e^{rt}$$

which may be rewritten as

$$(\underline{A} - r\underline{I})\underline{\xi} e^{rt} = 0 \tag{8.44}$$

where $\underline{I}$ is the *identity matrix*

$$\underline{I} = \begin{pmatrix} 1 & 0 \\ 0 & 1 \end{pmatrix}$$

Since $e^{rt}$ is never zero we may divide Equation 8.44 through by it and obtain

$$(\underline{A} - r\underline{I})\underline{\xi} = \underline{0} \tag{8.45}$$

Under what circumstances is Equation 8.45 satisfied? Recall from our review of matrix algebra in Section 8.4 that the solution to

$$\underline{C}\underline{y} = \underline{z}$$

is

$$\underline{y} = \underline{C}^{-1}\underline{z}$$

as long as

$$\det \underline{C} \neq 0$$

Returning to Equation 8.45, if the matrix $\underline{A} - r\underline{I}$ is invertible then the only solution is

$$\underline{\xi} = (\underline{A} - r\underline{I})^{-1}(\underline{0}) = \underline{0}$$

which will not satisfy the interesting cases, that is, those with nonzero initial conditions. Hence the first condition that Equation 8.45 must satisfy is

$$\det(\underline{A} - r\underline{I}) = 0 \tag{8.46}$$

Substituting Equation 8.42 into Equation 8.46,

$$\det\begin{pmatrix} -a-b-r & b \\ b & -a-b-r \end{pmatrix} = (a+b+r)^2 - b^2 = 0 \tag{8.47}$$

Equation 8.47 has solutions $r = -(a+b) \pm b$:

$$r_1 = -a$$

$$r_2 = -a - 2b$$

These solutions are called *eigenvalues*. Real-valued eigenvalues provide the rates of exponential decay or growth in the motion, which we will henceforth call *damping rate*. The damping rate is the inverse of the *time constant* introduced in Section 2.1.1.

Now we see that a second condition for a solution to Equation 8.45 is

$$(\underline{A} - r_i \underline{I})\underline{\xi}^{(i)} = \underline{0} \tag{8.48}$$

For a given eigenvalue $r_i$ (solution to Equation 8.46), the vector $\underline{\xi}^{(i)}$ satisfying Equation 8.48 is called the *eigenvector* associated with the eigenvalue.

Substituting $r_1 = -a$ in Equation 8.48 and again using Equation 8.42 we have

$$\begin{pmatrix} -a-b-(-a) & b \\ b & -a-b-(-a) \end{pmatrix} \begin{pmatrix} \xi_1^{(1)} \\ \xi_2^{(1)} \end{pmatrix} = \begin{pmatrix} 0 \\ 0 \end{pmatrix}$$

which, in component form, is

$$-b\xi_1^{(1)} + b\xi_2^{(1)} = 0$$
$$b\xi_1^{(1)} - b\xi_2^{(1)} = 0 \tag{8.49}$$

The matrix equation 8.49 is not two independent equations. It cannot be; its determinant is zero. It is essentially one equation:

$$-\xi_1^{(1)} + \xi_2^{(1)} = 0 \tag{8.50}$$

To proceed we *set* a value for one or the other of the unknowns. At this point we can only determine the ratios of the eigenvector's components. Their magnitudes will be determined when we apply initial conditions.

In setting a value for one or the other of the unknowns it is simplest to choose the value one. We set $\xi_1^{(1)} = 1$ and then, by Equation 8.50, $\xi_2^{(1)} = 1$, and so the eigenvector associated with the first eigenvalue, $r_1 = -a$, is

$$\underline{\xi}^{(1)} = \begin{pmatrix} 1 \\ 1 \end{pmatrix} \tag{8.51}$$

To determine the second eigenvector we return to Equation 8.48, substitute the second eigenvalue $r_2 = -a - 2b$ and solve for the eigenvector $\underline{\xi}^{(2)}$ associated with it.

$$\begin{pmatrix} -a-b-r_2 & b \\ b & -a-b-r_2 \end{pmatrix} \underline{\xi}^{(2)} = 0$$

$$\begin{pmatrix} -a-b-(-a-2b) & b \\ b & -a-b-(-a-2b) \end{pmatrix} \underline{\xi}^{(2)} = 0$$

We have

$$\begin{pmatrix} b & b \\ b & b \end{pmatrix} \underline{\xi}^{(2)} = 0 \tag{8.52}$$

The two equations in Equation 8.52 are not independent. (Again, they cannot be.) They are both equivalent to

$$\xi_1^{(2)} + \xi_2^{(2)} = 0 \tag{8.53}$$

Again we have one equation and two unknowns. We set $\xi_1^{(2)} = 1$ and then, from Equation 8.53, $\xi_2^{(2)} = -1$ and so the eigenvector associated with the second eigenvalue, $r_2 = -a - 2b$, is

$$\underline{\xi}^{(2)} = \begin{pmatrix} 1 \\ -1 \end{pmatrix} \tag{8.54}$$

Thus the general solution to Equations 8.40 and 8.41 has the form

$$\underline{x}(t) = c_1 \begin{pmatrix} 1 \\ 1 \end{pmatrix} e^{-at} + c_2 \begin{pmatrix} 1 \\ -1 \end{pmatrix} e^{-(a+2b)t} \tag{8.55}$$

We determine the constants $c_1$ and $c_2$ from the initial condition

$$\underline{x}(0) = \begin{pmatrix} T_{10} - T_\infty \\ T_{20} - T_\infty \end{pmatrix}$$

Setting $t = 0$ in Equation 8.55 and applying the initial condition we have

$$\underline{x}(0) = c_1 \begin{pmatrix} 1 \\ 1 \end{pmatrix} e^{-a(0)} + c_2 \begin{pmatrix} 1 \\ -1 \end{pmatrix} e^{-(a+2b)(0)} = c_1 \begin{pmatrix} 1 \\ 1 \end{pmatrix} + c_2 \begin{pmatrix} 1 \\ -1 \end{pmatrix} = \begin{pmatrix} T_{10} - T_\infty \\ T_{20} - T_\infty \end{pmatrix} \tag{8.56}$$

The vector Equation 8.56 represents two scalar equations:

$$c_1 \cdot 1 + c_2 \cdot 1 = T_{10} - T_\infty$$

$$c_1 \cdot 1 + c_2 \cdot (-1) = T_{20} - T_\infty$$

or

$$c_1 + c_2 = T_{10} - T_\infty$$

$$c_1 - c_2 = T_{20} - T_\infty$$

which has solution

$$c_1 = \frac{(T_{10} + T_{20} - 2T_\infty)}{2}$$

$$c_2 = \frac{(T_{10} - T_{20})}{2} \tag{8.57}$$

From Equations 8.55 and 8.57, the solution to Equation 8.39 is

$$\underline{x}(t) = \frac{(T_{10} + T_{20} - 2T_\infty)}{2}\begin{pmatrix} 1 \\ 1 \end{pmatrix}e^{-at} + \frac{(T_{10} - T_{20})}{2}\begin{pmatrix} 1 \\ -1 \end{pmatrix}e^{-(a+2b)t} \tag{8.58}$$

For real-valued eigenvalues, eigenvectors are the ratios in the magnitude of the state variables in each mode of motion the system can demonstrate. In other words, they are the ratios in the magnitude of the initial conditions that lead solely to the mode of motion associated with a given eigenvalue.

To what physical behavior do the eigenvalues and eigenvectors correspond in this example? The first eigenvalue is the damping rate each wafer experiences as its temperature approaches that of the airstream *when the temperatures of the two wafers are initially the same*. In this mode the wafers exchange no heat by conduction across the separating wall. All heat is gained or lost to the airstream. The second mode of behavior occurs *when one wafer is hotter than the airstream and the other colder*, each by the same amount. This maximizes conduction across the separating wall. Each wafer then approaches its equilibrium temperature at a faster rate because heat is exchanged with both the airstream and the other wafer.

### 8.5.1.2 Complex Eigenvalues: Electric Circuit Example

Because engineering students sometimes find the mathematical concepts of eigenvalues and eigenvectors abstract, we will dwell for a time on several other engineering examples to give concrete meaning to these important ideas. This section and the next discuss complex eigenvalues and eigenvectors.

Figure 8.5 depicts an electrical circuit that, we will discover, is modeled by a vector of two states, as was the heat transfer example in Section 8.5.1.1. We will concentrate in this section on learning the mathematics associated with complex eigenvalues. The next section addresses a system modeled by a vector of four states, which has more interesting dynamics. In that section we will concentrate on the physical interpretation of the system's eigenvalues and eigenvectors.

In the circuit shown in Figure 8.5, we assume that there is initially no current flowing in the perimeter and that the capacitor carries an initial voltage $V_0$. At time $t = 0$ the switch is thrown and we can analyze the circuit as two connected loops, as shown in Figure 8.6.

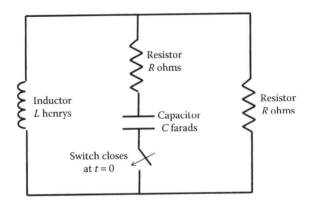

**FIGURE 8.5** Electrical circuit example.

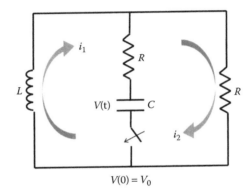

**FIGURE 8.6** Defining circuit variables.

Using Kirchhoff's voltage law, the sum of voltage drops around the left-hand loop equals zero:

$$L\frac{di_1}{dt} + R(i_1 - i_2) + V = 0 \tag{8.59}$$

and the sum of voltage drops around the right-hand loop equals zero:

$$Ri_2 - R(i_1 - i_2) - V = 0 \tag{8.60}$$

By Kirchhoff's current law, the time rate of change of the charge over the capacitor equals the current flowing from it:

$$\frac{dq}{dt} = C\frac{dV}{dt} = i_1 - i_2 \tag{8.61}$$

From Equation 8.60,

$$i_2 = \left(\frac{1}{2}\right)\left(i_1 + \frac{V}{R}\right) \tag{8.62}$$

Using Equation 8.62 in Equations 8.59 and 8.61,

$$\frac{dV}{dt} = \frac{1}{C}i_1 - \frac{1}{2C}(i_1 + V/R) = -\frac{1}{2RC}V + \frac{1}{2C}i_1 \tag{8.63}$$

$$\frac{di_1}{dt} = -\frac{R}{L}i_1 - \frac{1}{L}V + \frac{R}{2L}(i_1 + V/R) = -\frac{1}{2L}V - \frac{R}{2L}i_1 \tag{8.64}$$

To put this into the state space representation, let

$$\underline{x} = \begin{pmatrix} V \\ i_1 \end{pmatrix}$$

Then Equations 8.63 and 8.64 become

$$\underline{\dot{x}} = \underline{A}\underline{x}$$

where

$$\underline{A} = \begin{pmatrix} -1/(2RC) & 1/(2C) \\ -1/(2L) & -R/(2L) \end{pmatrix}$$

The initial condition is

$$\underline{x}(0) = \begin{pmatrix} V(0) \\ i_1(0) \end{pmatrix} = \begin{pmatrix} V_0 \\ 0 \end{pmatrix} \tag{8.65}$$

Let's assume that $R = 4$, $L = 1$, and $C = 1/16$. (The values of circuit elements are typically orders of magnitude different from these. However, with appropriate scaling our assumptions can be representative. We will not concern ourselves with scaling issues here.) With the assumed values,

$$\underline{A} = \begin{pmatrix} -2 & 8 \\ -1/2 & -2 \end{pmatrix} \tag{8.66}$$

$$\begin{pmatrix} \dot{x}_1 \\ \dot{x}_2 \end{pmatrix} = \begin{pmatrix} -2 & 8 \\ -1/2 & -2 \end{pmatrix} \begin{pmatrix} x_1 \\ x_2 \end{pmatrix} \tag{8.67}$$

To solve Equation 8.67 we try (as before) $\underline{x} = \underline{\xi}e^{rt}$ and derive the equation

$$(\underline{A} - r\underline{I})\underline{\xi} = 0 \tag{8.68}$$

where, again, $\underline{I}$ is the identity matrix, in this case

$$\underline{I} = \begin{pmatrix} 1 & 0 \\ 0 & 1 \end{pmatrix}$$

As always, Equation 8.68 results in two conditions. The first is

$$\det(\underline{A} - r\underline{I}) = 0 \qquad (8.69)$$

Equation 8.69 tells us that only certain values $r$ in our trial solution $\underline{x} = \underline{\xi}e^{rt}$ will work. Again, solutions $r_i$ to Equation 8.68 are called *eigenvalues*. The second condition is

$$(\underline{A} - r_i\underline{I})\underline{\xi}^{(i)} = 0 \qquad (8.70)$$

For each eigenvalue $r_i$ satisfying Equation 8.69 there is a corresponding vector of constants $\underline{\xi}^{(i)}$ called the *eigenvector* that satisfies Equation 8.70. Executing Equation 8.69:

$$\det(\underline{A} - r\underline{I}) = \det\begin{pmatrix} -2 - r & 8 \\ -1/2 & -2 - r \end{pmatrix} =$$

$$(-2 - r)(-2 - r) - (-1/2)8 = (r + 2)^2 + 4 = 0$$

$$r + 2 = \pm\sqrt{-4}$$

Hence the eigenvalues for this problem are

$$r_1 = -2 + 2i$$

$$r_2 = -2 - 2i$$

The next step is to substitute $r_1 = -2 + 2i$ and Equation 8.66 (for $\underline{A}$) into Equation 8.70. The result is

$$(\underline{A} - r_1\underline{I})\underline{\xi}^{(1)} = 0$$

$$\begin{pmatrix} -2 - (-2 + 2i) & 8 \\ -1/2 & -2 - (-2 + 2i) \end{pmatrix}\begin{pmatrix} \xi_1^{(1)} \\ \xi_2^{(1)} \end{pmatrix} = \begin{pmatrix} 0 \\ 0 \end{pmatrix}$$

$$\begin{pmatrix} -2i & 8 \\ -1/2 & -2i \end{pmatrix}\begin{pmatrix} \xi_1^{(1)} \\ \xi_2^{(1)} \end{pmatrix} = \begin{pmatrix} 0 \\ 0 \end{pmatrix} \qquad (8.71)$$

Writing out Equation 8.71 component by component we have

$$-2i\xi_1^{(1)} + 8\xi_2^{(1)} = 0$$
$$-(1/2)\xi_1^{(1)} - 2i\xi_2^{(1)} = 0 \qquad (8.72)$$

Equations 8.72 represent just one equation (as they must, since the determinant is zero). That equation can be written

$$-i\xi_1^{(1)} + 4\xi_2^{(1)} = 0$$

As always, there is an undetermined component. We *set* $\xi_1^{(1)} = 1$ and then $\xi_2^{(1)} = (1/4)i$. Hence the first eigenvector is

$$\underline{\xi}^{(1)} = \begin{pmatrix} 1 \\ i/4 \end{pmatrix}$$

We could go through the whole procedure again for the second eigenvalue $r_2 = -2 - 2i$ but we need not. We will always find (when the components of $\underline{A}$ are real) that when two eigenvalues are complex conjugates then their eigenvectors are complex conjugates also. Hence we can immediately write

$$\underline{\xi}^{(2)} = \begin{pmatrix} 1 \\ -i/4 \end{pmatrix}$$

The fundamental solutions associated with complex eigenvalues $r_1 = -\lambda + i\omega$ and $r_2 = -\lambda - i\omega$ can always be written

$$\underline{x}^{(1)} = (\underline{a} + i\underline{b})e^{-\lambda t + i\omega t}$$

$$\underline{x}^{(2)} = (\underline{a} - i\underline{b})e^{-\lambda t - i\omega t} \tag{8.73}$$

where $\underline{a}$ and $\underline{b}$ are vectors of real-valued constants. In this case

$$\underline{a} = \begin{pmatrix} 1 \\ 0 \end{pmatrix}, \qquad \underline{b} = \begin{pmatrix} 0 \\ 1/4 \end{pmatrix} \tag{8.74}$$

What do Equations 8.73 mean in real terms? As we did for second-order LTI equations with characteristic polynomials having complex roots, we invoke Euler's equation

$$e^{i\omega t} = \cos \omega t + i \sin \omega t \tag{8.75}$$

Substituting Equation 8.75 into Equations 8.73 results in

$$\underline{x}^{(1)} = e^{-\lambda t}(\underline{a}\cos \omega t - \underline{b}\sin \omega t) + ie^{-\lambda t}(\underline{a}\sin \omega t + \underline{b}\cos \omega t)$$

$$\underline{x}^{(2)} = e^{-\lambda t}(\underline{a}\cos \omega t - \underline{b}\sin \omega t) - ie^{-\lambda t}(\underline{a}\sin \omega t + \underline{b}\cos \omega t)$$

A general solution must have the form

$$\underline{x} = c_1 \underline{x}^{(1)} + c_2 \underline{x}^{(2)} =$$

$$(c_1 + c_2)e^{-\lambda t}(\underline{a}\cos \omega t - \underline{b}\sin \omega t) +$$

$$i(c_1 - c_2)e^{-\lambda t}(\underline{a}\sin \omega t + \underline{b}\cos \omega t)$$

If $\underline{x}$ is to be real then the coefficients $c_1, c_2$ must be complex. If we let

$$c_1 = (c_3 + ic_4)/2$$

$$c_2 = (c_3 - ic_4)/2$$

where $c_3, c_4$ are real-valued constants, then

$$\underline{x} = c_3 e^{-\lambda t}(\underline{a}\cos\omega t - \underline{b}\sin\omega t) + c_4 e^{-\lambda t}(\underline{a}\sin\omega t + \underline{b}\cos\omega t) \tag{8.76}$$

and all terms in Equation 8.76 are real. Therefore we see that the two *real-valued* fundamental solutions associated with complex conjugate eigenvalues $r_{1,2} = -\lambda \pm i\omega$ are

$$e^{-\lambda t}(\underline{a}\cos\omega t - \underline{b}\sin\omega t) \quad \text{and} \quad e^{-\lambda t}(\underline{a}\sin\omega t + \underline{b}\cos\omega t)$$

It is important to note that in solving a new problem we do not have to go through the intermediate steps represented here by Equations 8.73 and invocation of Euler's equation. We can go directly from identification of eigenvectors to the real solution represented here by Equation 8.76.

Complex eigenvalues represent the natural frequencies and damping factors in the motions. A two-state system such as this example has just one frequency and one damping factor. Eigenvectors give the relative magnitude of the state vector components and their relative phase. In this example, the ratio of the first and second components of the complex eigenvector is purely imaginary, indicating that the two components of the state vector are ninety degrees out of phase, meaning one is zero when the other is at its maxima or minima.

Using Equation 8.74, we determine the coefficients $c_3, c_4$ from the initial condition

$$\underline{x}(0) = c_3 e^{-\lambda(0)}(\underline{a}\cos\omega(0) - \underline{b}\sin\omega(0)) +$$

$$c_4 e^{-\lambda(0)}(\underline{a}\sin\omega(0) + \underline{b}\cos\omega(0)) = c_3\underline{a} + c_4\underline{b}$$

Using Equations 8.65 and 8.74

$$\begin{pmatrix} V_0 \\ 0 \end{pmatrix} = c_3 \begin{pmatrix} 1 \\ 0 \end{pmatrix} + c_4 \begin{pmatrix} 0 \\ 1/4 \end{pmatrix}$$

$$c_3 = V_0$$

$$c_4 = 0$$

In summary, the state vector for the circuit illustrated in Figures 8.5 and 8.6 is given by

$$\underline{x}(t) = \begin{pmatrix} V \\ i_1 \end{pmatrix} = V_0 e^{-2t}\left( \begin{pmatrix} 1 \\ 0 \end{pmatrix}\cos 2t - \begin{pmatrix} 0 \\ 1/4 \end{pmatrix}\sin 2t \right)$$

### 8.5.1.3 Complex Eigenvalues: Aircraft Dynamics Example

We now move on to a second example featuring complex eigenvalues that demonstrates more interesting behavior because it is a four-state system.

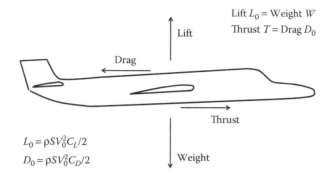

**FIGURE 8.7**   Aircraft in straight and level flight.

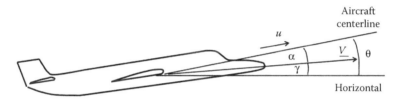

$V$ = Velocity vector
$u$ = Fractional increase in airspeed
$\alpha$ = Increase in angle of attack (angle between aircraft centerline and velocity vector)
$\theta$ = Increase in angle between aircraft centerline and horizontal
$\gamma$ = Increase in angle between aircraft velocity vector and horizontal

**FIGURE 8.8**   Perturbations about straight and level flight.

Figure 8.7 illustrates an aircraft in straight and level flight. In equilibrium, the aircraft's lift balances its weight and its thrust counteracts air resistance (drag).

Figure 8.8 defines variables capturing perturbations about straight and level flight. Aircraft demonstrate interesting lateral motions also (see Challenge Problem CP8.2), but our example will concentrate on longitudinal dynamics.

Challenge Problem CP8.5 guides the student through derivation of the linear equations, which turn out to be

$$\frac{du}{dt} = -2\frac{D_0}{L_0} \cdot \frac{g}{V_0} u + \left( D_\alpha - \frac{g}{V_0} \right)\gamma - D_\alpha\theta$$

$$\frac{d\gamma}{dt} = \frac{2g}{V_0} \cdot u - L_\alpha\gamma + L_\alpha\theta$$

$$\frac{d\theta}{dt} = q \qquad\qquad (8.77)$$

$$\frac{dq}{dt} = M_\alpha\gamma - M_\alpha\theta - M_q q$$

Figures 8.7 and 8.8 define $L_0, D_0$, and $V_0$, $g$ = acceleration of gravity, and the remaining terms are constants called "aerodynamic derivatives" defined in Challenge Problem CP8.5. Note that the angle of attack has disappeared from our equations, since it is given by

$$\alpha = \gamma - \theta \tag{8.78}$$

and a new variable, the angular rotation rate

$$q = \frac{d\theta}{dt} \tag{8.79}$$

has been added.

The four-dimensional vector

$$\underline{x} = \begin{pmatrix} x_1 \\ x_2 \\ x_3 \\ x_4 \end{pmatrix} = \begin{pmatrix} u \\ \gamma \\ \theta \\ q \end{pmatrix} \tag{8.80}$$

is the *state vector* for the aircraft's linearized longitudinal dynamics. In matrix form, Equation 8.77 can be written

$$\begin{pmatrix} \dot{u} \\ \dot{\gamma} \\ \dot{\theta} \\ \dot{q} \end{pmatrix} = \begin{pmatrix} -2(D_0/L_0)(g/V_0) & D_\alpha - (g/V_0) & -D_\alpha & 0 \\ 2g/V_0 & -L_\alpha & L_\alpha & 0 \\ 0 & 0 & 0 & 1 \\ 0 & M_\alpha & -M_\alpha & -M_q \end{pmatrix} \begin{pmatrix} u \\ \gamma \\ \theta \\ q \end{pmatrix} \tag{8.81}$$

Aircraft longitudinal dynamics captured in these equations exhibit two distinctively different behaviors, which engineers call the *phugoid* and *short–period* modes. We will see that these modes relate to eigenvalues and eigenvectors of the system of ODEs.

To enhance the student's understanding of these modes we will first make some assumptions in Equation 8.81 and solve the simplified ODEs. These solutions will be approximate. We will then solve Equation 8.81 exactly and determine the accuracy of our approximations.

To examine the phugoid mode we *set* $\gamma = \theta$ (i.e., set the angle of attack to zero.) Then the first two rows in Equations 8.77 become

$$\frac{du}{dt} = -2\frac{D_0}{L_0} \cdot \frac{g}{V_0} u - \frac{g}{V_0}\gamma$$

$$\frac{d\gamma}{dt} = \frac{2g}{V_0} \cdot u$$

Differentiating the first equation and substituting the second into it we arrive at

$$\frac{d^2u}{dt^2} + 2\frac{D_0}{L_0} \cdot \frac{g}{V_0}\frac{du}{dt} + 2\frac{g^2}{V_0^2}u = 0 \tag{8.82}$$

The *lift-to-drag ratio* L/D is an important one in flight vehicle design. One of the vehicle designer's main objectives is to make L/D as large as possible. In subsonic aircraft L/D is typically approximately 10. Then

$$\frac{D_0}{L_0} = \frac{1}{10}$$

in Equation 8.82 and from Chapter 5 we can determine that Equation 8.82 represents underdamped oscillations and has solution

$$u(t) = c_1 e^{-\lambda_1 t} \sin \omega_1 t + c_2 e^{-\lambda_1 t} \cos \omega_1 t \tag{8.83}$$

where

$$\lambda_1 = \frac{D_0}{L_0} \cdot \frac{g}{V_0}$$

and, approximately,

$$\omega_1 = \sqrt{2} \frac{g}{V_0}$$

The natural frequency for an aircraft traveling 350 nautical miles/h is approximately 0.077/s, which translates to a period of 81 s. The *damping ratio*

$$\varsigma_1 \cong \frac{\lambda_1}{\omega_1} = .07$$

which is quite low. From

$$\frac{d\gamma}{dt} = \frac{2g}{V_0} \cdot u$$

the flight path angle is given by

$$\gamma(t) = c_3 e^{-\lambda_1 t} \sin \omega_1 t + c_4 e^{-\lambda_1 t} \cos \omega_1 t$$

The phugoid motion is a sinusoidal flight path, with speed trading off with altitude (i.e., speed is high when altitude is low and vice versa). Figure 8.9 depicts it.

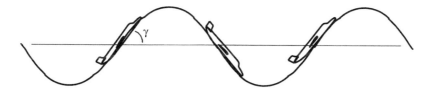

Note the zero angle of attack.

**FIGURE 8.9** Motion of the phugoid mode (vertical scale exaggerated).

To examine the short-period mode we return to Equations 8.81 and make different simplifying assumptions: we set $u = 0$ and $\gamma = 0$. The resulting equations are

$$\frac{d\theta}{dt} = q$$
$$\frac{dq}{dt} = -M_\alpha \theta - M_q q \tag{8.84}$$

Differentiating the first equation and substituting it into the second, we arrive at

$$\frac{d^2\theta}{dt^2} + M_q \frac{d\theta}{dt} + M_\alpha \theta = 0 \tag{8.85}$$

This has solution

$$\theta(t) = c_5 e^{-\lambda_2 t} \sin \omega_2 t + c_6 e^{-\lambda_2 t} \cos \omega_2 t$$

where, since $M_q$ is small,

$$\omega_2 \cong \sqrt{M_\alpha}$$

$$\lambda_2 = \frac{M_q}{2}$$

In keeping with its name, the short-period mode has a period much shorter than the phugoid mode. For subsonic aircraft it is typically 1 to 3 sec. In our approximation of the short-period mode the aircraft oscillates in the pitch plane without changing flight path, as shown in Figure 8.10.

Why do we not experience these motions when we travel in a commercial aircraft? Because we have been analyzing the natural motions of the aircraft "hands off"; that is, when its control surfaces and throttle are unchanged. Our analysis has omitted the actions of the pilot or autopilot to counteract these motions, which they will do, because the motions are after all deviations from the desired straight and level flight.

Now we return to Equation 8.81 and solve it exactly. For clarity we substitute numbers, estimating the value of each term in the matrix $\underline{A}$. It is beyond the scope of this text to pursue details, but assuming the parameters of a small corporate jet in a typical flight condition, one finds that the resulting equations are

$$\begin{pmatrix} \dot{u} \\ \dot{\gamma} \\ \dot{\theta} \\ \dot{q} \end{pmatrix} = \begin{pmatrix} -0.0101 & -0.310 & -0.0235 & 0 \\ 0.109 & -1.10 & 1.10 & 0 \\ 0 & 0 & 0 & 1 \\ 0 & 12.2 & -12.2 & -0.755 \end{pmatrix} \begin{pmatrix} u \\ \gamma \\ \theta \\ q \end{pmatrix} \tag{8.86}$$

Note the zero flight path angle.

**FIGURE 8.10**  Approximate motion of the short-period mode.

Equation 8.86 is of the form

$$\frac{d\underline{x}}{dt} = \underline{A}\underline{x} \tag{8.87}$$

We try

$$\underline{x}(t) = \underline{\xi}e^{rt} \tag{8.88}$$

where $\underline{\xi}$ is a vector of constants, as we will for all LTI systems. Substituting Equation 8.88 into Equation 8.87 we obtain

$$(\underline{A} - r\underline{I})\underline{\xi} = \underline{0} \tag{8.89}$$

where, again, $\underline{I}$ is the identity matrix, this time 4 by 4.

Again, the first condition that Equation 8.89 must satisfy is

$$\det(\underline{A} - r\underline{I}) = 0 \tag{8.90}$$

For our aircraft example,

$$\underline{A} - r\underline{I} = \begin{pmatrix} -0.0101 - r & -0.031 & -0.0235 & 0 \\ 0.109 & -1.10 - r & 1.10 & 0 \\ 0 & 0 & -r & 1 \\ 0 & 12.2 & -12.2 & -0.755 - r \end{pmatrix}$$

We find, after considerable algebra,

$$\det(\underline{A} - r\underline{I}) = r^4 + 1.8651r^3 + 13.0830r^2 + 0.134158r + 0.072474.$$

Therefore the values of $r$ that allow a solution to Equation 8.87 in the form of Equation 8.88 must satisfy the quartic equation

$$r^4 + 1.8651r^3 + 13.0830r^2 + 0.134158r + 0.072474 = 0 \tag{8.91}$$

Web-based "root calculators" make it easy to solve Equation 8.91. The results are

$$r_{1,2} = -0.00474 \pm i0.0743$$
$$r_{3,4} = -0.928 \pm i3.49 \tag{8.92}$$

Referring back to our approximations, we can compare the first two roots to the phugoid approximation and the second two to the short-period approximation. Table 8.2 shows that our approximate results are quite satisfactory except for the damping factor in the short-period mode.

The roots in Equations 8.92 are the *eigenvalues* for the aircraft's linearized longitudinal dynamics. As before, the second condition for a solution to Equation 8.89 is

$$(\underline{A} - r_i\underline{I})\underline{\xi}^{(i)} = \underline{0} \tag{8.93}$$

## TABLE 8.2

### Comparing Approximate and Actual Eigenvalues for Aircraft Dynamics Example

| Phugoid Mode | Approximate | Actual |
|---|---|---|
| Natural frequency $\omega$ (1/sec) | 0.0771 | 0.0743 |
| Damping factor $\lambda$ (1/sec) | 0.00509 | 0.00474 |

| Short-Period Mode | Approximate | Actual |
|---|---|---|
| Natural frequency $\omega$ (1/sec) | 3.49 | 3.49 |
| Damping factor $\lambda$ (1/sec) | 0.378 | 0.928 |

For a given eigenvalue (solution to Equation 8.91), we solve Equation 8.93 to find the eigenvector $\underline{\xi}^{(i)}$ associated with that eigenvalue. We have

$$\begin{pmatrix} -0.0101 - r_i & -0.031 & -0.0235 & 0 \\ 0.109 & -1.10 - r_i & 1.10 & 0 \\ 0 & 0 & -r_i & 1 \\ 0 & 12.2 & -12.2 & -0.755 - r_i \end{pmatrix} \begin{pmatrix} \xi_1^{(i)} \\ \xi_2^{(i)} \\ \xi_3^{(i)} \\ \xi_4^{(i)} \end{pmatrix} = \begin{pmatrix} 0 \\ 0 \\ 0 \\ 0 \end{pmatrix} \qquad (8.94)$$

Equations 8.92 show us that the four eigenvalues are two sets of complex conjugate pairs. Hence, recalling the discussion in the previous section, we need only solve Equation 8.94 once for each complex conjugate pair, say for $r_1$ and $r_3$. As in the previous section, the eigenvectors will be in the form

$$\underline{\xi}^{(i)} = \underline{a}^{(i)} + i\underline{b}^{(i)}$$

where $\underline{a}^{(i)}$ and $\underline{b}^{(i)}$ are vectors of real constants. With $\underline{a}^{(i)}$ and $\underline{b}^{(i)}$ in hand the real fundamental solutions are then given by

$$\underline{x}^{(1)}(t) = \exp(-\lambda_1 t)(\underline{a}^{(1)} \cos \omega_1 t - \underline{b}^{(1)} \sin \omega_1 t)$$

$$\underline{x}^{(2)}(t) = \exp(-\lambda_1 t)(\underline{a}^{(1)} \sin \omega_1 t + \underline{b}^{(1)} \cos \omega_1 t) \qquad (8.95)$$

$$\underline{x}^{(3)}(t) = \exp(-\lambda_3 t)(\underline{a}^{(3)} \cos \omega_3 t - \underline{b}^{(3)} \sin \omega_3 t)$$

$$\underline{x}^{(4)}(t) = \exp(-\lambda_3 t)(\underline{a}^{(3)} \sin \omega_3 t + \underline{b}^{(3)} \cos \omega_3 t)$$

where, again,

$$\underline{x}(t) = \begin{pmatrix} u(t) \\ \gamma(t) \\ \theta(t) \\ q(t) \end{pmatrix}$$

Recall from Equation 8.92 that $r_1 = -.00474 + i0.0743$. As always, there will be one undetermined vector component and the others will be given as a constant times the undetermined component. We elect to let $\xi_3$, the component for $\theta$, be the undetermined component because $\theta$ is active in both modes. The calculations are tedious because of four dimensions and complex arithmetic, but they are straightforward. The eigenvector associated with the first eigenvalue is found to be $\underline{\xi}^{(1)} = \underline{a}^{(1)} + i\underline{b}^{(1)}$ where

$$\underline{a}^{(1)} = \begin{pmatrix} -0.052 \\ 0.9993 \\ 1 \\ -0.0047 \end{pmatrix}, \qquad \underline{b}^{(1)} = \begin{pmatrix} 0.73 \\ 0.0045 \\ 0 \\ 0.074 \end{pmatrix}$$

When we substitute $r_3 = -0.928 + i3.49$ into Equation 8.94 we find that $\underline{\xi}^{(3)} = \underline{a}^{(3)} + i\underline{b}^{(3)}$ where

$$\underline{a}^{(3)} = \begin{pmatrix} 0.003 \\ 0.015 \\ 1 \\ -0.928 \end{pmatrix}, \qquad \underline{b}^{(3)} = \begin{pmatrix} 0.021 \\ -0.31 \\ 0 \\ 3.49 \end{pmatrix}$$

Table 8.3 compares these results to the approximate ones we derived earlier. The approximate results are satisfactory for the phugoid mode, less so for the short period mode.

The general solution to our original ODE, Equation 8.86, can be written

$$\underline{x}(t) = c_7 \underline{x}^{(1)}(t) + c_8 \underline{x}^{(2)}(t) + c_9 \underline{x}^{(3)}(t) + c_{10} \underline{x}^{(4)}(t) \tag{8.96}$$

where the fundamental solutions $\underline{x}^{(i)}(t)$ are given by Equations 8.95. Of course the coefficients $c_i$ depend on initial conditions.

---

**TABLE 8.3**

**Comparing Approximate and Actual Eigenvectors for Aircraft Dynamics Example**

| Phugoid Mode (Relative Magnitudes) | Approximate | Actual |
|---|---|---|
| $u$ | $-0.047 + 0.71i$ | $-0.052 + 0.73i$ |
| $\gamma$ | 1 | $0.9993 + 0.0045i$ |
| $\theta$ | 1 | 1 |
| $q(1/\text{sec})$ | $-0.0051 + 0.077i$ | $-0.0046 + 0.074i$ |

| Short-Period Mode (Relative Magnitudes) | Approximate | Actual |
|---|---|---|
| $u$ | 0 | $0.003 + 0.021i$ |
| $\gamma$ | 0 | $0.015 - 0.31i$ |
| $\theta$ | 1 | 1 |
| $q$ (1/sec) | $-0.38 + 3.49i$ | $-0.928 + 3.49i$ |

Note that Equation 8.96 can be written in the form

$$\underline{x}(t) = \underline{\Psi}(t) \cdot \underline{c} \tag{8.97}$$

where the *ith* column of $\underline{\Psi}(t)$ is the fundamental solution $\underline{x}^{(i)}(t)$. The matrix $\underline{\Psi}(t)$ is called the *fundamental matrix* for the system of Equations 8.86.

### 8.5.1.4 Repeated Eigenvalues

We begin this section with an example to illustrate how repeated eigenvalues can arise. The example will also demonstrate that the engineer's analyses may result in equations not in state space form, and he or she may have to do some work to put them in that form. And it will be useful for illustrating once more the physical behavior that eigenvalues and their associated eigenvectors represent. This section will then proceed to explore different forms that solutions can take when the system has repeated eigenvalues.

Consider the electrical circuit in Figures 8.11 and 8.12. Figure 8.11 shows the circuit prior to a switch being thrown at time $t = 0$. We assume that at $t = 0$ the circuit has been in

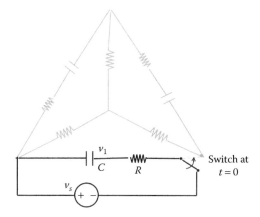

**FIGURE 8.11**    Electrical circuit prior to switch being thrown.

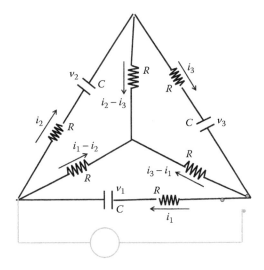

**FIGURE 8.12**    Electrical circuit after switch.

the configuration shown in Figure 8.11 long enough for the circuit to reach steady state. In the steady state the voltage over the capacitor $v_1$ equals the constant input voltage $v_S$.

Figure 8.12 pictures the circuit after $t = 0$ and shows the important circuit variables. All the resistors have the same value $R$ and all the capacitors the same value $C$. We want to determine the behavior of the circuit after $t = 0$.

Because voltage over a capacitor cannot change instantaneously, initial conditions at $t = 0$ are $v_1 = v_S$, $v_2 = v_3 = 0$.

Applying Kirchhoff's voltage law around the three internal loops:

$$v_1 + (3i_1 - i_2 - i_3)R = 0$$
$$v_2 + (3i_2 - i_1 - i_3)R = 0 \qquad (8.98)$$
$$v_3 + (3i_3 - i_1 - i_2)R = 0$$

Applying Kirchhoff's current law at the capacitors:

$$\frac{dq_1}{dt} = C\frac{dv_1}{dt} = i_1$$

$$\frac{dq_2}{dt} = C\frac{dv_2}{dt} = i_2 \qquad (8.99)$$

$$\frac{dq_3}{dt} = C\frac{dv_3}{dt} = i_3$$

Substituting Equations 8.99 into Equations 8.98:

$$v_1 = RC\left(-3\frac{dv_1}{dt} + \frac{dv_2}{dt} + \frac{dv_3}{dt}\right)$$

$$v_2 = RC\left(-3\frac{dv_2}{dt} + \frac{dv_1}{dt} + \frac{dv_3}{dt}\right) \qquad (8.100)$$

$$v_3 = RC\left(-3\frac{dv_3}{dt} + \frac{dv_1}{dt} + \frac{dv_2}{dt}\right)$$

Equations 8.100 are of the form

$$\underline{x} = \underline{A}^{-1}\frac{d\underline{x}}{dt}$$

Using an online matrix calculator we determine that

$$\begin{pmatrix} 3 & -1 & -1 \\ -1 & 3 & -1 \\ -1 & -1 & 3 \end{pmatrix}^{-1} = \begin{pmatrix} 1/2 & 1/4 & 1/4 \\ 1/4 & 1/2 & 1/4 \\ 1/4 & 1/4 & 1/2 \end{pmatrix}$$

Hence we can invert Equation 8.100 and obtain the state space formulation

$$\frac{d\underline{x}}{dt} = \underline{A}\underline{x} \tag{8.101}$$

where

$$\underline{x} = \begin{pmatrix} v_1 \\ v_2 \\ v_3 \end{pmatrix} \tag{8.102}$$

and

$$\underline{A} = \frac{-1}{4RC} \begin{pmatrix} 2 & 1 & 1 \\ 1 & 2 & 1 \\ 1 & 1 & 2 \end{pmatrix} \tag{8.103}$$

Note that the matrix $\underline{A}$ is symmetric, meaning that $A_{ij} = A_{ji}$. This is important in what follows.

Solved Problem 8.5 works this example and finds that the eigenvalues are $r_1 = -1/(RC)$, $r_2 = -1/(4RC)$, $r_3 = -1/(4RC)$ and the eigenvectors are:

$$\underline{\xi}^{(1)} = \begin{pmatrix} 1 \\ 1 \\ 1 \end{pmatrix}, \qquad \underline{\xi}^{(2)} = \begin{pmatrix} 1 \\ 0 \\ -1 \end{pmatrix}, \qquad \underline{\xi}^{(3)} = \begin{pmatrix} 1 \\ -1 \\ 0 \end{pmatrix}$$

Let us first explore the physical behavior that the three eigenvalues and their associated eigenvectors represent. The dynamics are easier to visualize if we examine the currents. The eigenvector

$$\underline{\xi}^{(1)} = \begin{pmatrix} 1 \\ 1 \\ 1 \end{pmatrix}$$

implies equal voltages $v_1(t), v_2(t), v_3(t)$ and hence, via Equations 8.99, equal currents $i_1(t)$, $i_2(t)$, and $i_3(t)$. Looking at the circuit, Figure 8.12, we see that in this mode all the current flows *around the perimeter* of the triangle. The inside paths have zero current. In the second and third dynamic modes current flows into the center of the triangle. In the second mode, current flows through sides one and three of the triangle but not through side two. In the third mode, current flows through sides one and two but not through side three.

But the most important observation here is that this example reveals a major difference between an nth-order LTI ODE in the form

$$\frac{d^n y}{dt^n} + a_1 \frac{d^{n-1}y}{dt^{n-1}} + \dots + a_n y = 0 \tag{8.104}$$

and a system of n first-order LTI ODEs in the form

$$\frac{dx}{dt} = Ax \tag{8.105}$$

As discussed in Section 6.2, if $\lambda$ is a root of multiplicity m in the characteristic equation for the nth order ODE (Equation 8.104) the m fundamental solutions $y_i$ must take the form $y_1 = e^{\lambda t}, y_2 = te^{\lambda t}, ..., y_m = t^{m-1}e^{\lambda t}$. In contrast, if $\lambda$ is a root of multiplicity m in the equation $\det(\underline{A} - r\underline{I}) = 0$ for the system of $n$ first-order ODEs (Equation 8.105), the system may have more than one and as many as m fundamental solutions of the form $\underline{x}^{(i)}(t) = \underline{\xi}^{(i)}e^{\lambda t}$. This example has repeated eigenvalues but linearly independent eigenvectors nevertheless. The key fact is that the matrix $\underline{A}$ in Equation 8.103 is symmetric.

**Theorem**

In the LTI system of first order ODEs

$$\frac{dx}{dt} = \underline{A}x$$

if the $n$ by $n$ matrix $\underline{A}$ is symmetric then the system will always have $n$ linearly independent eigenvectors, no matter the multiplicity of its eigenvalues. The fundamental solutions to the ODE will have the form

$$\underline{x}(t) = \underline{\xi}e^{rt}$$

where

$$\det(\underline{A} - r\underline{I}) = 0$$

$$(\underline{A} - r\underline{I})\underline{\xi} = 0$$

We state the theorem without proof. Note that this is just as was the case with non-repeated eigenvalues.

If $\underline{A}$ is *not* symmetric and it has a repeated eigenvalue then it may not have $n$ linearly independent eigenvalues and the fundamental solutions can be more complicated. We will satisfy ourselves by summarizing two examples, the first of which is worked out in detail in Solved Problem 8.6 and the second of which we leave to the student to solve by an easier method (Laplace transforms) in Word Problem WP8.6.

**Example 1:**

The matrix

$$A = \begin{pmatrix} 1 & -5 \\ 5 & -9 \end{pmatrix}$$

has the double eigenvalue $r_1 = r_2 = -4$ and the sole eigenvector

$$\underline{\xi} = \begin{pmatrix} 1 \\ 1 \end{pmatrix}$$

The associated fundamental solutions have the form

$$\underline{x}^{(1)}(t) = \underline{\xi} e^{rt}$$

$$\underline{x}^{(2)}(t) = \underline{\xi} t e^{rt} + \underline{\eta} e^{rt}$$

where

$$(\underline{A} - rI)\underline{\xi} = 0$$

$$(\underline{A} - rI)\underline{\eta} = \underline{\xi}$$

We find that

$$\underline{\eta} = \begin{pmatrix} 0 \\ -1/5 \end{pmatrix}$$

**Example 2 (from [9]):**

The matrix

$$\underline{A} = \begin{pmatrix} 1 & 1 & 1 \\ 2 & 1 & -1 \\ -3 & 2 & 4 \end{pmatrix}$$

has a triple eigenvalue $r = 2$ and one eigenvector

$$\underline{\xi} = \begin{pmatrix} 0 \\ 1 \\ -1 \end{pmatrix}$$

The associated fundamental solutions have the form

$$\underline{x}^{(1)}(t) = \underline{\xi} e^{rt}$$

$$\underline{x}^{(2)}(t) = \underline{\xi} t e^{rt} + \underline{\eta} e^{rt}$$

$$\underline{x}^{(3)}(t) = \underline{\xi}(t^2/2)e^{rt} + \underline{\eta} t e^{rt} + \underline{\varsigma} e^{rt}$$

where

$$(\underline{A} - rI)\underline{\xi} = 0$$

$$(\underline{A} - rI)\underline{\eta} = \underline{\xi}$$

$$(\underline{A} - rI)\underline{\varsigma} = \underline{\eta}$$

We find

$$\underline{\eta} = \begin{pmatrix} 1 \\ 1 \\ 0 \end{pmatrix} \quad \underline{\varsigma} = \begin{pmatrix} 2 \\ 0 \\ 3 \end{pmatrix}$$

The examples in this section all have repeated real roots. Repeated complex roots can also occur, in which case solutions can include terms such as $t^m e^{-\lambda t} \cos \omega t$ and $t^m e^{-\lambda t} \sin \omega t$.

The pattern evident in the solution of these examples holds in all $n$ dimensional time-invariant systems with repeated eigenvalues and less than $n$ independent eigenvectors. The pattern is probably best understood by investigating advanced properties of matrices that are beyond the scope of this text. See, for example, [10] or [11]. However, we will give in Section 8.5.1.9 a system-oriented interpretation of the pattern and in Section 8.5.2.2 an easier method (Laplace transforms) of solving such problems.

### 8.5.1.5   State Transition Matrix

We return now to the general linear time-varying homogeneous ODE

$$\dot{\underline{x}} = \underline{A}(t)\underline{x}$$

$$\underline{x}(t_0) = \underline{x}_0 \tag{8.106}$$

Equation 8.97 introduced the fundamental matrix $\underline{\Psi}(t)$ as it applied to the aircraft dynamics system. More generally, the solution to any linear ODE of the form in Equations 8.106 can be written as

$$\underline{x}(t) = \underline{\Psi}(t)\underline{c} \tag{8.107}$$

where the columns of $\underline{\Psi}(t)$ are the fundamental solutions of Equations 8.106.

A more convenient matrix is the *state transition matrix* $\underline{\Phi}(t,t_0)$, which describes how the state vector transitions from one point in time to another:

$$\underline{x}(t) = \underline{\Phi}(t,t_0)\underline{x}(t_0) \tag{8.108}$$

By definition

$$\underline{\Phi}(t_0,t_0) = \underline{I} \tag{8.109}$$

where, again, $I$ is the identity matrix. From Equation 8.107,

$$\underline{x}(t_0) = \underline{\Psi}(t_0)\underline{c} \tag{8.110}$$

Now the columns of $\underline{\Psi}(t_0)$ consist of the fundamental solutions $\underline{x}^{(i)}(t_0)$. By definition these solutions are linearly independent, so $\underline{\Psi}(t_0)$ is invertible. Therefore

$$\underline{c} = \underline{\Psi}^{-1}(t_0)\underline{x}(t_0) \tag{8.111}$$

and then, from Equations 8.107 and 8.111,

$$\underline{x}(t) = \underline{\Psi}(t)\underline{\Psi}^{-1}(t_0)\underline{x}(t_0) \tag{8.112}$$

From Equations 8.108 and 8.112

$$\underline{\Phi}(t,t_0) = \underline{\Psi}(t)\underline{\Psi}^{-1}(t_0) \tag{8.113}$$

The state transition matrix is convenient for homogeneous equations and we will discover in Section 8.5.2 that it plays a fundamental role in the solution of non-homogeneous equations.

For LTI systems the state transition matrix takes the form

$$\underline{\Phi}(t,t_0) = \underline{\Phi}(t - t_0)$$

### 8.5.1.6 Linear System Stability in State Space

In state space, a *linear* system in the form of Equation 8.106 is said to be stable if every component of the vector $\underline{x}(t)$ is bounded as time goes to infinity when each component of $\underline{x}(t_0)$ is bounded. (Recall from Chapter 4 that nonlinear systems require a more nuanced definition of stability.) Similarly, a linear system is said to be asymptotically stable if every component of the vector $\underline{x}(t)$ goes to zero as time goes to infinity when each component of $\underline{x}(t_0)$ is bounded. For linear systems, from $\underline{x}(t) = \underline{\Phi}(t,t_0)\underline{x}(t_0)$ it is easy to see that a sufficient condition for a system to be stable is that each component of $\underline{\Phi}(t,t_0)$ is bounded as time goes to infinity and a sufficient condition that the system is asymptotically stable is that each component of $\underline{\Phi}(t,t_0)$ goes to zero. For LTI systems, Sections 8.5.1.1 through 8.5.1.5 show that each component of the state transition matrix can be expressed as a linear combination of terms in the form

$$e^{\lambda t}((a_0 + a_1 t + \ldots + a_k t^k)\cos \omega t + (b_0 + b_1 t + \ldots + b_k t^k)\sin \omega t)$$

where $k \leq n$, the dimension of the system. In this expression, any of the terms $\lambda, a_i, b_i, \omega$ could be zero. From this we can see that an LTI system is stable and damped if $\lambda < 0$, because then

$$\lim_{t \to \infty} e^{\lambda t} t^m = 0$$

for any $m$. Hence an LTI system is stable and damped if the real part of every eigenvalue is negative. Similarly, a system is unstable if it has one or more eigenvalues with a positive real part. If a system has one or more eigenvalues with real part equal to zero and the remaining eigenvalues with negative real parts then nothing general can be said; each case must be considered on its own.

### 8.5.1.7   Coordinate Systems: Road Vehicle Dynamics

The next several sections explore the possibility of simplifying solution of a system of ODEs by a more convenient choice of state variables, which we can think of as selection of an appropriate coordinate system. We begin with an illustrative example.

Figure 8.13 depicts a road vehicle and its suspension system.

Model abstractions: Suspension springs can be coils (as shown), leaf springs, torsional bars or other realization. We will model the spring stiffness constant $k$ ($F = -kx$). Shock absorbers can be dashpots (as shown) or many other realizations. We will model the damping constant $\mu$ ($F = -\mu\dot{x}$).

Model simplifications: We assume that tires, wheels and undercarriage have negligible weight. Also, that tires are rigid; that is, have no compliance ("give"). And finally, that the vehicle center of gravity is midway between wheels. This last assumption makes our model more appropriate to a wagon towed behind a motor vehicle than to a motor vehicle itself, because weight of a motor places the center of gravity forward.

Figure 8.14 and Table 8.4 define parameters and variables for our model.

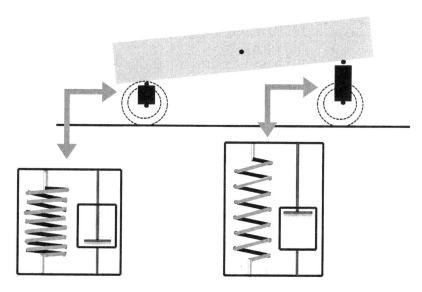

**FIGURE 8.13**   Road vehicle and its suspension system.

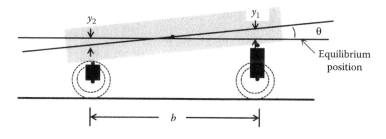

**FIGURE 8.14**   Variable definition for road vehicle model.

## TABLE 8.4

### Variable and Parameter Definition for Road Vehicle Model

$y_1$     Position of vehicle centerline at right wheel relative to equilibrium position

$y_2$     Position of vehicle centerline at left wheel relative to equilibrium position

$y_{cg}$   Position of vehicle center of gravity relative to equilibrium position

$\theta$     Angle of vehicle centerline relative to horizontal

$m$     Mass of vehicle

$b$     Wheelbase of vehicle

$I$     Moment of inertia of vehicle

$k$     Suspension spring stiffness constant

$\mu$     Suspension damping constant

The equilibrium position in Figure 8.14 is the vehicle's steady state in the presence of gravity.

From geometry, we have

$$y_{cg} = (y_1 + y_2)/2 \tag{8.114}$$

$$\theta = (y_1 - y_2)/b \tag{8.115}$$

The force on the car at each wheel is given by:

$$F_1 = -ky_1 - \mu\dot{y}_1 \tag{8.116}$$

$$F_2 = -ky_2 - \mu\dot{y}_2 \tag{8.117}$$

Translational and rotational equations of motion are:

$$m\ddot{y}_{cg} = F_1 + F_2 \tag{8.118}$$

$$I\ddot{\theta} = (F_1 - F_2)b/2 \tag{8.119}$$

Combining Equations 8.114 through 8.119, we arrive at

$$\ddot{y}_1 + \ddot{y}_2 = -(2k/m)(y_1 + y_2) - (2\mu/m)(\dot{y}_1 + \dot{y}_2) \tag{8.120}$$

$$\ddot{y}_1 - \ddot{y}_2 = -\left(b^2 k/2I\right)(y_1 - y_2) - \left(b^2\mu/2I\right)(\dot{y}_1 - \dot{y}_2) \tag{8.121}$$

One scheme to solve Equations 8.120 and 8.121 is to let

$$\underline{x} = \begin{pmatrix} x_1 \\ x_2 \\ x_3 \\ x_4 \end{pmatrix} = \begin{pmatrix} y_1 \\ y_2 \\ \dot{y}_1 \\ \dot{y}_2 \end{pmatrix}$$

However, from Equations 8.120 and 8.121 it is apparent that a better choice is

$$x_1 = y_1 + y_2$$

$$x_2 = y_1 - y_2$$

Then Equations 8.120 and 8.121 become

$$\ddot{x}_1 = -(2k/m)x_1 - (2\mu/m)\dot{x}_1 \tag{8.122}$$

$$\ddot{x}_2 = -(b^2k/2I)x_2 - (b^2\mu/2I)\dot{x}_2 \tag{8.123}$$

Equations 8.122 and 8.123, while not in state space form, are uncoupled and easy to solve.

Before we solve Equations 8.122 and 8.123 it is useful to compute the moment of inertia $I$. We assume that vehicle mass is uniformly distributed and that vehicle length is 10% greater than its wheelbase as illustrated in Figure 8.15.

Then

$$I = \int_{-l/2}^{l/2} \rho z^2 dz = 2\rho(l/2)^3/3 = \rho l^3/12 = (\rho l)l^2/12 \cong mb^2/10 \tag{8.124}$$

Using Equation 8.124, introducing the dimensionless ratio $v = \mu/\sqrt{km}$, and assuming that the system is underdamped in both translation and rotation, we find that the solution to Equations 8.122 and 8.123 can be written as

$$x_1 = e^{-\lambda_1 t}(c_1 \cos \omega_1 t + c_2 \sin \omega_1 t) \tag{8.125}$$

$$x_2 = e^{-\lambda_2 t}(c_3 \cos \omega_2 t + c_4 \sin \omega_2 t) \tag{8.126}$$

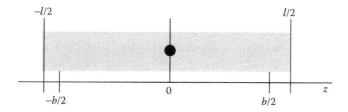

**FIGURE 8.15** Geometry for calculation of moment of inertia.

where

$$\lambda_1 = v\sqrt{k/m} \tag{8.127}$$

$$\omega_1 = \sqrt{k/m}\sqrt{2-v^2} \tag{8.128}$$

$$\lambda_2 = (5v/2)\sqrt{k/m} \tag{8.129}$$

$$\omega_2 = \sqrt{(k/m)}\sqrt{5-(5v/2)^2} \tag{8.130}$$

Having solved the equations with the variables we did, recognizing the system's dynamic modes is easy. In mode 1, the vehicle translates up and down without rotating, as shown in Figure 8.16. In mode 2, the vehicle rotates while keeping its center of gravity still, as shown in Figure 8.17.

We note as an aside that Equations 8.125 through 8.130 tell us that, as a function of $m$, $k$, and $\mu$: (a) vehicle oscillations always damp out more quickly in the rotational mode (mode 2) than in the translational mode (mode 1) and (b) the frequency of rotational oscillations can be higher or lower than the frequency of translational oscillations. The vehicle designer may desire critical damping as a compromise between rough ride and undesirable oscillation but he or she cannot achieve critical damping in both modes simultaneously with this system configuration.

The example shows that if the choice of dependent variable uncouples the system equations, solution is greatly eased. In this example the appropriate choice of dependent variables was apparent when we wrote down the basic equations of motion. In general

**FIGURE 8.16**   Translational mode.

**FIGURE 8.17**   Rotational mode.

it is not so easy to find a set of variables that uncouples the equations. The next section addresses that issue.

## 8.5.1.8    Coordinate Systems: A General Analytic Approach

This section explores the possibility of simplifying solution of a general LTI system of ODEs by a more convenient choice of state variables. We will find that we can often choose a state vector $\underline{z}(t)$ for which the state equations have the form

$$\underline{\dot{z}} = \underline{R}\underline{z} \tag{8.131}$$

where

$$\underline{R} = \begin{pmatrix} r_1 & 0 & ... & 0 & 0 \\ 0 & r_2 & ... & 0 & 0 \\ ... & ... & ... & ... & ... \\ 0 & 0 & ... & r_{n-1} & 0 \\ 0 & 0 & ... & 0 & r_n \end{pmatrix} \tag{8.132}$$

Then the solution is easy:

$$z_i = z_i(0)e^{r_i t} \quad i = 1,...,n \tag{8.133}$$

We must, of course, admit the possibility of complex eigenvalues.

**Theorem**

Let $\underline{\dot{x}} = \underline{A}\underline{x}$ and where $\underline{A}$ is an $n$ by $n$ matrix with $n$ linearly independent eigenvectors. Then the state vector

$$\underline{z}(t) = \underline{\Psi}^{-1}(0)\underline{x}(t) \tag{8.134}$$

satisfies $\underline{\dot{z}} = \underline{R}\underline{z}$ where $\underline{R}$ is the diagonalized matrix given by Equation 8.132, $r_i$ $(i = 1,...,n)$ are the eigenvalues of $\underline{A}$ and $\underline{\Psi}(0)$ is the fundamental matrix evaluated at $t = 0$.
    (Stated without proof.)
    Remark: In this theorem, when eigenvalues are complex, we are using the complex form of the fundamental solutions: $\underline{x}(t) = \underline{\xi}e^{rt}$.
    Given that (a) we can often diagonalize a system's $\underline{A}$ matrix and (b) solving a problem in that form is simple, why do engineers not use the method more often? The answer lies in two parts. First, the procedure to diagonalize the $\underline{A}$ matrix is about as much work as that involved in solving the equations in coupled form. Note that one had to determine the system's eigenvalues and eigenvectors to find the state vector for the diagonalized $\underline{A}$ matrix. Second, and perhaps more important, diagonalization often hides physical significance. Engineers want to use state variables that have physical meaning. A state vector for a diagonalized system is typically a linear combination of physically meaningful variables, and for engineers a meaningless variable in itself. But this is not always the case. The road vehicle example in Section 8.5.1.7 shows that mathematically convenient variables can sometimes be physically relevant as well. See Word Problem WP8.4 for another such example.

### 8.5.1.9   A System Viewpoint on Coordinate Systems

In the previous section we saw that when an $n$-dimensional coupled LTI system has $n$ independent eigenvectors its dynamics are equivalent to $n$ first-order systems operating *in parallel*, completely independent of one another. Although typically not useful for computations the equivalence is conceptually valuable, helping us keep in mind that in such coupled systems there are $n$ independent simultaneous motions.

But what about those $n$-dimensional systems with less than $n$ independent eigenvectors? Is there a similarly valuable construct? The answer is yes. It can be shown that when a system has an eigenvalue of multiplicity $m$ with only $m - p$ independent eigenvectors then part of the system can be transformed into equations equivalent to those for identical first-order systems operating *in series*; that is, cascaded identical systems. There may be more than one series operating in parallel, but the total number of output-to-input connections in all the series together will be $p$ [12]. Such series have their own unique dynamics. This is one interpretation of the pattern noted in the solutions to Examples 1 and 2 in Section 8.5.1.4 on repeated eigenvalues. Challenge Problem CP8.6 guides the student through an analysis of the dynamics of cascaded identical first-order systems.

### 8.5.2   Nonhomogeneous Linear Systems

This section explores methods for solving problems in the form

$$\underline{\dot{x}} = \underline{A}(t)\underline{x} + \underline{B}(t)\underline{u}(t), \qquad \underline{x}(t_0) = \underline{x}_0$$

where $\underline{u}(t)$ is nonzero. As was the case for second- and higher-order linear systems, we examine three different methods. First, for the general time-varying case, we derive the matrix kernel (variation of parameters) method. We then develop the equations for the matrix Laplace transform and matrix undetermined coefficient methods, which are applicable only to the time-invariant case.

### 8.5.2.1   Matrix Kernel Method

Section 8.5.1.5 introduced the state transition matrix $\underline{\Phi}(t,t_0)$. This matrix is central to the method we are about to derive. Among useful properties of the state transition matrix are the following:

$$\frac{d}{dt}\underline{\Phi}(t,t_0) = \underline{\dot{\Psi}}(t)\underline{\Psi}^{-1}(t_0) = \underline{A}(t)\underline{\Psi}(t)\underline{\Psi}^{-1}(t_0)$$

$$\frac{d}{dt}\underline{\Phi}(t,t_0) = \underline{A}(t)\underline{\Phi}(t,t_0) \tag{8.135}$$

Also,

$$\underline{\Phi}^{-1}(t,t_0) = \underline{\Phi}(t_0,t) \tag{8.136}$$

Proof:

$$\underline{\Phi}^{-1}(t,t_0) = \left(\underline{\Psi}(t)\underline{\Psi}^{-1}(t_0)\right)^{-1} = \underline{\Psi}(t_0)\underline{\Psi}^{-1}(t) = \underline{\Phi}(t_0,t)$$

Finally,

$$\underline{\Phi}(t,t_0) = \underline{\Phi}(t,t_1)\underline{\Phi}(t_1,t_0) \tag{8.137}$$

Proof:

$$\underline{\Phi}(t,t_0) = \underline{\Psi}(t)\underline{\Psi}^{-1}(t_0) = \underline{\Psi}(t)\underline{I}\underline{\Psi}^{-1}(t_0)$$

$$\underline{\Phi}(t,t_0) = \underline{\Psi}(t)(\underline{\Psi}^{-1}(t_1)\underline{\Psi}(t_1))\underline{\Psi}^{-1}(t_0)$$

$$\underline{\Phi}(t,t_0) = (\underline{\Psi}(t)\underline{\Psi}^{-1}(t_1))(\underline{\Psi}(t_1)\underline{\Psi}^{-1}(t_0)) = \underline{\Phi}(t,t_1)\underline{\Phi}(t_1,t_0)$$

We will need Equations 8.136 and 8.137 in the proof of the following important theorem, which states that the state transition matrix is the matrix kernel.

**Theorem**

The equation

$$\underline{\dot{x}} = \underline{A}(t)\underline{x} + \underline{B}(t)\underline{u}(t), \qquad \underline{x}(t_0) = \underline{x}_0 \tag{8.138}$$

has solution

$$\underline{x}(t) = \underline{\Phi}(t,t_0)\underline{x}(t_0) + \int_{t_0}^{t} \underline{\Phi}(t,\tau)\underline{B}(\tau)\underline{u}(\tau)d\tau \tag{8.139}$$

where

$$\frac{d}{dt}\underline{\Phi}(t,t_0) = \underline{A}(t)\underline{\Phi}(t,t_0) \tag{8.140}$$

$$\underline{\Phi}(t_0,t_0) = \underline{I} \tag{8.141}$$

**Proof**

Recall that Section 2.2 solved the nonhomogeneous first-order linear ODE

$$\frac{dy}{dt} + p(t)y = g(t), \qquad y(t_0) = y_0$$

by finding an integrating factor $\mu(t)$ such that

$$\mu(t)\left(\frac{dy}{dt} + p(t)y\right) = \frac{d}{dt}(\mu(t)y)$$

In an exactly analogous way, we seek here a matrix integrating factor $\underline{M}(t)$ such that

$$\underline{M}(t)\left(\underline{\dot{x}} - \underline{A}(t)\underline{x}\right) = \frac{d}{dt}\left(\underline{M}\underline{x}\right) \tag{8.142}$$

Expanding the left-hand side of Equation 8.142,

$$\underline{M}(t)\left(\underline{\dot{x}} - \underline{A}(t)\underline{x}\right) = \underline{M}\underline{\dot{x}} - \underline{M}\underline{A}\underline{x} \tag{8.143}$$

Expanding the right-hand side of Equation 8.142,

$$\frac{d}{dt}\left(\underline{M}\underline{x}\right) = \underline{M}\underline{\dot{x}} + \underline{\dot{M}}\underline{x} \tag{8.144}$$

Equating Equations 8.143 and 8.144, we see that the matrix integrating factor we seek must satisfy

$$\underline{\dot{M}} = -\underline{M}\underline{A} \tag{8.145}$$

Now consider

$$\underline{\Phi}(t,t_0) \cdot \underline{\Phi}^{-1}(t,t_0) = \underline{I}$$

$$\frac{d}{dt}\left(\underline{\Phi}(t,t_0) \cdot \underline{\Phi}^{-1}(t,t_0)\right) = \frac{d}{dt}\underline{I} = \underline{0}$$

$$\underline{\dot{\Phi}}\underline{\Phi}^{-1} + \underline{\Phi}\underline{\dot{\Phi}}^{-1} = \underline{0}$$

$$\underline{A}\underline{\Phi}\underline{\Phi}^{-1} + \underline{\Phi}\underline{\dot{\Phi}}^{-1} = \underline{A}\underline{I} + \underline{\Phi}\underline{\dot{\Phi}}^{-1} = \underline{A} + \underline{\Phi}\underline{\dot{\Phi}}^{-1} = \underline{0}$$

$$\underline{\dot{\Phi}}^{-1} = -\underline{\Phi}^{-1}\underline{A}$$

Hence the matrix integrating factor we seek is

$$\underline{M} = \underline{\Phi}^{-1}(t,t_0) \tag{8.146}$$

Multiplying both sides of the equation

$$\underline{\dot{x}} - \underline{A}(t)\underline{x} = \underline{B}(t)\underline{u}(t)$$

by $\underline{\Phi}^{-1}(t,t_0)$ and using Equations 8.142 and 8.146 we have

$$\frac{d}{dt}\left(\underline{\Phi}^{-1}(t,t_0)\underline{x}\right) = \underline{\Phi}^{-1}(t,t_0)\underline{B}(t)\underline{u}(t)$$

$$\underline{\Phi}^{-1}(t,t_0)\underline{x} - x_0 = \int_{t_0}^{t} \underline{\Phi}^{-1}(\tau,t_0)\underline{B}(\tau)\underline{u}(\tau)d\tau$$

$$\underline{\Phi}^{-1}(t,t_0)\underline{x}(t) = \underline{x}_0 + \int_{t_0}^{t} \underline{\Phi}^{-1}(\tau,t_0)\underline{B}(\tau)\underline{u}(\tau)d\tau$$

$$\underline{x}(t) = \underline{\Phi}(t,t_0)\underline{x}_0 + \int_{t_0}^{t} \underline{\Phi}(t,t_0)\underline{\Phi}^{-1}(\tau,t_0)\underline{B}(\tau)\underline{u}(\tau)d\tau$$

Now, from the properties of the state transition matrix, Equations 8.136 and 8.137,

$$\underline{\Phi}(t,t_0)\underline{\Phi}^{-1}(\tau,t_0) = \underline{\Phi}(t,t_0)\underline{\Phi}(t_0,\tau) = \underline{\Phi}(t,\tau)$$

Hence

$$\underline{x}(t) = \underline{\Phi}(t,t_0)\underline{x}_0 + \int_{t_0}^t \underline{\Phi}(t,\tau)\underline{B}(\tau)\underline{u}(\tau)d\tau$$

as was to be proved.

**Theorem**

The LTI system of equations

$$\underline{\dot{x}} = \underline{A}\underline{x} + \underline{B}\underline{u}(t)$$

has solution

$$\underline{x}(t) = \underline{\Phi}(t-t_0)\underline{x}(t_0) + \int_{t_0}^t \underline{\Phi}(t-\tau)\underline{B}\underline{u}(\tau)d\tau \tag{8.147}$$

where $\underline{\Phi}(t) = e^{\underline{A}t}$ and $e^{\underline{A}t}$ satisfies all the usual properties of the exponential function. (Stated without proof.)

The Matrix Kernel method uses Equation 8.139 or, if $\underline{A}$ is constant, Equation 8.147. The procedure for the time-invariant case is as follows:

To solve:

$$\underline{\dot{x}} = \underline{A}\underline{x} + \underline{B}\underline{u}(t)$$

$$\underline{x}(0) = \underline{x}_0 \tag{8.148}$$

$$\underline{A},\underline{B},\underline{u}(t),\underline{x}_0 = \text{given}$$

Step 1: Find the eigenvalues of the solutions to the homogeneous equation $\underline{\dot{x}} = \underline{A}\underline{x}$. That is, try $\underline{x} = \underline{\xi}e^{rt}$ and find the values $r_i$ satisfying $\det(r\underline{I} - \underline{A}) = 0$.

Step 2: Find the eigenvectors associated with the eigenvalues via $(r_i\underline{I} - \underline{A})\underline{\xi}^{(i)} = \underline{0}$.

Step 3: Determine the fundamental solutions and form the fundamental matrix. This is often, but not always, given by $\underline{\Psi}(t) = \left(\underline{\xi}^{(1)}e^{rt} \ldots \underline{\xi}^{(n)}e^{r_n t}\right)$. We can use the complex form of fundamental solutions in this step.

Step 4: Calculate the inverse of the fundamental matrix evaluated at zero. Use the defining equation $\underline{\Psi}(0) \cdot \underline{\Psi}^{-1}(0) = \underline{I}$.

Step 5: Calculate the state transition matrix from $\underline{\Phi}(t) = \underline{\Psi}(t) \cdot \underline{\Psi}^{-1}(0)$.

Step 6: Calculate the solution $\underline{x}(t) = \underline{\Phi}(t)\underline{x}_0 + \int_0^t \underline{\Phi}(t-\tau)\underline{B}\underline{u}(\tau)d\tau$.

### Example: Electrical Circuit with Time-Varying Voltage Input

Figure 8.18 depicts an electrical circuit in which the voltage input has the form $v_{in}(t) = 10e^{-t}$ for $t > 0$ and all voltages and currents are zero for $t < 0$. Our object is to find the voltage $v_{out}$ over the capacitor as a function of time.

Step 0: We must first derive the state equations. Figure 8.19 identifies circuit variables for us to work with.

From Kirchhoff's voltage law we have $v_{in}(t) = v_{out} + (i_1 - i_2)R$ and $L di_2/dt = (i_1 - i_2)R$. From Kirchhoff's current law (or conservation of charge) and noting that the charge on the capacitor is given by $q = C v_{out}$ we also have $dq/dt = C dv_{out}/dt = i_1$. Reducing the number of variables: $i_1 = i_2 + (v_{in}(t) - v_{out})/R$ and then

$$\frac{dv_{out}}{dt} = (v_{in} - v_{out})/RC + i_2/C \tag{8.149}$$

$$\frac{di_2}{dt} = (v_{in} - v_{out})/L \tag{8.150}$$

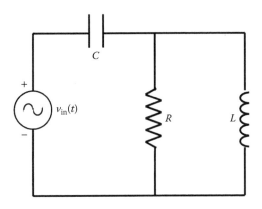

**FIGURE 8.18**   Electrical circuit with time-varying voltage input.

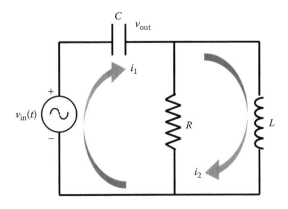

**FIGURE 8.19**   Electrical circuit with identified circuit variables.

Choosing $v_{out}$ and $i_2$ as state variables, Equations 8.149 and 8.150 can be written in state space form as

$$\frac{d}{dt}\begin{pmatrix} v_{out} \\ i_2 \end{pmatrix} = \begin{pmatrix} -1/RC & 1/C \\ -1/L & 0 \end{pmatrix}\begin{pmatrix} v_{out} \\ i_2 \end{pmatrix} + \begin{pmatrix} 1/RC \\ 1/L \end{pmatrix} v_{in}(t) \qquad (8.151)$$

We are now ready to proceed with the steps defined earlier to solve nonhomogenous time-invariant systems of first-order ODEs.

Step 1: Find the eigenvalues of the solutions to the homogeneous equation

$$\frac{d}{dt}\begin{pmatrix} v_{out} \\ i_2 \end{pmatrix} = \begin{pmatrix} -1/RC & 1/C \\ -1/L & 0 \end{pmatrix}\begin{pmatrix} v_{out} \\ i_2 \end{pmatrix}$$

using

$$\det(r\underline{I} - \underline{A}) = \det\begin{pmatrix} r+1/RC & -1/C \\ 1/L & r \end{pmatrix} = r^2 + (1/RC)r + (1/LC) = 0 \qquad (8.152)$$

Suppose $C = 1$, $R = 1/5$, and $L = 1/6$. Then Equation 8.152 becomes

$$r^2 + 5r + 6 = (r+2)(r+3) = 0$$

and the eigenvalues are $r_1 = -2$ and $r_2 = -3$.

Step 2: Find the eigenvectors associated with the eigenvalues via $(r_i\underline{I} - \underline{A})\underline{\xi}^{(i)} = \underline{0}$.

$$\begin{pmatrix} -2+5 & -1 \\ 6 & -2 \end{pmatrix}\begin{pmatrix} \xi_1^{(1)} \\ \xi_2^{(1)} \end{pmatrix} = \begin{pmatrix} 0 \\ 0 \end{pmatrix} \qquad \begin{pmatrix} -3+5 & -1 \\ 6 & -3 \end{pmatrix}\begin{pmatrix} \xi_1^{(2)} \\ \xi_2^{(2)} \end{pmatrix} = \begin{pmatrix} 0 \\ 0 \end{pmatrix}$$

$$3\xi_1^{(1)} - \xi_2^{(1)} = 0, \qquad 2\xi_1^{(2)} - \xi_2^{(2)} = 0$$

In each case we let the first component $\xi^{(1)} = 1$ and then the eigenvectors are

$$\underline{\xi}^{(1)} = \begin{pmatrix} 1 \\ 3 \end{pmatrix}, \qquad \underline{\xi}^{(2)} = \begin{pmatrix} 1 \\ 2 \end{pmatrix}$$

Step 3: Determine the fundamental solutions and form the fundamental matrix. The fundamental solutions are

$$\underline{\xi}^{(1)}e^{r_1 t} = \begin{pmatrix} 1 \\ 3 \end{pmatrix}e^{-2t}, \qquad \underline{\xi}^{(2)}e^{r_2 t} = \begin{pmatrix} 1 \\ 2 \end{pmatrix}e^{-3t}$$

These vectors form the columns of the fundamental matrix:

$$\underline{\Psi}(t) = \begin{pmatrix} e^{-2t} & e^{-3t} \\ 3e^{-2t} & 2e^{-3t} \end{pmatrix}$$

Step 4: Calculate the inverse of the fundamental matrix evaluated at zero. Use the defining equation $\underline{\Psi}(0) \cdot \underline{\Psi}^{-1}(0) = I$.

$$\begin{pmatrix} 1 & 1 \\ 3 & 2 \end{pmatrix}\begin{pmatrix} a & b \\ c & d \end{pmatrix} = \begin{pmatrix} 1 & 0 \\ 0 & 1 \end{pmatrix}$$

$$a + c = 1 \qquad\qquad b + d = 0$$

$$3a + 2c = 0 \qquad\quad 3b + 2d = 1$$

$$c = 1 - a \qquad\qquad b = -d$$

$$3a + 2(1-a) = 0 \quad 3b + 2(-b) = 1$$

$$a = -2 \qquad\qquad b = 1$$

$$c = 3 \qquad\qquad d = -1$$

$$\underline{\Psi}^{-1}(0) = \begin{pmatrix} -2 & 1 \\ 3 & -1 \end{pmatrix}$$

Step 5: Calculate the state transition matrix from $\underline{\Phi}(t) = \underline{\Psi}(t) \cdot \underline{\Psi}^{-1}(0)$.

$$\underline{\Phi}(t) = \underline{\Psi}(t) \cdot \underline{\Psi}^{-1}(0) = \begin{pmatrix} e^{-2t} & e^{-3t} \\ 3e^{-2t} & 2e^{-3t} \end{pmatrix}\begin{pmatrix} -2 & 1 \\ 3 & -1 \end{pmatrix} = \begin{pmatrix} -2e^{-2t} + 3e^{-3t} & e^{-2t} - e^{-3t} \\ -6e^{-2t} + 6e^{-3t} & 3e^{-2t} - 2e^{-3t} \end{pmatrix}$$

Here is a place to check. We have done a fair amount of algebra already. We can (partially) check our results so far by ensuring that, using the equation for $\underline{\Phi}(t)$,

$$\underline{\Phi}(0) = I = \begin{pmatrix} 1 & 0 \\ 0 & 1 \end{pmatrix}$$

It does.

Step 6: Calculate the solution $\underline{x}(t) = \underline{\Phi}(t)\underline{x}_0 + \int_0^t \underline{\Phi}(t - \tau)\underline{B}\underline{u}(\tau)d\tau$. Here $\underline{x}_0 = \underline{0}$,

$$\underline{B} = \begin{pmatrix} 1/RC \\ 1/L \end{pmatrix} = \begin{pmatrix} 5 \\ 6 \end{pmatrix}$$

and $\underline{u}(t) = v(t) = 10e^{-t}$. Thus

$$\underline{x}(t) = \int_0^t \begin{pmatrix} -2e^{-2(t-\tau)} + 3e^{-3(t-\tau)} & e^{-2(t-\tau)} - e^{-3(t-\tau)} \\ -6e^{-2(t-\tau)} + 6e^{-3(t-\tau)} & 3e^{-2(t-\tau)} - 2e^{-3(t-\tau)} \end{pmatrix} \begin{pmatrix} 5 \\ 6 \end{pmatrix} (10e^{-\tau}) d\tau$$

$$x_1(t) = 50 \int_0^t (-2e^{-2(t-\tau)} + 3e^{-3(t-\tau)})e^{-\tau}d\tau + 60 \int_0^t (e^{-2(t-\tau)} - e^{-3(t-\tau)})e^{-\tau}d\tau$$

$$x_1(t) = \int_0^t (-40e^{-2(t-\tau)} + 90e^{-3(t-\tau)})e^{-\tau}d\tau = -40e^{-2t} \int_0^t e^{\tau}d\tau + 90e^{-3t} \int_0^t e^{2\tau}d\tau$$

$$x_1(t) = -40e^{-2t}(e^t - 1) + 90e^{-3t}(e^{2t} - 1)/2 = 5e^{-t} + 40e^{-2t} - 45e^{-3t}$$

In summary, the voltage over the capacitor is

$$V_{out} = x_1(t) = 5e^{-t} + 40e^{-2t} - 45e^{-3t}$$

### 8.5.2.2   Matrix Laplace Transform Method

In this section we turn our attention to solving

$$\dot{\underline{x}} = \underline{A}\underline{x} + \underline{B}\underline{u}(t), \quad \underline{x}(0) = \underline{x}_0 \tag{8.153}$$

by Laplace transforms. We should note that the method of course applies to homogeneous ODEs as well. For notational simplicity we define

$$L\{\underline{x}(t)\} = \underline{x}(s), \quad L\{\underline{u}(t)\} = \underline{u}(s)$$

An elemental point is that, for any vector function $\underline{v}(t)$, based on the fundamental definition of the Laplace transform and the integral of a vector function,

$$L\{\underline{v}(t)\} = L\left\{ \begin{pmatrix} v_1(t) \\ \ldots \\ v_n(t) \end{pmatrix} \right\} = \begin{pmatrix} L\{v_1(t)\} \\ \ldots \\ L\{v_n(t)\} \end{pmatrix} = \begin{pmatrix} v_1(s) \\ \ldots \\ v_n(s) \end{pmatrix} = \underline{v}(s)$$

Similarly, for any matrix function $\underline{M}(t)$,

$$L\{\underline{M}(t)\} = L\left\{ \begin{pmatrix} M_{11}(t) & \ldots & M_{1n}(t) \\ \ldots & \ldots & \ldots \\ M_{n1}(t) & \ldots & M_{nn}(t) \end{pmatrix} \right\} = \begin{pmatrix} L\{M_{11}(t)\} & \ldots & L\{M_{1n}(t)\} \\ \ldots & \ldots & \ldots \\ L\{M_{n1}(t)\} & \ldots & L\{M_{nn}(t)\} \end{pmatrix} = \underline{M}(s)$$

Transforming Equation 8.153, we have

$$L\{\dot{\underline{x}}\} = sL\{\underline{x}\} - \underline{x}(0) = s\underline{x}(s) - \underline{x}_0$$

$$L\{\underline{A}\underline{x}(t) + \underline{B}\underline{u}(t)\} = L\{\underline{A}\underline{x}(t)\} + L\{\underline{B}\underline{u}(t)\} = \underline{A}L\{\underline{x}(t)\} + \underline{B}L\{\underline{u}(t)\} = \underline{A}\underline{x}(s) + \underline{B}\underline{u}(s)$$

$$(s\underline{I} - \underline{A})\underline{x}(s) - \underline{x}_0 = \underline{B}\underline{u}(s)$$

$$\underline{x}(s) = (s\underline{I} - \underline{A})^{-1}\underline{x}_0 + (s\underline{I} - \underline{A})^{-1}\underline{B}\underline{u}(s) \tag{8.154}$$

Now consider the time-domain solution of Equation 8.154, which we have seen from Section 8.5.2.1 is

$$\underline{x}(t) = \underline{\Phi}(t)\underline{x}_0 + \int_0^t \underline{\Phi}(t - \tau)\underline{B}\underline{u}(\tau)d\tau \tag{8.155}$$

Transforming Equation 8.155 and remembering from Section 7.6.2 that the Laplace transform of the convolution of $h(t)$ and $g(t)$ is the product $H(s)G(s)$ we find

$$\underline{x}(s) = L\{\underline{\Phi}(t)\}\underline{x}_0 + L\{\underline{\Phi}(t)\}\underline{B}\underline{u}(s) \tag{8.156}$$

Equating Equations 8.154 and 8.156 we see that

$$L\{\underline{\Phi}(t)\} = \underline{\Phi}(s) = (s\underline{I} - \underline{A})^{-1} \tag{8.157}$$

The formula

$$\underline{\Phi}(t) = L^{-1}\left\{(s\underline{I} - \underline{A})^{-1}\right\}$$

can be a handy way to compute the transition matrix.

**Example: Solve**

$$\underline{\dot{x}} = \begin{pmatrix} -3 & 2 \\ 3 & -4 \end{pmatrix}\underline{x} + \begin{pmatrix} 0 \\ 1 \end{pmatrix}u_0(t) \quad \underline{x}(0) = \begin{pmatrix} 2 \\ 0 \end{pmatrix} \tag{8.158}$$

where $u_0(t)$ is the unit step function.

Step 1: Transform each element of the time-domain equation:

$$L\{\underline{x}(t)\} = \underline{x}(s)$$

$$L\{\underline{\dot{x}}(t)\} = s\underline{x}(s) - \underline{x}(0) = s\underline{x}(s) - \begin{pmatrix} 2 \\ 0 \end{pmatrix}$$

$$L\{u_0(t)\} = \frac{1}{s}$$

$$L\{\underline{A}\underline{x} + \underline{B}\underline{u}\} = \begin{pmatrix} -3 & 2 \\ 3 & -4 \end{pmatrix}\underline{x}(s) + \begin{pmatrix} 0 \\ 1 \end{pmatrix}\frac{1}{s}$$

Step 2: Collect terms involving $\underline{x}(s)$ on left side, everything else on right:

$$\begin{pmatrix} s+3 & -2 \\ -3 & s+4 \end{pmatrix} \underline{x}(s) = \begin{pmatrix} 2 \\ 0 \end{pmatrix} + \begin{pmatrix} 0 \\ 1 \end{pmatrix} \frac{1}{s}$$

Step 3: Solve for $\underline{x}(s)$:

$$(s+3)x_1 - 2x_2 = 2$$

$$-3x_1 + (s+4)x_2 = 1/s$$

$$x_2 = -1 + (s+3)x_1/2$$

$$-3x_1 + (s+4)(-1 + (s+3)x_1/2) = 1/s$$

$$((s+4)(s+3)/2 - 3)x_1 = (s+4) + 1/s$$

$$x_1 = \frac{2(s+4)}{s^2 + 7s + 6} + \frac{2}{s(s^2 + 7s + 6)} = \frac{2s^2 + 8s + 2}{s(s^2 + 7s + 6)}$$

$$x_2 = -1 + \frac{(s+3)}{2} \frac{2s^2 + 8s + 2}{s(s^2 + 7s + 6)} = \frac{7s + 3}{s(s^2 + 7s + 6)}$$

Step 4: Manipulate each component to match our short table of Laplace transforms, as we did previously for scalar ODEs:

$$x_1 = \frac{2s^2 + 8s + 2}{s(s^2 + 7s + 6)} = \frac{c_1}{s} + \frac{c_2}{s+1} + \frac{c_3}{s+6} = \frac{c_1(s+1)(s+6) + c_2 s(s+6) + c_3 s(s+1)}{s(s^2 + 7s + 6)}$$

$$\frac{2s^2 + 8s + 2}{s(s^2 + 7s + 6)} = \frac{(c_1 + c_2 + c_3)s^2 + (7c_1 + 6c_2 + c_3)s + 6c_1}{s(s^2 + 7s + 6)}$$

Equating coefficients of like powers of $s$:

$$c_1 + c_2 + c_3 = 2$$

$$7c_1 + 6c_2 + c_3 = 8$$

$$6c_1 = 2$$

The results are $c_1 = 1/3$, $c_2 = 4/5$, and $c_3 = 13/15$. Hence

$$x_1(s) = \frac{(1/3)}{s} + \frac{(4/5)}{s+1} + \frac{(13/15)}{s+6}$$

and then

$$x_1(t) = \left(\frac{1}{3}\right) + \left(\frac{4}{5}\right)e^{-t} + \left(\frac{13}{15}\right)e^{-6t}$$

In similar fashion

$$x_2(s) = \frac{7s+3}{s(s^2+7s+6)} = \frac{c_4}{s} + \frac{c_5}{s+1} + \frac{c_6}{s+6} = \frac{(1/2)}{s} + \frac{(4/5)}{s+1} - \frac{(13/10)}{s+6}$$

and

$$x_2(t) = \left(\frac{1}{2}\right) + \left(\frac{4}{5}\right)e^{-t} - \left(\frac{13}{10}\right)e^{-6t}$$

Chapter 7 demonstrated many attractive attributes of the Laplace transform method for solving ODEs. In the state space representation we can add another: sidestepping calculation of eigenvectors and especially the complexities that arise in the eigenvalue/eigenvector approach when the number of independent eigenvectors is less than the dimension of the system. Word Problem WP8.6 is an example.

### 8.5.2.3  Matrix Undetermined Coefficients Method

We first note that an equation of the form $\dot{\underline{x}} = \underline{A}\underline{x} + \underline{B}\underline{u}(t)$ where

$$\underline{u}(t) = \begin{pmatrix} u_1(t) \\ \dots \\ u_m(t) \end{pmatrix}$$

can be written as $\dot{\underline{x}} = \underline{A}\underline{x} + \sum_{k=1}^{m} \underline{b}_k u_k(t)$. Then, since the equation is linear, the problem can be broken into $m$ parts, each solved individually via

$$\dot{\underline{x}}^{(k)} = \underline{A}\underline{x}^{(k)} + \underline{b}_k u_k(t)$$

and then the desired result obtained from

$$\underline{x}(t) = \sum_{k=1}^{m} \underline{x}^{(k)}(t)$$

Therefore it suffices in this section to consider ODEs of the form

$$\dot{\underline{x}} = \underline{A}\underline{x} + \underline{b}u(t) \tag{8.159}$$

where $u(t)$ is a *scalar* function of time and $\underline{b}$ is a vector of constants.

By direct analogy with the method we applied to second-order LTI ODEs in Chapter 5, Table 8.5 lists particular solutions we should try for a given form of input function.

We found for second-order linear ODEs that when the input $u(t)$ is itself a solution to the differential equation, the trial solution does not work. A similar caution applies here. For example, if $u(t) = e^{kt}$ and $k$ is an eigenvalue of $\underline{A}$, then the solution of 8.159 is more complicated than given in Table 8.5.

---

**TABLE 8.5**

**Trial Particular Solutions for Given Input Functions Using the Matrix Undetermined Coefficients Method**

| Input $u(t)$ | Trial Solution |
|---|---|
| $e^{kt}$ | $\underline{c}e^{kt}$ |
| $t$ | $\underline{c}t + \underline{d}$ |
| $te^{kt}$ | $(\underline{c}t + \underline{d})e^{kt}$ |
| $t^2$ | $\underline{c}t^2 + \underline{d}t + \underline{e}$ |
| $\sin\omega t$ | $\underline{c}\sin\omega t + \underline{d}\cos\omega t$ |
| $e^{kt}\sin\omega t$ | $(\underline{c}\sin\omega t + \underline{d}\cos\omega t)e^{kt}$ |
| $t\sin\omega t$ | $(\underline{c}t + \underline{d})\sin\omega t + (\underline{e}t + \underline{f})\cos\omega t$ |

---

We can derive the formal particular solutions for the given inputs in Table 8.5. For example, suppose we need to solve

$$\underline{\dot{x}} = \underline{A}\underline{x} + \underline{b}te^{kt} \tag{8.160}$$

where $\underline{A}$, $\underline{b}$, and $k$ are given. We substitute the trial solution from Table 8.5 on the right-hand side of Equation 8.160 and obtain

$$\underline{A}\underline{x} + \underline{b}te^{kt} = \underline{A}(\underline{c}t + \underline{d})e^{kt} + \underline{b}te^{kt} \tag{8.161}$$

Differentiating the trial solution and substituting it in the left-hand side of Equation 8.160 we have

$$\underline{\dot{x}} = \frac{d}{dt}\left((\underline{c}t + \underline{d})e^{kt}\right) = \underline{c}e^{kt} + k\underline{c}te^{kt} + k\underline{d}e^{kt} \tag{8.162}$$

Equating Equations 8.161 and 8.162 we find

$$\underline{c}e^{kt} + k\underline{c}te^{kt} + k\underline{d}e^{kt} = \underline{A}(\underline{c}t + \underline{d})e^{kt} + \underline{b}te^{kt} \tag{8.163}$$

Dividing through by $e^{kt}$ and rearranging we have

$$(\underline{c} + (k\underline{I} - \underline{A})\underline{d}) + ((k\underline{I} - \underline{A})\underline{c} - \underline{b})t = \underline{0} \tag{8.164}$$

Each component of the vector Equation 8.164 is of the form

$$g_i + h_i t = 0 \tag{8.165}$$

For Equation 8.165 to be true for all $t$ we must have $g_i = 0$ and $h_i = 0$, or

$$\underline{c} + (k\underline{I} - \underline{A})\underline{d} = \underline{0} \tag{8.166}$$

and

$$(k\underline{I} - \underline{A})\underline{c} - \underline{b} = \underline{0} \tag{8.167}$$

Recalling that $\underline{A}$, $\underline{b}$, and $k$ are given, we first solve Equation 8.167 for $\underline{c}$. Formally, we write

$$\underline{c} = (k\underline{I} - \underline{A})^{-1}\underline{b} \tag{8.168}$$

but in practice we use the process of row reduction instead of actually inverting the matrix. With $\underline{c}$ in hand we calculate $\underline{d}$ from Equation 8.166. Again, although we use row reduction, we can write the formal solution as

$$\underline{d} = -(k\underline{I} - \underline{A})^{-1}\underline{c} = -(k\underline{I} - \underline{A})^{-2}\underline{b} \tag{8.169}$$

We see from Equations 8.168 and 8.169 that this method only works when the matrix $k\underline{I} - \underline{A}$ is invertible, which we know means that $k$ is not an eigenvalue of $\underline{A}$.

We could follow a process similar to that in Equations 8.160 through 8.169 and formally solve for the unknown vectors in each of the trial solutions in Table 8.5.

Suppose the particular solution determined by the process just described is $\underline{x}_P(t)$. To satisfy the initial condition $\underline{x}(0) = \underline{x}_0$ we use the methods of Section 8.5.1 and compute

$$\underline{x}(t) = \underline{\Psi}(t)\underline{c} + \underline{x}_p(t)$$

where $\underline{c}$ satisfies

$$\underline{x}_0 = \underline{\Psi}(0)\underline{c} + \underline{x}_p(0)$$

### 8.5.2.4 Worth of the Three Methods, in Summary

We have now seen three different methods for solving nonhomogeneous ODEs in state space form. As we have said, the matrix kernel (variation of parameters) method is the most general, applying in principle to time-varying as well as time-invariant systems, though the method depends on finding fundamental solutions, which can be difficult for time-varying ODEs. (We will return to this in the next chapter.) The matrix kernel method is also usually the most time-consuming.

The matrix Laplace transform method applies only for LTI ODEs, but it is relatively straightforward and sidesteps some of the complications of the eigenvalue/eigenvector method for solving homogeneous ODEs and the matrix kernel method for solving nonhomogeneous ODEs.

The matrix undetermined coefficient method also applies only to LTI ODEs and has a number of other limitations which we summarized in Chapter 5, Section 5.3.3.6. When initial conditions must be met one is drawn into consideration of eigenvalues and eigenvectors, as in the matrix kernel method, but even so the undetermined coefficient method is typically faster than computing the convolution. When only steady-state solutions are needed, the matrix undetermined coefficient method is normally the simplest of the three.

For many problems of engineering interest the best choice may be one of personal preference. Using different methods of solution on the same problem is an excellent check on one's own work, but as an alternative one can always directly substitute a supposed solution into the ODE and see if the ODE is satisfied.

## 8.6 A BRIEF LOOK AT N-DIMENSIONAL NONLINEAR SYSTEMS

We end our discussion of systems of first-order ODEs with a short examination of nonlinear systems. At the beginning of Section 8.5 we noted that many systems of engineering interest are accurately modeled as linear, within limits on the magnitude of the state vector. But beyond those limits the behavior of these systems is nonlinear, and many other systems are inherently nonlinear for any magnitude of the state. What are the implications of nonlinearity? How to solve nonlinear ODEs is not an issue. We accept that we will usually be unable to solve them analytically and must therefore turn to numerical integration. What we will endeavor to do here is to convey (a) an elementary understanding of qualitative similarities and differences in the behavior of linear and nonlinear systems and (b) a (necessarily limited) sense of mathematical tools available for preliminary design and analysis of nonlinear systems. Problems relating to this section at the end of this chapter are devoted to computer-based explorations of nonlinear dynamics.

### 8.6.1 Equilibrium Points

Linear systems can have but one equilibrium point. We saw in Chapter 4 that first-order nonlinear ODEs can have multiple equilibrium points. An example of this in $n$-dimensional systems is the three-body gravitational problem. Consider the Earth in orbit around the Sun, as shown in Figure 8.20. There are five locations in the Earth's orbital plane where a small object such as a spacecraft can be in equilibrium relative to the Sun and Earth. These equilibrium points are called the Lagrangian points, L1 through L5, though the first three were discovered by Leonhard Euler. Joseph-Louis Lagrange discovered the last two

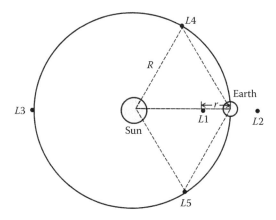

**FIGURE 8.20**   The five Lagrangian points, equilibrium locations for a spacecraft in the plane of the Earth's orbit around the Sun.

a few years later. The first three are unstable, though only slightly so. Spacecraft have been placed at L1 and maintained there with modest station-keeping. L4 and L5 are stable. L1 and L2 are approximately the same distance $r$ from the Earth:

$$r = R \left( \frac{M_E}{3M_S} \right)^{1/3}$$

where $R$ is the distance from Earth to Sun, $M_E$ is the mass of the Earth and $M_S$ is the mass of the Sun. L1 and L2 are approximately four times more distant from the Earth than the Moon is. See, for example, [13] for details.

### 8.6.2 STABILITY

Consider again the nonlinear system

$$\dot{\underline{w}} = \underline{g}(t, \underline{w}), \qquad \underline{w}(t_0) = \underline{w}_0$$

we linearized at the beginning of Section 8.5. If $\underline{g}(t, \underline{w})$ is not a function of time, then we have

$$\dot{\underline{w}} = \underline{g}(\underline{w}), \qquad \underline{w}(t_0) = \underline{w}_0 \tag{8.170}$$

Let $\underline{x} = \underline{w} - \underline{w}_0$. If $\underline{g}(\underline{w})$ is differentiable at $\underline{w}_0$ then

$$\dot{\underline{x}} = \dot{\underline{w}} = \underline{g}(\underline{w}) = \underline{g}(\underline{w}_0 + \underline{x}) = \underline{g}(\underline{w}_0) + \left( \frac{\partial \underline{g}}{\partial \underline{x}} \right) \underline{x} + \underline{r}(\underline{x})$$

where

$$\left( \frac{\partial \underline{g}}{\partial \underline{x}} \right)_{ij} = \frac{\partial g_i}{\partial w_j}(\underline{w}_0)$$

and $\underline{r}(\underline{x})$ is a remainder term. If

$$\lim_{\underline{x} \to 0} \frac{|\underline{r}(\underline{x})|}{|\underline{x}|} = 0$$

and if $\underline{w}_0$ is an equilibrium point of the system, that is,

$$\underline{g}(\underline{w}_0) = 0$$

then the resulting linear system is

$$\dot{\underline{x}} = \underline{A}\underline{x} \tag{8.171}$$

where

$$A_{ij} = \frac{\partial g_i}{\partial w_j}(\underline{w}_0)$$

is easily computed.

Under these conditions the following facts can be proved, which are comforting and certainly important enough to be stated in a theorem.

### *Theorem*
If the linear system (Equation 8.171) is damped (asymptotically stable) then the nonlinear system (Equation 8.170) is asymptotically stable at $\underline{w}_0$. Likewise, if the linear system is unstable, then the nonlinear system is unstable at $\underline{w}_0$.

(Stated without proof.)

It is fortunate that our ability to analyze linear systems can be so useful in understanding nonlinear systems. However, we cannot say how far the region of stability extends around each equilibrium point.

### 8.6.3  IMPOSITION OF LIMITS

We discussed the properties of linearity in Chapter 1. Restating them here in state space representation, for a system in the form

$$\underline{\dot{x}} = \underline{A}(t)\underline{x} + \underline{B}(t)\underline{u}$$

where $\underline{x}_0 = 0$, if the output is $\underline{x}_\alpha(t)$ when the input is $\underline{u}_\alpha(t)$, and the output is $\underline{x}_\beta(t)$ when the input is $\underline{u}_\beta(t)$, then for any constants $a$ and $b$, when the input is $a\underline{u}_\alpha(t) + b\underline{u}_\beta(t)$ then the output is $a\underline{x}_\alpha(t) + b\underline{x}_\beta(t)$. A similar statement can be made with respect to initial conditions.

This handy property is not true for nonlinear systems. One major implication of this is that nonlinear systems impose limits. Consider, for example, the aircraft performance calculations in Computer Project B, defined at the end of Chapters 3 and 6. Figure 8.21 (repeating Figure 3.6) reminds us of the fundamental variables in aircraft translational motion. Recall Figure 8.7 also for additional definition.

Figure 8.22 shows the result of programming the aircraft to achieve a commanded altitude $h_C$. The desired flight path angle is held to a steady 0.2 radians until the aircraft altitude is within 1000 ft of $h_C$ and then it adopts a control law to bring the aircraft

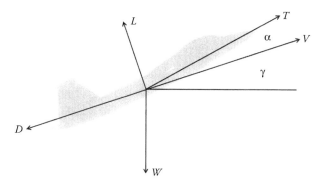

**FIGURE 8.21**   Variables in aircraft longitudinal flight.

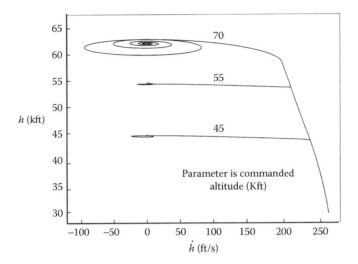

**FIGURE 8.22**   Phase plane for aircraft trajectory to commanded altitude.

smoothly to $h_C$. Figure 8.22 is in the form of a *phase plane*. The value of $\dot{h}$ at any time is plotted against the value of $h$ at that same instant. We see that when the aircraft is commanded to 45 Kft, the aircraft reaches the point ($\dot{h} = 0$, $h = 45$ Kft) and remains there. We cannot see the other variables but they reach steady values as well, so the point becomes an equilibrium point for the system. Similarly, the point ($\dot{h} = 0$, $h = 55$ Kft) becomes an equilibrium point when $h_C = 55$ Kft. But when the aircraft is commanded to 70 Kft the equilibrium point is ($\dot{h} = 0$, $h = 62$ Kft). The aircraft cannot achieve the commanded altitude but rather falls into a phugoid oscillation (recall section 8.5.1.3) that settles out at 62 Kft. The aircraft "ceiling" is 62 Kft (approximately 18.9 km).

Two nonlinearities impose this limit. First, air density diminishes approximately exponentially as altitude increases. The air density at 62 Kft is less than one eleventh the density at sea level. Aircraft lift and thrust depend on air density while, obviously, aircraft weight does not. Eventually lift and thrust cannot counter weight. The other nonlinearity contributing to the limit is the dependence of lift on the angle of attack. For angles of attack less than some value (call it $\alpha_{MAX}$) the dependence is linear; lift is proportional to angle of attack. In Computer Project B's model, $\alpha_{MAX} = 8$ degrees. For angles of attack greater than $\alpha_{MAX}$ airflow separates from the wing and lift is greatly reduced. So the aircraft is controlled so that its angle of attack never exceeds $\alpha_{MAX}$. The result is a limit on lift.

In a linear model anything seems possible. Double the input to double the output. But nonlinearities impose limits.

We have already seen another example of limits set by nonlinearities, of a type different from that just described. Chapter 4 showed that, in contrast to linear LTI systems, which have solutions valid for all time, nonlinear systems may be governed by equations that have solution only for a limited time.

### 8.6.4   Uniquely Nonlinear Dynamics

We finish our brief essay comparing and contrasting linear and nonlinear systems with remarks about several unique behaviors nonlinear systems can exhibit. The first of these we encountered in Chapter 3: harmonic distortion. When a stable and damped LTI system is excited by a sine-wave input of frequency $\omega$, its steady-state output is purely a sine wave of that frequency. When a nonlinear system is excited by a similar input, its steady-state output usually includes sine waves of frequencies that are multiples of $\omega$. This is a phenomenon of considerable interest to engineers.

The remaining behaviors we shall discuss are not commonly encountered in engineering practice today but you should be aware that physical phenomena exist that can exhibit them.

Consider van der Pol's equation, in state space form:

$$\dot{x}_1 = x_2$$
$$\dot{x}_2 = -x_1 + \mu(1 - x_1^2)x_2$$
(8.172)

This equation originated in analysis of early radio circuits and has since been used to model phenomena in biology and geology.

Figures 8.23 through 8.26 graph the result of numerically integrating the van der Pol equation under four conditions: two different initial conditions and two different values for the parameter $\mu$. Figures 8.23 and 8.24 present $x_1$ as a function of time for the two different values of $\mu$. Figures 8.25 and 8.26 are phase plane plots. The remarkable fact about the equation is that, for any fixed value of $\mu > 0$, all solutions lead to a *limit cycle*; a periodic trajectory that is the destination of all solutions, independent of initial conditions. Many systems have trajectories that are periodic from the start, for example,

$$\dot{x}_1 = x_2$$
$$\dot{x}_2 = -x_1$$

a system you should recognize, but in those systems different initial conditions produce different trajectories. Limit cycles are unique to nonlinear equations.

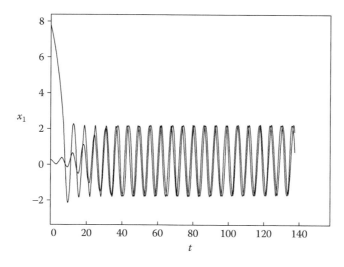

**FIGURE 8.23** Graph of $x_1(t)$ for the van der Pol Equation 8.172 with $\mu = 0.25$, $x_1(0) = 8$, and $x_1(0) = 0.1$.

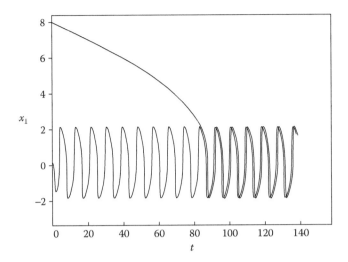

**FIGURE 8.24**   Graph of $x_1(t)$ for the van der Pol Equation 8.172 with $\mu = 3$, $x_1(0) = 8$, and $x_1(0) = 0.1$.

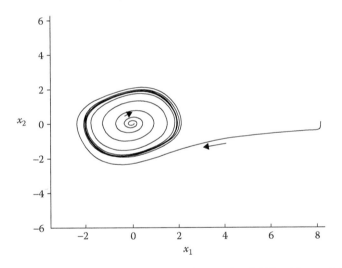

**FIGURE 8.25**   Phase plane graph for the van der Pol Equation 8.172 with $\mu = 0.25$.

The final uniquely nonlinear behavior we will mention became well known through study of the innocent-looking equations

$$\dot{x}_1 = \sigma(-x_1 + x_2)$$

$$\dot{x}_2 = rx_1 - x_2 - x_1 x_3 \tag{8.173}$$

$$\dot{x}_3 = -bx_3 + x_1 x_2$$

where $\sigma, r$, and $b$ are constants. E.N. Lorenz, a meteorologist, developed these equations as a model for convective flow in the atmosphere and used $\sigma = 10$, $b = 8/3$, and $r = 28$.

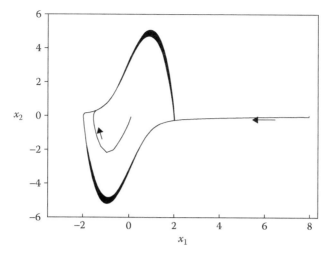

**FIGURE 8.26**   Phase plane graph for the van der Pol Equation 8.172 with $\mu = 3$.

In the abstract of the paper [14] announcing his results, Lorenz said that "... slightly differing initial states can evolve into considerably different states." This is what has become known as *chaos*, essentially unpredictable behavior in deterministic systems. With the initial conditions Lorenz used,

$$\underline{x}(0) = \begin{pmatrix} 0 \\ 1 \\ 0 \end{pmatrix}$$

we can attempt to reproduce his results.

The dark solid line in Figure 8.27 graphs $x_2$ as a function of time for the aforementioned case. We see that, after an initial spike, the motion is a slowly growing oscillation until about $t = 17$. Then the character of the motion changes to something like the initial

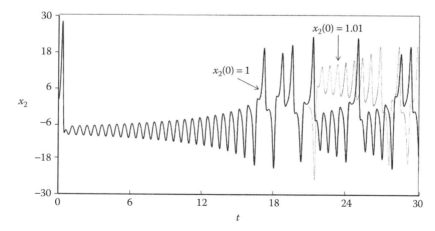

**FIGURE 8.27**   Graph of $x_2(t)$ for the Lorenz Equations 8.173 for $x_2(0) = 1.0$ and $x_2(0) = 1.01$.

spike, then switches back and forth between these two modes. The light, dashed line in Figure 8.27 is the result of changing the initial conditions very slightly, to

$$\underline{x}(0) = \begin{pmatrix} 0 \\ 1.01 \\ 0 \end{pmatrix}$$

We see that, after about $t = 21$ the motion is indeed quite different, as Lorenz reported. The phase plane plot of $x_1(t)$ versus $x_2(t)$ (Figure 8.28) conveys more clearly the character of the motion. There are two periodic orbits that the trajectory approaches but eventually diverges from, again and again. Graphs very similar to Figure 8.27 (for the $x_2(0) = 1$ case) and Figure 8.28 appeared in Lorenz' paper.

However, we have not reproduced Lorenz' results perfectly. The differences are due to differences in the numerical integration algorithm employed. In fact, when we use an algorithm other than the Runge–Kutta algorithm that produced Figures 8.27 and 8.28, we begin to see significantly different results at about the time the divergence occurs for the slight deviation in initial conditions. The equations are sensitive to differences in method of calculation as well as initial conditions.

It is now recognized that chaotic behavior is not rare in nature. In fact, a system as simple as the double pendulum (recall WP6.1, Chapter 6) exhibits it, given large enough initial conditions.

Engineers strive to avoid chaos. However, you should be aware that it may lurk in innocent-looking *nonlinear* equations, which may govern a real system.

### 8.6.5   MATHEMATICAL MODELING TOOLS FOR NONLINEAR SYSTEMS

As a practical matter engineers have to deal with nonlinear systems frequently and, as we have said, ultimately rely on simulation via numerical integration for performance prediction prior to testing the actual system itself. But additional mathematical modeling tools are

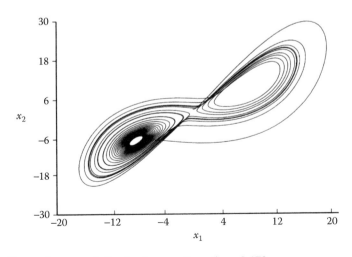

**FIGURE 8.28**   Phase plane graph for the Lorenz Equations 8.173.

needed to facilitate preliminary design. A number of effective modeling tools have been developed, but they are typically limited to specific applications and fundamentally heuristic; that is, based on experiment to determine what works. It is difficult to generalize and this brief survey cannot be comprehensive but we deem it worthwhile to give one example.

For a system that has one nonlinear element the concept of *equivalent gain* is often employed. In the analysis model of the overall system, the nonlinear element is replaced with a linear one, a constant gain. The great advantage of the method is that it permits engineers to bring to bear the powerful methods for design and analysis of linear systems. The worth of the concept depends on choosing the right value for the equivalent gain.

As a specific example, consider the saturation nonlinearity shown in Figure 8.29, the most common nonlinearity in control systems [15]. If the input to the element is $x$, the output is

$$y(x) = L \quad \text{if } x > L$$

$$y(x) = kx \quad \text{if } -L < x < L$$

$$y(x) = -L \quad \text{if } x < -L$$

The right value to choose for the equivalent gain of the saturation element depends on the nature and magnitude of the inputs it will experience over its operational life. For example, if the input to the element in Figure 8.29 is $a u_T(t)$, where $u_T(t)$ is the unit step function stepping up at time $T$, the equivalent gain, the ratio of output to input, is $k$ if $k|a| < L$ and $L / k|a|$ if $k|a| > L$. If the input to the element is $a \sin \omega t$ the output is $ka \sin \omega t$ if $k|a| < L$ and the equivalent gain is $k$. If $k|a| > L$ the output includes higher harmonics of $\omega$ but these are typically ignored and the equivalent gain is based on the magnitude of the fundamental harmonic; that is, the magnitude of the $\sin \omega t$ term in the output. It can be shown that the equivalent gain in this case is (from [15])

$$G_{EQ} = \frac{2}{\pi} \left( k \arcsin\left(\frac{L}{ka}\right) + \frac{L}{a} \sqrt{1 - \left(\frac{L}{ka}\right)^2} \right)$$

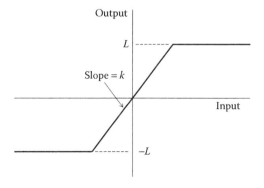

**FIGURE 8.29**    Saturation nonlinearity.

With the equivalent gain replacing the nonlinear element in the system model, using the tools available for design and analysis of linear systems, the engineer tests the model against the expected range of signals and their magnitudes and determines if the equivalent gain of the nonlinear element yields satisfactory system performance in all cases.

## 8.7   SUMMARY

This chapter has dealt with systems of first-order ODEs, also known as the state space representation. The general form is

$$\frac{d\underline{x}}{dt} = \underline{f}(t,\underline{x})$$

$$\underline{x}(t_0) = \underline{x}_0$$

The chapter presented an existence and uniqueness theorem very similar to the one for scalar first-order ODEs, showing that systems of equations in state space form have at least a local solution provided $\underline{f}(t,\underline{x})$ is sufficiently smooth. An extensive variety of computer programs to numerically integrate ODEs in state space form are used by practicing engineers and we were introduced to several of them, including the Runge–Kutta and Adams–Moulton algorithms. Most of the chapter was devoted to linear systems. For homogeneous LTI systems in the form

$$\underline{\dot{x}} = \underline{A}\underline{x}$$

$$\underline{x}(0) = \underline{x}_0$$

finding fundamental solutions begins with trying $\underline{x}(t) = \underline{\xi}e^{rt}$, where for each eigenvalue $r_i$ satisfying

$$\det(r\underline{I} - \underline{A}) = 0$$

there is an eigenvector $\underline{\xi}^{(i)}$ satisfying

$$(r_i\underline{I} - \underline{A})\underline{\xi} = 0$$

When an eigenvalue is complex ($r = \lambda + i\omega$) its eigenvector is also ($\underline{\xi} = \underline{a} + i\underline{b}$) and the real fundamental solutions are

$$\underline{x}^{(1)}(t) = e^{\lambda t}(\underline{a}\cos\omega t - \underline{b}\sin\omega t)$$

$$\underline{x}^{(2)}(t) = e^{\lambda t}(\underline{a}\sin\omega t + \underline{b}\cos\omega t)$$

If the $n$ by $n$ matrix $\underline{A}$ is symmetric, it will have $n$ independent eigenvectors even if it has repeated eigenvalues and then all fundamental solutions will be in the aforementioned forms. If $\underline{A}$ is not symmetric and it has repeated eigenvalues it may not have $n$ independent eigenvectors, in which case some of its fundamental solutions will have more complicated forms involving $t^k e^{rt}$, including where $r$ is complex.

An LTI system is stable and damped if the real part of every eigenvalue is negative. It is unstable if it has one or more eigenvalues with a positive real part.

We have a choice of state variables in writing the state space description of a system and our choice can sometimes ease our solution of the ODEs or enhance our understanding of the dynamics of the system.

The chapter determined that the solution of the nonhomogeneous linear time-varying system

$$\dot{\underline{x}} = \underline{A}(t)\underline{x} + \underline{B}(t)\underline{u}(t) \quad \underline{x}(t_0) = \underline{x}_0$$

can be written

$$\underline{x}(t) = \underline{\Phi}(t,t_0)\underline{x}(t_0) + \int_{t_0}^{t} \underline{\Phi}(t,\tau)\underline{B}(\tau)\underline{u}(\tau)d\tau$$

where the state transition matrix $\underline{\Phi}(t,\tau)$ is given by

$$\underline{\Phi}(t,\tau) = \underline{\Psi}(t)\underline{\Psi}^{-1}(\tau)$$

and the columns of $\underline{\Psi}(t)$ are the fundamental solutions of

$$\dot{\underline{x}} = \underline{A}(t)\underline{x}$$

We learned three methods to solve the nonhomogenous LTI system

$$\dot{\underline{x}} = \underline{A}\underline{x} + \underline{B}\underline{u}(t) \quad \underline{x}(t_0) = \underline{x}_0$$

the matrix kernel method

$$\underline{x}(t) = \underline{\Phi}(t-t_0)\underline{x}(t_0) + \int_{t_0}^{t} \underline{\Phi}(t-\tau)\underline{B}\underline{u}(\tau)d\tau$$

the matrix Laplace transform method, which can be formally written as

$$\underline{x}(t) = L^{-1}\{\underline{x}(s)\}$$

$$\underline{x}(s) = (s\underline{I} - \underline{A})^{-1}(\underline{x}_0 + \underline{B}\underline{u}(s))$$

$$\underline{u}(s) = L\{\underline{u}(t)\}$$

and the matrix undetermined coefficients method, which can be formally written as

$$\underline{x}(t) = \underline{\Phi}(t)(\underline{x}_0 - \underline{x}_P(0)) + \underline{x}_P(t)$$

$$\underline{x}_P(t) = \sum_{i=1}^{m} \underline{c}^{(i)} f_i(t)$$

where the scalar functions $f_i(t)$ are suggested by the scalar functions found in $\underline{u}(t)$ and each $\underline{c}^{(i)}$ is a vector of constants that one solves for by substituting the trial solution in the matrix ODE.

This chapter explored nonlinear systems briefly and compared and contrasted their behavior with that of linear systems. Nonlinear systems can have multiple equilibrium points where linear systems have at most one. Under certain not very stringent conditions a nonlinear system

$$\underline{\dot{w}} = \underline{g}(\underline{w}), \quad \underline{w}(t_0) = \underline{w}_0$$

is asymptotically stable or unstable at an equilibrium point $\underline{w}_0$ depending on whether the linearized system

$$\underline{\dot{x}} = \underline{Ax}, \quad A_{ij} = \frac{\partial g_i}{\partial w_j}(\underline{w}_0)$$

is asymptotically stable (damped) or unstable. Nonlinearities impose limits where linear systems do not. And finally, nonlinear systems can exhibit behavior such as harmonic distortion, limit cycles and virtual unpredictability (chaos) that linear systems do not. Engineers have developed effective methods for designing and analyzing nonlinear systems but they tend to be ad hoc. This chapter gave one example, the concept of equivalent gain, which enables engineers to employ on nonlinear systems the powerful methods available for design and analysis of linear systems.

## Solved Problems

**SP8.1** Use the Euler method to compute to five decimal places the first three time steps ($t = 0.025, 0.05, 0.075$) in the numerical integration of

$$\frac{dz}{dt} = \ln(vz)\cos(4t) \quad z(0) = 6$$

$$\frac{dv}{dt} = -\left(\frac{z}{v}\right)(1+8t)^2 \quad v(0) = 4$$

*Solution:*
The Euler method applied to this problem yields

$$t_{n+1} = t_n + \Delta t$$

$$z_{n+1} = z_n + \ln(v_n z_n)\cos(4t_n) \cdot \Delta t$$

$$v_{n+1} = v_n - \left(\frac{z_n}{v_n}\right)(1+8t_n)^2 \cdot \Delta t$$

where $\Delta t = 0.025$.

For $n = 0$:

$$t_1 = 0 + 0.025 = 0.025$$

$$z_1 = 6 + \ln(6 \cdot 4)\cos(4(0))(0.025)$$

$$z_1 = 6 + (3.178)(1)(0.025) = 6.07945$$

$$v_1 = 4 - \left(\frac{6}{4}\right)(1 + 8(0))^2(0.025)$$

$$v_1 = 4 - (1.5)(1)(0.025) = 3.96250$$

For $n = 1$:

$$t_2 = 0.025 + 0.025 = 0.05$$

$$z_2 = 6.07945 + \ln((6.07945) \cdot (3.9625))\cos(4(0.025)))(0.025)$$

$$z_2 = 6.07945 + (3.18179)(0.9950)(0.025) = 6.15860$$

$$v_2 = 3.96250 - \left(\frac{6.07945}{3.9625}\right)(1 + 8(0.025))^2(0.025)$$

$$v_2 = 3.96250 - (1.5342)(1.440)(0.025) = 3.90727$$

For $n = 2$:

$$t_3 = 0.05 + 0.025 = 0.075$$

$$z_3 = 6.15860 + \ln((6.15860) \cdot (3.90727))\cos(4(0.05)))(0.025)$$

$$z_3 = 6.15860 + (3.18069)(0.98007)(0.025) = 6.23653$$

$$v_3 = 3.90727 - \left(\frac{6.15860}{3.90727}\right)(1 + 8(0.5))^2(0.025)$$

$$v_3 = 3.90727 - (1.57619)(1.96)(0.025) = 3.83004$$

**SP8.2** Use the Euler method to compute to four decimal places the first three time steps ($t = 0.05, 0.10, 0.15$) in the numerical integration of

$$\frac{dz}{dt} = \frac{z}{v^2} \qquad z(0) = 2$$

$$\frac{dv}{dt} = \frac{\sqrt{v(t + 0.1)}}{z^2} \qquad v(0) = 4$$

*Solution*:
The Euler method applied to this problem yields

$$t_{n+1} = t_n + \Delta t$$

$$z_{n+1} = z_n + \left(z_n / v_n^2\right) \cdot \Delta t$$

$$v_{n+1} = v_n + \left(\sqrt{v_n(t_n + 0.1)} / z_n^2\right) \cdot \Delta t$$

where $\Delta t = 0.05$.
For $n = 0$:

$$t_1 = 0 + 0.05 = 0.05$$

$$z_1 = 2 + \left(2/4^2\right)(0.05)$$

$$z_1 = 2 + (0.125)(0.05) = 2.0063$$

$$v_1 = 4 + \left(\sqrt{4(0 + 0.1)} / 2^2\right)(0.05)$$

$$v_1 = 4 + (0.1581)(0.05) = 4.0079$$

For $n = 1$:

$$t_2 = 0.05 + 0.05 = 0.1$$

$$z_2 = 2.0063 + \left(2.0063/(4.0079)^2\right)(0.05)$$

$$z_2 = 2.0063 + (0.1250)(0.05) = 2.0125$$

$$v_2 = 4.0079 + \left(\sqrt{(4.0079)(0.05 + 0.1)} / (2.0063)^2\right)(0.05)$$

$$v_2 = 4.0079 + (0.1926)(0.05) = 4.0175$$

For $n = 2$:

$$t_3 = 0.1 + 0.05 = 0.15$$

$$z_3 = 2.0125 + \left(2.0125/(4.0175)^2\right)(0.05)$$

$$z_3 = 2.0125 + (0.1247)(0.05) = 2.0187$$

$$v_3 = 4.0175 + \left(\sqrt{(4.0175)(0.1 + 0.1)} / (2.0125)^2\right)(0.05)$$

$$v_3 = 4.0175 + (0.2213)(0.05) = 4.0286$$

**SP8.3** Solve

$$\frac{d\underline{x}}{dt} = \begin{pmatrix} 1 & 1 \\ 4 & -2 \end{pmatrix} \underline{x}, \quad \underline{x}(0) = \begin{pmatrix} 1 \\ 2 \end{pmatrix} \tag{8.174}$$

*Solution*: Trying $\underline{x}(t) = \underline{\xi}e^{rt}$ in Equation 8.174,

$$r\underline{\xi}e^{rt} = \begin{pmatrix} 1 & 1 \\ 4 & -2 \end{pmatrix} \underline{\xi}e^{rt}$$

$$(\underline{A} - r\underline{I})\underline{\xi} = \begin{pmatrix} 1-r & 1 \\ 4 & -2-r \end{pmatrix} \underline{\xi} = 0 \tag{8.175}$$

The first condition for Equation 8.175 to be satisfied for nonzero $\underline{\xi}$ is

$$\det \begin{pmatrix} 1-r & 1 \\ 4 & -2-r \end{pmatrix} = 0$$

$$(1-r)(-2-r) - (4)(1) = 0$$

$$r^2 + r - 6 = 0 \tag{8.176}$$

The quadratic in Equation 8.176 factors into

$$(r+3)(r-2) = 0$$

Hence the eigenvalues for Equation 8.174 are

$$r_1 = -3$$

$$r_2 = +2$$

Substituting $r_1 = -3$ into Equation 8.175,

$$\begin{pmatrix} r_1-1 & -1 \\ -4 & r_1+2 \end{pmatrix} \underline{\xi}^{(1)} = 0$$

$$\begin{pmatrix} -3-1 & -1 \\ -4 & -3+2 \end{pmatrix} \underline{\xi}^{(1)} = 0$$

$$\begin{pmatrix} -4 & -1 \\ -4 & -1 \end{pmatrix} \underline{\xi}^{(1)} = 0 \tag{8.177}$$

The matrix equation 8.177 is not two independent equations. It cannot be; its determinant is zero. It is one equation:

$$-4\xi_1^{(1)} - \xi_2^{(1)} = 0 \tag{8.178}$$

To proceed we *set* a value for one or the other of the unknowns. It is simplest to choose the value one. We set $\xi_1^{(1)} = 1$ and then, by Equation 8.178, $\xi_2^{(1)} = -4$, and so the eigenvector associated with the first eigenvalue, $r_1 = -3$, is

$$\underline{\xi}^{(1)} = \begin{pmatrix} 1 \\ -4 \end{pmatrix} \tag{8.179}$$

Next we return to Equation 8.175, substitute the second eigenvalue $r_2 = 2$ and solve for the eigenvector $\underline{\xi}^{(2)}$ associated with it.

$$\begin{pmatrix} r_2 - 1 & -1 \\ -4 & r_2 + 2 \end{pmatrix} \underline{\xi}^{(2)} = 0$$

$$\begin{pmatrix} 2 - 1 & -1 \\ -4 & 2 + 2 \end{pmatrix} \underline{\xi}^{(2)} = 0$$

We have

$$\begin{pmatrix} 1 & -1 \\ -4 & 4 \end{pmatrix} \underline{\xi}^{(2)} = 0 \tag{8.180}$$

The two equations in Equation 8.180 are not exactly the same, as in Equation 8.177, but they are not independent. (Again, they cannot be.) They are both equivalent to

$$\xi_1^{(2)} - \xi_2^{(2)} = 0 \tag{8.181}$$

Again we have one equation and two unknowns. We set $\xi_1^{(2)} = 1$ and then, from Equation 8.181, $\xi_2^{(2)} = 1$ and so the eigenvector associated with the second eigenvalue, $r_2 = 2$, is

$$\underline{\xi}^{(2)} = \begin{pmatrix} 1 \\ 1 \end{pmatrix} \tag{8.182}$$

Thus the general solution to Equation 8.174 has the form

$$\underline{x}(t) = c_1 \begin{pmatrix} 1 \\ -4 \end{pmatrix} e^{-3t} + c_2 \begin{pmatrix} 1 \\ 1 \end{pmatrix} e^{2t} \tag{8.183}$$

We determine the constants $c_1$ and $c_2$ from the initial condition

$$\underline{x}(0) = \begin{pmatrix} 1 \\ 2 \end{pmatrix}$$

Setting $t = 0$ in Equation 8.183 and applying the initial condition we have

$$\underline{x}(0) = c_1 \begin{pmatrix} 1 \\ -4 \end{pmatrix} e^{-3(0)} + c_2 \begin{pmatrix} 1 \\ 1 \end{pmatrix} e^{2(0)} = c_1 \begin{pmatrix} 1 \\ -4 \end{pmatrix} + c_2 \begin{pmatrix} 1 \\ 1 \end{pmatrix} = \begin{pmatrix} 1 \\ 2 \end{pmatrix} \qquad (8.184)$$

The vector Equation 8.184 represents two scalar equations:

$$c_1 \cdot 1 + c_2 \cdot 1 = 1$$

$$c_1 \cdot (-4) + c_2 \cdot 1 = 2$$

or

$$c_1 + c_2 = 1$$

$$-4c_1 + c_2 = 2$$

which has solution

$$c_1 = -\frac{1}{5}$$

$$c_2 = \frac{6}{5} \qquad (8.185)$$

From Equations 8.183 and 8.185, the solution to Equation 8.174 is

$$\underline{x}(t) = (-1/5) \begin{pmatrix} 1 \\ -4 \end{pmatrix} e^{-3t} + (6/5) \begin{pmatrix} 1 \\ 1 \end{pmatrix} e^{2t} \qquad (8.186)$$

**SP 8.4** Solve

$$\begin{pmatrix} \dot{x}_1 \\ \dot{x}_2 \end{pmatrix} = \begin{pmatrix} -5 & -1 \\ 5 & -3 \end{pmatrix} \begin{pmatrix} x_1 \\ x_2 \end{pmatrix}, \qquad \begin{pmatrix} x_1(0) \\ x_2(0) \end{pmatrix} = \begin{pmatrix} 3 \\ -4 \end{pmatrix} \qquad (8.187)$$

*Solution*: As always, for LTI homogeneous matrix ODEs such as this, we try

$$\underline{x}(t) = \underline{\xi} e^{rt} \qquad (8.188)$$

Substituting Equation 8.188 into the matrix ODE in Equations 8.187 leads in short order to

$$(\underline{A} - r\underline{I})\underline{\xi} = \begin{pmatrix} -5 - r & -1 \\ 5 & -3 - r \end{pmatrix} \begin{pmatrix} \xi_1 \\ \xi_2 \end{pmatrix} = \begin{pmatrix} 0 \\ 0 \end{pmatrix} \qquad (8.189)$$

which, for nonzero $\underline{\xi}$, implies that $r$ must satisfy

$$\det(\underline{A} - r\underline{I}) = \det \begin{pmatrix} -5 - r & -1 \\ 5 & -3 - r \end{pmatrix} = (-5 - r)(-3 - r) - (5)(-1) = r^2 + 8r + 20 = 0 \quad (8.190)$$

Completing the square:

$$r^2 + 8r + 20 = (r+4)^2 - 4^2 + 20 = (r+4)^2 + 4 = 0 \qquad (8.191)$$

Hence the two eigenvalues of the system in Equation 8.183 are

$$r_1 = -4 + 2i$$
$$r_2 = -4 - 2i \qquad (8.192)$$

where $i = \sqrt{-1}$.

We must next find the eigenvector $\underline{\xi}^{(1)}$ associated with the eigenvalue $r_1$. Substituting $r_1 = -4 + 2i$ into Equation 8.189:

$$(r_1 \underline{I} - \underline{A})\underline{\xi}^{(1)} = \begin{pmatrix} r_1 + 5 & 1 \\ -5 & r_1 + 3 \end{pmatrix}\begin{pmatrix} \xi_1^{(1)} \\ \xi_2^{(1)} \end{pmatrix} = \begin{pmatrix} -4 + 2i + 5 & 1 \\ -5 & -4 + 2i + 3 \end{pmatrix}\begin{pmatrix} \xi_1^{(1)} \\ \xi_2^{(1)} \end{pmatrix} = \begin{pmatrix} 0 \\ 0 \end{pmatrix}$$

$$\begin{pmatrix} 1 + 2i & 1 \\ -5 & -1 + 2i \end{pmatrix}\begin{pmatrix} \xi_1^{(1)} \\ \xi_2^{(1)} \end{pmatrix} = \begin{pmatrix} 0 \\ 0 \end{pmatrix} \qquad (8.193)$$

The matrix Equation 8.193 is equivalent to two scalar equations:

$$(1 + 2i)\xi_1^{(1)} + \xi_2^{(1)} = 0$$
$$(-5)\xi_1^{(1)} + (-1 + 2i)\xi_2^{(1)} = 0 \qquad (8.194)$$

Although it may not be apparent looking at Equations 8.194, the two scalar equations are not independent. This is as it must be, since the determinant is zero. It is really one equation in two unknowns. Therefore we are free to set the value of one of the unknowns. We choose to set $\xi_1^{(1)} = 1$. Then $\xi_2^{(1)} = -1 - 2i$. Therefore the eigenvector associated with the first eigenvalue $r_1 = -4 + 2i$ is

$$\underline{\xi}^{(1)} = \begin{pmatrix} 1 \\ -1 - 2i \end{pmatrix}$$

This can be written as

$$\underline{\xi}^{(1)} = \underline{a} + \underline{b}i = \begin{pmatrix} 1 \\ -1 \end{pmatrix} + \begin{pmatrix} 0 \\ -2 \end{pmatrix}i \qquad (8.195)$$

From Section 8.5.1.2 we learned that if the first eigenvalue is $r_1 = -\lambda + i\omega$ and its associated eigenvector is $\underline{\xi}^{(1)} = \underline{a} + \underline{b}i$ then the two real fundamental solutions associated with $r_1$ and its complex conjugate are

$$\underline{x}^{(1)}(t) = e^{\lambda t}(\underline{a}\cos\omega t - \underline{b}\sin\omega t)$$

$$\underline{x}^{(2)}(t) = e^{\lambda t}(\underline{a}\sin\omega t + \underline{b}\cos\omega t)$$

Here, from Equation 8.192, $\lambda = -4$ and $\omega = 2$, and from Equation 8.195,

$$\underline{a} = \begin{pmatrix} 1 \\ -1 \end{pmatrix}, \qquad \underline{b} = \begin{pmatrix} 0 \\ -2 \end{pmatrix}$$

Then

$$\underline{x}^{(1)}(t) = e^{-4t}\left( \begin{pmatrix} 1 \\ -1 \end{pmatrix} \cos 2t - \begin{pmatrix} 0 \\ -2 \end{pmatrix} \sin 2t \right)$$

$$\underline{x}^{(2)}(t) = e^{-4t}\left( \begin{pmatrix} 1 \\ -1 \end{pmatrix} \sin 2t + \begin{pmatrix} 0 \\ -2 \end{pmatrix} \cos 2t \right)$$

(8.196)

Any solution of Equations 8.187 must be of the form

$$\underline{x}(t) = c_1 \underline{x}^{(1)}(t) + c_2 \underline{x}^{(2)}(t)$$

(8.197)

The coefficients $c_1, c_2$ are determined by the initial conditions. From Equations 8.196 and 8.197 and the second part of Equations 8.187 we have

$$\underline{x}(0) = c_1 \begin{pmatrix} 1 \\ -1 \end{pmatrix} + c_2 \begin{pmatrix} 0 \\ -2 \end{pmatrix} = \begin{pmatrix} 3 \\ -4 \end{pmatrix}$$

(8.198)

Equation 8.198 is equivalent to the two scalar equations

$$c_1 \cdot (1) + c_2 \cdot (0) = 3$$

$$c_1 \cdot (-1) + c_2 \cdot (-2) = -4$$

or

$$c_1 = 3$$

$$-c_1 - 2c_2 = -4$$

which have solution $c_1 = 3$ and $c_2 = 1/2$. Then, from Equations 8.196 and 8.197,

$$\underline{x}(t) = 3e^{-4t}\left( \begin{pmatrix} 1 \\ -1 \end{pmatrix} \cos 2t - \begin{pmatrix} 0 \\ -2 \end{pmatrix} \sin 2t \right)$$

$$+ \left( \frac{1}{2} \right) e^{-4t}\left( \begin{pmatrix} 1 \\ -1 \end{pmatrix} \sin 2t + \begin{pmatrix} 0 \\ -2 \end{pmatrix} \cos 2t \right)$$

which can be simplified to

$$\underline{x}(t) = e^{-4t}\left( \begin{pmatrix} 3 \\ -4 \end{pmatrix} \cos 2t + \begin{pmatrix} 1/2 \\ 11/2 \end{pmatrix} \sin 2t \right)$$

(8.199)

**SP 8.5** This problem is from Section 8.5.1.4.
Solve

$$\frac{d\underline{x}}{dt} = \underline{A}\underline{x} \qquad (8.200)$$

where

$$\underline{x}(0) = \begin{pmatrix} v_S \\ 0 \\ 0 \end{pmatrix} \qquad (8.201)$$

and

$$\underline{A} = \frac{-1}{4RC}\begin{pmatrix} 2 & 1 & 1 \\ 1 & 2 & 1 \\ 1 & 1 & 2 \end{pmatrix} \qquad (8.202)$$

Trying $\underline{x}(t) = \underline{\xi}e^{rt}$ we find as before that Equation 8.200 implies $r\underline{\xi}e^{rt} = \underline{A}\underline{\xi}e^{rt}$, which in turn requires

$$\det(\underline{A} - r\underline{I}) = 0 \qquad (8.203)$$

For simplicity define

$$r_* = (4RC)r \qquad (8.204)$$

Then substituting Equations 8.202 and 8.204 into Equation 8.203, we arrive at

$$\det\begin{pmatrix} 2+r_* & 1 & 1 \\ 1 & 2+r_* & 1 \\ 1 & 1 & 2+r_* \end{pmatrix} = 0 \qquad (8.205)$$

Equation 8.205 becomes

$$(2+r_*)\left[(2+r_*)^2 - 1\right] - \left[(2+r_*) - 1\right] + \left[1 - (2+r_*)\right] =$$
$$(2+r_*)(2+r_*)(2+r_*) - 3(2+r_*) + 2 = \qquad (8.206)$$
$$r_*^3 + 6r_*^2 + 9r_* + 4 = 0$$

Using the hand-calculation method we applied in Chapter 6, or using a web-based root calculator, we find that the cubic in Equation 8.206 factors into

$$(r_* + 4)(r_* + 1)^2 = 0 \qquad (8.207)$$

Hence the nondimensional eigenvalues for the system represented by Equations 8.200, 8.201, and 8.202 are

$$r_{*1} = -4$$

$$r_{*2} = r_{*3} = -1$$

The eigenvectors for this system are given for $i = 1, 2, 3$ by

$$\begin{pmatrix} 2 + r_{*i} & 1 & 1 \\ 1 & 2 + r_{*i} & 1 \\ 1 & 1 & 2 + r_{*i} \end{pmatrix} \begin{pmatrix} \xi_1^{(i)} \\ \xi_2^{(i)} \\ \xi_3^{(i)} \end{pmatrix} = \begin{pmatrix} 0 \\ 0 \\ 0 \end{pmatrix} \tag{8.208}$$

We choose $\xi_1^{(i)}$ as the undetermined component and set $\xi_1^{(i)} = 1$. For the first eigenvalue $(r_{*1} = -4)$ the first and second rows of Equation 8.204 become

$$\xi_2^{(1)} + \xi_3^{(1)} = 2$$

$$-2\xi_2^{(1)} + \xi_3^{(1)} = -1 \tag{8.209}$$

The solutions to Equation 8.209 are

$$\xi_2^{(1)} = 1$$

$$\xi_3^{(1)} = 1$$

Hence the eigenvector associated with eigenvalue $r_{*1} = -4$ is

$$\underline{\xi}^{(1)} = \begin{pmatrix} 1 \\ 1 \\ 1 \end{pmatrix} \tag{8.210}$$

For the eigenvalues $r_{*2} = -1, r_{*3} = -1$ the first and second rows of Equation 8.208 become, for $i = 1, 2$

$$\xi_2^{(i)} + \xi_3^{(i)} = -1$$

$$\xi_2^{(i)} + \xi_3^{(i)} = -1 \tag{8.211}$$

The equations are identical. We can form two independent eigenvalues from them. In each we arbitrarily set one of the components equal to zero. Then together with the first component the eigenvectors for the repeated eigenvalues $r_{*2} = r_{*3} = -1$ are

$$\underline{\xi}^{(2)} = \begin{pmatrix} 1 \\ 0 \\ -1 \end{pmatrix} \quad \underline{\xi}^{(3)} = \begin{pmatrix} 1 \\ -1 \\ 0 \end{pmatrix} \tag{8.212}$$

Summarizing our results so far, returning to dimensional variables, the general solution to Equations 8.200 and 8.202 is

$$\underline{x}(t) = c_1 \begin{pmatrix} 1 \\ 1 \\ 1 \end{pmatrix} e^{-t/RC} + c_2 \begin{pmatrix} 1 \\ 0 \\ -1 \end{pmatrix} e^{-t/4RC} + c_3 \begin{pmatrix} 1 \\ -1 \\ 0 \end{pmatrix} e^{-t/4RC} \qquad (8.213)$$

We must determine the coefficients $c_i$ from initial conditions, Equation 8.201. Setting $t = 0$ in Equation 8.213 and applying initial conditions we have

$$\underline{x}(0) = \begin{pmatrix} v_S \\ 0 \\ 0 \end{pmatrix} = c_1 \begin{pmatrix} 1 \\ 1 \\ 1 \end{pmatrix} + c_2 \begin{pmatrix} 1 \\ 0 \\ -1 \end{pmatrix} + c_3 \begin{pmatrix} 1 \\ -1 \\ 0 \end{pmatrix}$$

or

$$c_1 + c_2 + c_3 = v_S$$
$$c_1 - c_3 = 0 \qquad (8.214)$$
$$c_1 - c_2 = 0$$

The solution to Equations 8.214 is

$$c_1 = c_2 = c_3 = \frac{v_S}{3} \qquad (8.215)$$

Then the solution to Equations 8.200 through 8.202 is

$$\underline{x}(t) = \frac{v_S}{3} \begin{pmatrix} e^{-t/RC} + 2e^{-t/4RC} \\ e^{-t/RC} - e^{-t/4RC} \\ e^{-t/RC} - e^{-t/4RC} \end{pmatrix} \qquad (8.216)$$

**SP 8.6** Solve

$$\underline{\dot{x}} = \begin{pmatrix} 1 & -5 \\ 5 & -9 \end{pmatrix} \underline{x}, \qquad \begin{pmatrix} x_1(0) \\ x_2(0) \end{pmatrix} = \begin{pmatrix} 2 \\ -1 \end{pmatrix} \qquad (8.217)$$

*Solution*: As before we try the solution $\underline{x}(t) = \underline{\xi} e^{rt}$ and arrive again at the equations

$$\det(\underline{A} - r\underline{I}) = 0 \qquad (8.218)$$

$$(\underline{A} - r\underline{I})\underline{\xi} = \underline{0} \qquad (8.219)$$

From Equations 8.217 and 8.218

$$\det\begin{pmatrix} 1-r & -5 \\ 5 & -9-r \end{pmatrix} = (1-r)(-9-r) + 25 = r^2 + 8r + 16 = (r+4)^2 = 0$$

$$r_1 = r_2 = -4 \tag{8.220}$$

Substituting Equation 8.220 into Equation 8.219

$$5\xi_1^{(1)} - 5\xi_2^{(1)} = 0 \tag{8.221}$$

We choose $\xi_1^{(1)} = 1$ and then from Equation 8.221 $\xi_2^{(1)} = 1$. Hence

$$\underline{\xi}^{(1)} = \begin{pmatrix} 1 \\ 1 \end{pmatrix} \tag{8.222}$$

This provides us

$$\underline{x}^{(1)}(t) = \underline{\xi}^{(1)} e^{r_1 t} = \begin{pmatrix} 1 \\ 1 \end{pmatrix} e^{-4t} \tag{8.223}$$

But what of $\underline{x}^{(2)}(t)$? Recall from Section 5.1.2 that when repeated roots occur in second-order systems the second solution has the form $te^{\lambda t}$. Extrapolating that result to the state-space case, we might expect the second solution

$$\underline{x}^{(2)}(t) = \underline{\gamma} t e^{rt} \tag{8.224}$$

Trying Equation 8.224 here:

$$\underline{\dot{x}}^{(2)} = \underline{\gamma} e^{rt} + r\underline{\gamma} t e^{rt}$$

$$\underline{\dot{x}}^{(2)} - \underline{A}\underline{x} = \underline{\gamma} e^{rt} + r\underline{\gamma} t e^{rt} - \underline{A}(\underline{\gamma} t e^{rt}) = \underline{\gamma} e^{rt} - t e^{rt}(\underline{A} - r\underline{I})\underline{\gamma} \tag{8.225}$$

Equation 8.225 is supposed to equal $\underline{0}$. We can make the vector $(\underline{A} - r\underline{I})\underline{\gamma}$ equal $\underline{0}$ by setting $\underline{\gamma} = \underline{\xi}^{(1)}$. (Recall Equations 8.219 and 8.222.) But then the first term on the right-hand side in Equation 8.221, $\underline{\gamma} e^{rt}$, is nonzero. So our first trial solution has failed. Given the nature of the failure we next try

$$\underline{x}^{(2)}(t) = \underline{\gamma} t e^{rt} + \underline{\eta} e^{rt} \tag{8.226}$$

Substituting Equation 8.226 into

$$\frac{d\underline{x}}{dt} = \underline{A}\underline{x}$$

we arrive at

$$\underline{\gamma}e^{rt} + r\underline{\gamma}te^{rt} + r\underline{\eta}e^{rt} = \underline{A}(\underline{\gamma}te^{rt} + \underline{\eta}e^{rt}) \tag{8.227}$$

Equating coefficients of $t^n$ in Equation 8.227 with $n = 0, 1$ we find

$$(\underline{A} - r\underline{I})\underline{\gamma} = 0 \tag{8.228}$$

$$(\underline{A} - r\underline{I})\underline{\eta} = \underline{\gamma} \tag{8.229}$$

Now Equation 8.228 is the same as Equation 8.219 and we found that the only solution to Equation 8.219 is

$$\underline{\gamma} = \underline{\xi}^{(1)} = \begin{pmatrix} 1 \\ 1 \end{pmatrix} \tag{8.230}$$

Hence Equation 8.229 becomes

$$(\underline{A} - r\underline{I})\underline{\eta} = \underline{\xi}^{(1)} \tag{8.231}$$

Using $r = -4$ we find that Equation 8.231 reduces to

$$5\eta_1 - 5\eta_2 = 1 \tag{8.232}$$

We have

$$\eta_2 = \frac{-1}{5} + \eta_1 \tag{8.233}$$

Then, using Equations 8.226, 8.230, and 8.233, the second solution can be written as

$$\underline{x}^{(2)}(t) = te^{-4t}\underline{\xi}^{(1)} + \begin{pmatrix} 0 \\ -1/5 \end{pmatrix}e^{-4t} + \eta_1\begin{pmatrix} 1 \\ 1 \end{pmatrix}e^{-4t} \tag{8.234}$$

where $\eta_1$ is an undetermined constant. But, recalling Equation 8.223, the last term in Equation 8.234 duplicates the first solution $\underline{x}^{(1)}(t)$, so we can set $\eta_1 = 0$ and then Equation 8.234 simplifies to

$$\underline{x}^{(2)}(t) = te^{-4t}\underline{\xi}^{(1)} + \begin{pmatrix} 0 \\ -1/5 \end{pmatrix}e^{-4t} \tag{8.235}$$

Comparing 8.226 and 8.235 we see that

$$\underline{\eta} = \begin{pmatrix} 0 \\ -1/5 \end{pmatrix} \tag{8.236}$$

Next, we write

$$\underline{x}(t) = c_1 \underline{x}^{(1)}(t) + c_2 \underline{x}^{(2)}(t) = c_1 \binom{1}{1} e^{-4t} + c_2 \left( \binom{1}{1} t e^{-4t} + \binom{0}{-1/5} e^{-4t} \right)$$

and solve for the coefficients via initial conditions. We must have

$$\underline{x}(0) = c_1 \binom{1}{1} + c_2 \binom{0}{-1/5} = \binom{2}{-1}$$

In scalar form, this is

$$c_1 = 2$$

$$c_1 - \frac{c_2}{5} = -1$$

The solutions are

$$c_1 = 2$$

$$c_2 = 15$$

So, the solution to Equations 8.217 is

$$\underline{x}(t) = 2 \binom{1}{1} e^{-4t} + 15 \left( \binom{1}{1} t e^{-4t} + \binom{0}{-1/5} e^{-4t} \right)$$

which can be simplified to

$$\underline{x}(t) = \binom{2}{-1} e^{-4t} + 15 \binom{1}{1} t e^{-4t}$$

**SP 8.7** Find the state transition matrix $\underline{\Phi}(t)$ for the system

$$\underline{\dot{x}} = \begin{pmatrix} -8 & 4 \\ -5 & -4 \end{pmatrix} \underline{x} \tag{8.237}$$

*Solution*:
Step 1 is to find the two fundamental solutions. We do this by trying $\underline{x}(t) = \underline{\xi} e^{rt}$, which leads to

$$(\underline{A} - rI)\underline{\xi} = \begin{pmatrix} -8-r & 4 \\ -5 & -4-r \end{pmatrix} \underline{\xi} = \underline{0} \tag{8.238}$$

which in turn implies

$$\det(\underline{A} - r\underline{I}) = \det\begin{pmatrix} -8-r & 4 \\ -5 & -4-r \end{pmatrix} = 0$$

$$(-8-r)(-4-r) - (4)(-5) = r^2 + 12r + 52 = 0$$

Completing the square:

$$r^2 + 12r + 52 = (r+6)^2 - 6^2 + 52 = (r+6)^2 + 4^2 = 0$$

Hence, the eigenvalues of the system are $r_{1,2} = -6 \pm 4i$. Substituting $r_1$ into Equation 8.238 results in

$$(r_1\underline{I} - \underline{A})\underline{\xi}^{(1)} = \begin{pmatrix} -6+4i+8 & -4 \\ 5 & -6+4i+4 \end{pmatrix}\underline{\xi}^{(1)} = \begin{pmatrix} 2+4i & -4 \\ 5 & -2+4i \end{pmatrix}\underline{\xi}^{(1)} = \underline{0} \quad (8.239)$$

Setting $\xi_1^{(1)} = 1$ in Equation 8.239 yields $\xi_2^{(1)} = 1/2 + i$. Writing $\underline{\xi} = \underline{a} + i\underline{b}$ we determine that

$$\underline{a} = \begin{pmatrix} 1 \\ 1/2 \end{pmatrix}, \qquad \underline{b} = \begin{pmatrix} 0 \\ 1 \end{pmatrix} \qquad\qquad (8.240)$$

From Section 8.5.1.2, we learned that if the first eigenvalue is $r_1 = -\lambda + i\omega$ and its associated eigenvector is $\underline{\xi}^{(1)} = \underline{a} + \underline{b}i$ then the *two real* fundamental solutions associated with $r_1$ and its complex conjugate are $\underline{x}^{(1)}(t) = e^{\lambda t}(\underline{a}\cos\omega t - \underline{b}\sin\omega t)$ and $\underline{x}^{(2)}(t) = e^{\lambda t}(\underline{a}\sin\omega t + \underline{b}\cos\omega t)$. Using that here:

$$\underline{x}^{(1)}(t) = e^{-6t}\left( \begin{pmatrix} 1 \\ 1/2 \end{pmatrix}\cos 4t - \begin{pmatrix} 0 \\ 1 \end{pmatrix}\sin 4t \right)$$

$$\underline{x}^{(2)}(t) = e^{-6t}\left( \begin{pmatrix} 1 \\ 1/2 \end{pmatrix}\sin 4t + \begin{pmatrix} 0 \\ 1 \end{pmatrix}\cos 4t \right)$$

$\qquad\qquad\qquad\qquad\qquad\qquad\qquad\qquad\qquad\qquad\qquad\qquad (8.241)$

Step 2 is to form the fundamental matrix $\underline{\Psi}(t)$, whose columns consist of the fundamental solutions:

$$\underline{\Psi}(t) = \begin{pmatrix} e^{-6t}\cos 4t & e^{-6t}\sin 4t \\ e^{-6t}\left((1/2)\cos 4t - \sin 4t\right) & e^{-6t}\left((1/2)\sin 4t + \cos 4t\right) \end{pmatrix} \qquad (8.242)$$

Step 3 is to determine $\underline{\Psi}^{-1}(0)$, which is defined by $\underline{\Psi}(0)\underline{\Psi}^{-1}(0) = \underline{I}$. Letting

$$\underline{\Psi}^{-1}(0) = \begin{pmatrix} a & b \\ c & d \end{pmatrix}$$

we have

$$\underline{\Psi}(0)\underline{\Psi}^{-1}(0) = \begin{pmatrix} 1 & 0 \\ 1/2 & 1 \end{pmatrix}\begin{pmatrix} a & b \\ c & d \end{pmatrix} = \begin{pmatrix} 1 & 0 \\ 0 & 1 \end{pmatrix}$$

$$1 \cdot a + (0) \cdot c = 1 \qquad 1 \cdot b + (0) \cdot d = 0$$

$$(1/2) \cdot a + 1 \cdot c = 0 \qquad (1/2) \cdot b + 1 \cdot d = 1$$

which has solution $a = 1, b = 0, c = -1/2, d = 1$, so that

$$\underline{\Psi}^{-1}(0) = \begin{pmatrix} 1 & 0 \\ -1/2 & 1 \end{pmatrix} \tag{8.243}$$

Step 4 is to compute the state transition matrix $\underline{\Phi}(t)$ from

$$\underline{\Phi}(t) = \underline{\Psi}(t)\underline{\Psi}^{-1}(0) \tag{8.244}$$

We have

$$\underline{\Phi}(t) = \begin{pmatrix} e^{-6t}\cos 4t & e^{-6t}\sin 4t \\ e^{-6t}\left((1/2)\cos 4t - \sin 4t\right) & e^{-6t}\left((1/2)\sin 4t + \cos 4t\right) \end{pmatrix}\begin{pmatrix} 1 & 0 \\ -1/2 & 1 \end{pmatrix}$$

$$\underline{\Phi}(t) = \begin{pmatrix} e^{-6t}\left(\cos 4t - (1/2)\sin 4t\right) & e^{-6t}\sin 4t \\ -e^{-6t}(5/4)\sin 4t & e^{-6t}\left(\cos 4t + (1/2)\sin 4t\right) \end{pmatrix}$$

We can (partially) check our calculations by finding out if $\underline{\Phi}(0) = \underline{I}$. It does.

**SP 8.8** Find a transformation $\underline{z} = \underline{T}\underline{x}$ to convert the system

$$\underline{\dot{x}} = \begin{pmatrix} -4 & 5 \\ -10 & -2 \end{pmatrix}\underline{x} \tag{8.245}$$

into a system in the form

$$\underline{\dot{z}} = \begin{pmatrix} r_1 & 0 \\ 0 & r_2 \end{pmatrix}\underline{z} \tag{8.246}$$

and show that the system matrix is indeed diagonalized with the transformation.

*Solution*:
Step 1: We must find the fundamental solutions to Equation 8.245. Trying $\underline{x}(t) = \underline{\xi}e^{rt}$ we quickly arrive at

$$\det(\underline{A} - r\underline{I}) = \det\begin{pmatrix} -4-r & 5 \\ -10 & -2-r \end{pmatrix} = (-4-r)(-2-r)-(5)(-10) = r^2 + 6r + 58 = 0$$

Completing the square,

$$r^2 + 6r + 58 = (r+3)^2 - 3^2 + 58 = (r+3)^2 + 7^2 = 0$$

so, $r_{1,2} = -3 \pm 7i$. Taking $r_1 = -3 + 7i$,

$$\begin{pmatrix} r_1 + 4 & -5 \\ 10 & r_1 + 2 \end{pmatrix}\begin{pmatrix} \xi_1^{(1)} \\ \xi_2^{(1)} \end{pmatrix} = \begin{pmatrix} -3 + 7i + 4 & -5 \\ 10 & -3 + 7i + 2 \end{pmatrix}\begin{pmatrix} \xi_1^{(1)} \\ \xi_2^{(1)} \end{pmatrix} = \tag{8.247}$$

$$\begin{pmatrix} 1 + 7i & -5 \\ 10 & -1 + 7i \end{pmatrix}\begin{pmatrix} \xi_1^{(1)} \\ \xi_2^{(1)} \end{pmatrix} = \begin{pmatrix} 0 \\ 0 \end{pmatrix}$$

From Equation 8.247, letting $\xi_1^{(1)} = 1$, we have $\xi_2^{(1)} = (1+7i)/5$. Then the first eigenvector is

$$\underline{\xi}^{(1)} = \begin{pmatrix} 1 \\ (1+7i)/5 \end{pmatrix}$$

We know that the second eigenvalue and second eigenvector are complex conjugates of the first, so the two (complex) fundamental solutions are

$$\underline{x}^{(1)} = \begin{pmatrix} 1 \\ (1+7i)/5 \end{pmatrix}e^{(-3+7i)t} \qquad \underline{x}^{(2)} = \begin{pmatrix} 1 \\ (1-7i)/5 \end{pmatrix}e^{(-3-7i)t}$$

The fundamental matrix is then

$$\underline{\Psi}(t) = \begin{pmatrix} e^{(-3+7i)t} & e^{(-3-7i)t} \\ (1+7i)e^{(-3+7i)t}/5 & (1-7i)e^{(-3-7i)t}/5 \end{pmatrix} \tag{8.248}$$

Step 2: From Section 8.5.1.8, the transformation that diagonalizes the system matrix is

$$\underline{z} = \underline{\Psi}^{-1}(0)\underline{x} \tag{8.249}$$

so, we must invert $\underline{\Psi}(0)$, which from Equation 8.248 is given by

$$\underline{\Psi}(0) = \begin{pmatrix} 1 & 1 \\ (1+7i)/5 & (1-7i)/5 \end{pmatrix} \tag{8.250}$$

Letting

$$\underline{\Psi}^{-1}(0) = \begin{pmatrix} a & b \\ c & d \end{pmatrix}$$

we have

$$\underline{\Psi}(0) \cdot \underline{\Psi}^{-1}(0) = \underline{I}$$

$$\begin{pmatrix} 1 & 1 \\ (1+7i)/5 & (1-7i)/5 \end{pmatrix}\begin{pmatrix} a & b \\ c & d \end{pmatrix} = \begin{pmatrix} 1 & 0 \\ 0 & 1 \end{pmatrix}$$

which represents the scalar equations

$$a + c = 1 \qquad\qquad\qquad b + d = 0$$

$$((1+7i)/5)a + ((1-7i)/5)c = 0 \qquad ((1+7i)/5)b + ((1-7i)/5)d = 1$$

The solutions are $a = 1/2 + i/14$, $b = -5i/14$, $c = 1/2 - i/14$, and $d = 5i/14$. Hence

$$\underline{\Psi}^{-1}(0) = \begin{pmatrix} 1/2 + i/14 & -5i/14 \\ 1/2 - i/14 & 5i/14 \end{pmatrix} \tag{8.251}$$

The transformed state vector is

$$\begin{pmatrix} z_1 \\ z_2 \end{pmatrix} = \begin{pmatrix} 1/2 + i/14 & -5i/14 \\ 1/2 - i/14 & 5i/14 \end{pmatrix}\begin{pmatrix} x_1 \\ x_2 \end{pmatrix} \tag{8.252}$$

Step 3: We must now show that the matrix $\underline{R}$ for the transformed system

$$\dot{\underline{z}} = \underline{R}\underline{z}$$

is indeed diagonalized. From Equation 8.249 and its inverse relation $\underline{x} = \underline{\Psi}(0)\underline{z}$,

$$\dot{\underline{z}} = \underline{\Psi}^{-1}(0)\dot{\underline{x}} = \underline{\Psi}^{-1}(0)\underline{A}\underline{x} = \underline{\Psi}^{-1}(0)\underline{A}\underline{\Psi}(0)\underline{z}$$

so,

$$\underline{R} = \underline{\Psi}^{-1}(0)\underline{A}\underline{\Psi}(0) \tag{8.253}$$

From Equations 8.245 and 8.250,

$$\underline{A}\underline{\Psi}(0) = \begin{pmatrix} -4 & 5 \\ -10 & -2 \end{pmatrix}\begin{pmatrix} 1 & 1 \\ (1+7i)/5 & (1-7i)/5 \end{pmatrix}$$

$$= \begin{pmatrix} -3+7i & -3-7i \\ -(52+14i)/5 & -(52-14i)/5 \end{pmatrix} \tag{8.254}$$

From Equations 8.251, 8.253, and 8.254,

$$\underline{R} = \begin{pmatrix} 1/2 + i/14 & -5i/14 \\ 1/2 - i/14 & 5i/14 \end{pmatrix}\begin{pmatrix} -3+7i & -3-7i \\ -(52+14i)/5 & -(52-14i)/5 \end{pmatrix}$$

and indeed

$$\underline{R} = \begin{pmatrix} -3+7i & 0 \\ 0 & -3-7i \end{pmatrix} \tag{8.255}$$

as was to be shown.

**SP 8.9** Solve

$$\begin{pmatrix} \dot{x}_1 \\ \dot{x}_2 \end{pmatrix} = \begin{pmatrix} -2 & 3 \\ -3 & -8 \end{pmatrix} \begin{pmatrix} x_1 \\ x_2 \end{pmatrix} + \begin{pmatrix} 2 \\ -3 \end{pmatrix} u_0(t) \tag{8.256}$$

$$\begin{pmatrix} x_1(0) \\ x_2(0) \end{pmatrix} = \begin{pmatrix} 0 \\ 0 \end{pmatrix} \tag{8.257}$$

where $u_0(t)$ is the unit step function that steps up at $t = 0$.

*Solution:*

Step 1: We must find the fundamental solutions, which means we must first solve the homogeneous equation

$$\begin{pmatrix} \dot{x}_1 \\ \dot{x}_2 \end{pmatrix} = \begin{pmatrix} -2 & 3 \\ -3 & -8 \end{pmatrix} \begin{pmatrix} x_1 \\ x_2 \end{pmatrix} \tag{8.258}$$

Trying $\underline{x}(t) = \underline{\xi} e^{rt}$ in Equation 8.258 leads quickly to

$$\det(\underline{A} - r\underline{I}) = \det \begin{pmatrix} -2 - r & 3 \\ -3 & -8 - r \end{pmatrix} = (-2 - r)(-8 - r) + 9 = r^2 + 10r + 25 = 0 \tag{8.259}$$

There is but one solution to Equation 8.259 and it is $r_1 = -5$. We next seek the eigenvectors associated with this eigenvalue through the equation

$$(\underline{A} - r_1\underline{I})\underline{\xi} = \begin{pmatrix} -2 - r_1 & 3 \\ -3 & 8 - r_1 \end{pmatrix} \begin{pmatrix} \xi_1 \\ \xi_2 \end{pmatrix} = \begin{pmatrix} 3 & 3 \\ -3 & -3 \end{pmatrix} \begin{pmatrix} \xi_1 \\ \xi_2 \end{pmatrix} = \begin{pmatrix} 0 \\ 0 \end{pmatrix} \tag{8.260}$$

We set $\xi_1 = 1$ and then from Equation 8.260 $\xi_2 = -1$. Hence, the first fundamental solution to Equation 8.254 is

$$\underline{x}^{(1)} = \underline{\xi}^{(1)} e^{rt} = \begin{pmatrix} 1 \\ -1 \end{pmatrix} e^{-5t} \tag{8.261}$$

Equation 8.260 has shown us this is *not* one of those problems where there is a root of multiplicity 2 and yet two independent eigenvectors; here there is only one. Hence, following Section 8.5.1.4, we seek a second fundamental solution in the form

$$\underline{x}^{(2)} = \underline{\xi}^{(1)} t e^{-5t} + \underline{\eta} e^{-5t} \tag{8.262}$$

We will use the general form

$$\underline{x}^{(2)} = \underline{\xi}^{(1)} t e^{rt} + \underline{\eta} e^{rt} \tag{8.263}$$

in what follows, to remind us of Example 1 in Section 8.5.1.4 and Solved Problem 8.6. Preparing to substitute Equation 8.263 in the homogeneous Equation 8.258, we first differentiate:

$$\dot{\underline{x}}^{(2)} = \underline{\xi}^{(1)}e^{r_1 t} + tr_1\underline{\xi}^{(1)}e^{r_1 t} + r_1\underline{\eta}e^{r_1 t} \tag{8.264}$$

Then we calculate

$$\underline{A}\underline{x}^{(2)} = \underline{A}\left(\underline{\xi}^{(1)}te^{r_1 t} + \underline{\eta}e^{r_1 t}\right) = t\underline{A}\underline{\xi}^{(1)}e^{r_1 t} + \underline{A}\underline{\eta}e^{r_1 t} \tag{8.265}$$

Now $\underline{A}\underline{\xi}^{(1)} = r_1\underline{\xi}^{(1)}$, so Equation 8.265 becomes

$$\underline{A}\underline{x}^{(2)} = tr_1\underline{\xi}^{(1)}e^{r_1 t} + \underline{A}\underline{\eta}e^{r_1 t} \tag{8.266}$$

From Equations 8.264 and 8.266

$$\underline{\xi}^{(1)}e^{r_1 t} + tr_1\underline{\xi}^{(1)}e^{r_1 t} + r_1\underline{\eta}e^{r_1 t} = tr_1\underline{\xi}^{(1)}e^{r_1 t} + \underline{A}\underline{\eta}e^{r_1 t}$$

Eliminating the term $tr_1\underline{\xi}^{(1)}e^{r_1 t}$ common to both sides and dividing through by $e^{r_1 t}$ we are left with

$$(\underline{A} - r_1\underline{I})\underline{\eta} = \underline{\xi} \tag{8.267}$$

In this specific problem Equation 8.267 becomes

$$\begin{pmatrix} 3 & 3 \\ -3 & -3 \end{pmatrix}\begin{pmatrix} \eta_1 \\ \eta_2 \end{pmatrix} = \begin{pmatrix} 1 \\ -1 \end{pmatrix} \tag{8.268}$$

or

$$\eta_1 + \eta_2 = \frac{1}{3} \tag{8.269}$$

By the same logic we employed in Solved Problem 8.6 (not duplicating $\underline{x}^{(1)}$ in $\underline{x}^{(2)}$), it suffices to set $\eta_1 = 0$, leaving $\eta_2 = 1/3$. See the discussion of Equations 8.234 and 8.235 in SP 8.6. Then

$$\underline{x}^{(2)}(t) = \begin{pmatrix} 1 \\ -1 \end{pmatrix}te^{-5t} + \begin{pmatrix} 0 \\ 1/3 \end{pmatrix}e^{-5t} \tag{8.270}$$

Step 2: With the two fundamental solutions in hand we form the fundamental matrix:

$$\underline{\Psi}(t) = e^{-5t}\begin{pmatrix} 1 & t \\ -1 & -t + 1/3 \end{pmatrix} \tag{8.271}$$

Step 3: Calculate the inverse of the fundamental matrix evaluated at zero. Use the defining equation $\underline{\Psi}(0) \cdot \underline{\Psi}^{-1}(0) = \underline{I}$. Let

$$\underline{\Psi}^{-1}(0) = \begin{pmatrix} a & b \\ c & d \end{pmatrix}$$

Then

$$\underline{\Psi}(0)\underline{\Psi}^{-1}(0) = \begin{pmatrix} 1 & 0 \\ -1 & 1/3 \end{pmatrix}\begin{pmatrix} a & b \\ c & d \end{pmatrix} = \underline{I} = \begin{pmatrix} 1 & 0 \\ 0 & 1 \end{pmatrix} \tag{8.272}$$

which is equivalent to the scalar equations

$$1 \cdot a + (0) \cdot c = 1 \qquad 1 \cdot b + (0) \cdot d = 0$$
$$-1 \cdot a + (1/3) \cdot c = 0 \qquad -1 \cdot b + (1/3) \cdot d = 1 \tag{8.273}$$

which have solution $a = 1$, $b = 0$, $c = 3$, and $d = 3$. Hence

$$\underline{\Psi}^{-1}(0) = \begin{pmatrix} 1 & 0 \\ 3 & 3 \end{pmatrix} \tag{8.274}$$

Step 4: Calculate the state transition matrix $\underline{\Phi}(t)$. From Equations 8.271 and 8.274,

$$\underline{\Phi}(t) = \underline{\Psi}(t)\underline{\Psi}^{-1}(0) = e^{-5t}\begin{pmatrix} 1 & t \\ -1 & -t+1/3 \end{pmatrix}\begin{pmatrix} 1 & 0 \\ 3 & 3 \end{pmatrix} = e^{-5t}\begin{pmatrix} 1+3t & 3t \\ -3t & 1-3t \end{pmatrix} \tag{8.275}$$

Step 5: Calculate the state vector $\underline{x}(t)$:

$$\underline{x}(t) = \underline{\Phi}(t)\underline{x}(0) + \int_0^t \underline{\Phi}(t-\tau)\underline{B}\underline{u}(\tau)d\tau \tag{8.276}$$

From Equations 8.256, 8.275, and 8.276 and the definition of the unit step function $u_0(t)$,

$$\underline{x}(t) = \int_0^t e^{-5(t-\tau)}\begin{pmatrix} 1+3(t-\tau) & 3(t-\tau) \\ -3(t-\tau) & 1-3(t-\tau) \end{pmatrix}\begin{pmatrix} 2 \\ -3 \end{pmatrix}(1)d\tau \tag{8.277}$$

$$x_1(t) = \int_0^t e^{-5(t-\tau)}\left[\left(1+3(t-\tau)\right)(2) + 3(t-\tau)(-3)\right]d\tau$$

$$x_2(t) = \int_0^t e^{-5(t-\tau)}\left[-3(t-\tau)(2) + \left(1-3(t-\tau)\right)(-3)\right]d\tau$$

$$x_1(t) = (2-3t)e^{-5t}\int_0^t e^{5\tau}d\tau + 3e^{-5t}\int_0^t \tau e^{5\tau}d\tau \tag{8.278}$$

$$x_2(t) = (-3+3t)e^{-5t}\int_0^t e^{5\tau}d\tau - 3e^{-5t}\int_0^t \tau e^{5\tau}d\tau$$

Now

$$\int_0^t e^{5\tau}d\tau = (e^{5t}-1)/5 \tag{8.279}$$

and, integrating by parts,

$$\int_0^t \tau e^{5\tau} d\tau = \frac{t e^{5t}}{5} - \frac{(e^{5t} - 1)}{25} \tag{8.280}$$

Substituting Equations 8.279 and 8.280 into Equations 8.278, there results, after some simplification,

$$x_1(t) = \left(\frac{7}{25}\right)(1 - e^{-5t}) + \left(\frac{3}{5}\right) t e^{-5t}$$

$$x_2(t) = -\left(\frac{12}{25}\right)(1 - e^{-5t}) - \left(\frac{3}{5}\right) t e^{-5t} \tag{8.281}$$

**SP 8.10** Using matrix Laplace transform methods, solve

$$\underline{\dot{x}} = \begin{pmatrix} -5 & -4 \\ 5 & -1 \end{pmatrix} \underline{x} + \begin{pmatrix} 1 \\ 3 \end{pmatrix} \cos 2t \tag{8.282}$$

$$\underline{x}(0) = \begin{pmatrix} 2 \\ -1 \end{pmatrix} \tag{8.283}$$

*Solution*:
We take the Laplace transform of Equation 8.282 and call $L\{\underline{x}(t)\} = \underline{x}(s)$. Now

$$L\{\underline{\dot{x}}(t)\} = sL\{\underline{x}(t)\} - \underline{x}(0) = s\underline{x}(s) - \underline{x}(0)$$

$$L\{\cos 2t\} = \frac{s}{s^2 + 2^2}$$

and there results, using Equations 8.282 and 8.283,

$$(s\underline{I} - \underline{A})x(s) = \underline{x}(0) + \underline{b}u(t)$$

$$\begin{pmatrix} s+5 & 4 \\ -5 & s+1 \end{pmatrix} \begin{pmatrix} x_1(s) \\ x_2(s) \end{pmatrix} = \begin{pmatrix} 2 \\ -1 \end{pmatrix} + \begin{pmatrix} 1 \\ 3 \end{pmatrix} \frac{s}{s^2 + 4}$$

Calling $\underline{x}(s) = \underline{x}$ for brevity, we have the scalar equations

$$(s+5)x_1 + 4x_2 = 2 + \frac{s}{(s^2 + 4)}$$

$$-5x_1 + (s+1)x_2 = -1 + \frac{3s}{(s^2 + 4)}$$

Multiplying the second equation by 4 and using the first to eliminate $x_2$ we have

$$4x_2 = -(s+5)x_1 + 2 + \frac{s}{(s^2 + 4)} \tag{8.284}$$

$$-20x_1 + (s+1)\left[-(s+5)x_1 + 2 + \frac{s}{(s^2 + 4)}\right] = -4 + \frac{12s}{(s^2 + 4)}$$

$$-\left((s+1)(s+5)+20\right)x_1 = -2(s+1)-\frac{s(s+1)}{(s^2+4)}-4+\frac{12s}{(s^2+4)}$$

$$x_1 = \frac{2s+6}{s^2+6s+25}+\frac{s^2-11s}{(s^2+6s+25)(s^2+4)} = \frac{2s^3+7s^2-3s+24}{(s^2+6s+25)(s^2+4)} \tag{8.285}$$

Using a partial fraction expansion:

$$x_1 = \frac{c_1 s+c_2}{s^2+6s+25}+\frac{c_3 s+c_4}{s^2+4} \tag{8.286}$$

$$x_1 = \frac{(c_1+c_3)s^3+(c_2+6c_3+c_4)s^2+(4c_1+25c_3+6c_4)s+(4c_2+25c_4)}{(s^2+6s+25)(s^2+4)} \tag{8.287}$$

Equating the coefficients of like powers of $s$ in the numerators of Equations 8.285 and 8.287, we have four simultaneous linear equations in four unknowns that can be written in matrix form as

$$\begin{pmatrix} 1 & 0 & 1 & 0 \\ 0 & 1 & 6 & 1 \\ 4 & 0 & 25 & 6 \\ 0 & 4 & 0 & 25 \end{pmatrix}\begin{pmatrix} c_1 \\ c_2 \\ c_3 \\ c_4 \end{pmatrix} = \begin{pmatrix} 2 \\ 7 \\ -3 \\ 24 \end{pmatrix} \tag{8.288}$$

Using an online calculator on the Internet we are given the result

$$\begin{pmatrix} c_1 \\ c_2 \\ c_3 \\ c_4 \end{pmatrix} = \begin{pmatrix} 2.3538 \\ 9.7179 \\ -0.3538 \\ -0.5949 \end{pmatrix} \tag{8.289}$$

which we can quickly check by substitution into Equation 8.288 and determine with a hand calculator to be correct. From Equations 8.286 and 8.289

$$x_1 = \frac{2.3538s+9.7179}{s^2+6s+25}-\frac{(0.3538s+0.5949)}{s^2+4} \tag{8.290}$$

We must manipulate Equation 8.290 to match the entries in our short table of Laplace transforms (Table 7.2):

$$x_1 = \frac{2.3538(s+3)-(2.3538)(3)+9.7179}{(s+3)^2+4^2}-\frac{0.3538s}{s^2+2^2}-\frac{0.5949}{s^2+2^2}$$

$$x_1 = 2.3538\frac{(s+3)}{(s+3)^2+4^2}+\frac{2.6565}{(s+3)^2+4^2}-\frac{0.3538s}{s^2+2^2}-\frac{0.5949}{s^2+2^2}$$

$$x_1 = 2.3538\frac{(s+3)}{(s+3)^2+4^2}+\left(\frac{2.6565}{4}\right)\frac{4}{(s+3)^2+4^2}-0.3538\frac{s}{s^2+2^2}-\left(\frac{0.5949}{2}\right)\frac{2}{s^2+2^2}$$

$$x_1(s) = 2.3538\frac{(s+3)}{(s+3)^2+4^2} + 0.6641\frac{4}{(s+3)^2+4^2} - 0.3538\frac{s}{s^2+2^2} - 0.2974\frac{2}{s^2+2^2} \quad (8.291)$$

From Equation 8.291 and Table 7.2,

$$x_1(t) = L^{-1}\{x_1(s)\}$$

$$= 2.3538e^{-3t}\cos 4t + 0.6641e^{-3t}\sin 4t - 0.3538\cos 2t - 0.2974\sin 2t$$

(8.292)

From Equations 8.284 and 8.285,

$$4x_2(s) = -(s+5)\left[\frac{2s^3+7s^2-3s+24}{(s^2+6s+25)(s^2+4)}\right] + 2 + \frac{s}{s^2+4} \quad (8.293)$$

After some algebra,

$$x_2(s) = \frac{-s^3+8s^2+16s+20}{(s^2+6s+25)(s^2+4)} \quad (8.294)$$

Again using a partial fraction expansion,

$$x_2 = \frac{c_5 s + c_6}{s^2+6s+25} + \frac{c_7 s + c_8}{s^2+4} \quad (8.295)$$

$$x_2 = \frac{(c_5+c_7)s^3 + (c_6+6c_7+c_8)s^2 + (4c_5+25c_7+6c_8)s + (4c_6+25c_8)}{(s^2+6s+25)(s^2+4)} \quad (8.296)$$

Again equating coefficients of like powers of $s$ in the numerators in Equations 8.294 and 8.296, we arrive at

$$\begin{pmatrix} 1 & 0 & 1 & 0 \\ 0 & 1 & 6 & 1 \\ 4 & 0 & 25 & 6 \\ 0 & 4 & 0 & 25 \end{pmatrix}\begin{pmatrix} c_5 \\ c_6 \\ c_7 \\ c_8 \end{pmatrix} = \begin{pmatrix} -1 \\ 8 \\ 16 \\ 20 \end{pmatrix} \quad (8.297)$$

Using an online calculator on the Internet we are given the result

$$\begin{pmatrix} c_5 \\ c_6 \\ c_7 \\ c_8 \end{pmatrix} = \begin{pmatrix} -1.8410 \\ 2.5641 \\ 0.8410 \\ 0.3897 \end{pmatrix} \quad (8.298)$$

From Equations 8.295 and 8.298,

$$x_2(s) = \frac{-1.8410s+2.5641}{s^2+6s+25} + \frac{0.8410s+0.3897}{s^2+4} \quad (8.299)$$

Manipulating Equation 8.299 into a form matching entries in Table 7.2, we arrive at

$$x_2(s) = -1.8410\frac{(s+3)}{(s+3)^2 + 4^2} + 2.0218\frac{4}{(s+3)^2 + 4^2} + 0.8410\frac{s}{s^2 + 2^2} + 0.1949\frac{2}{s^2 + 2^2}$$

and then

$$x_2(t) = L^{-1}\{x_2(s)\} =$$

$$-1.8410e^{-3t}\cos 4t + 2.0218e^{-3t}\sin 4t + 0.8410\cos 2t + 0.1949\sin 2t$$

(8.300)

In summary, from Equations 8.292 and 8.300, the solution to Equations 8.282 and 8.283 is

$$\underline{x}(t) = \begin{pmatrix} 2.3538 \\ -1.8410 \end{pmatrix} e^{-3t}\cos 4t + \begin{pmatrix} 0.6641 \\ 2.0218 \end{pmatrix} e^{-3t}\sin 4t$$

$$+ \begin{pmatrix} -0.3538 \\ 0.8410 \end{pmatrix}\cos 2t + \begin{pmatrix} -0.2974 \\ 0.1949 \end{pmatrix}\sin 2t$$

(8.301)

**SP8.11** Find the steady-state solution to

$$\begin{pmatrix} \dot{x}_1 \\ \dot{x}_2 \end{pmatrix} = \begin{pmatrix} 1 & -1 \\ 3 & -2 \end{pmatrix}\begin{pmatrix} x_1 \\ x_2 \end{pmatrix} + \begin{pmatrix} 2 \\ -1 \end{pmatrix} t\sin 2t$$

(8.302)

*Solution*:
Step 0 is to ensure that the system is stable and damped, so that the system's fundamental solutions are not part of the steady-state response. The characteristic equation is

$$r^2 + r + 1 = 0$$

The roots of this have negative real parts, so the system's fundamental solutions die away, leaving only the solution we are about to compute.
Equation 8.302 is of the form

$$\underline{\dot{x}} = \underline{A}\underline{x} + \underline{b}t\sin\omega t$$

(8.303)

We will develop a strategy for solution using this general form. Then, once we have a strategy, we will employ the given values and execute it.
Step 1: Based on Table 8.5, we try the solution

$$\underline{x}(t) = (\underline{c}t + \underline{d})\sin\omega t + (\underline{e}t + \underline{f})\cos\omega t$$

(8.304)

in Equation 8.303. Now

$$\underline{\dot{x}}(t) = \underline{c}\sin\omega t + \omega(\underline{c}t + \underline{d})\cos\omega t + \underline{e}\cos\omega t - \omega(\underline{e}t + \underline{f})\sin\omega t$$

(8.305)

Substituting Equations 8.304 and 8.305 into Equation 8.303 and combining terms
we have

$$(\omega\underline{e} + \underline{A}\underline{c} + \underline{b})t\sin\omega t + (\omega\underline{f} + \underline{A}\underline{d} - \underline{c})\sin\omega t +$$
$$(-\omega\underline{c} + \underline{A}\underline{e})t\cos\omega t + (-\omega\underline{d} + \underline{A}\underline{f} - \underline{e})\cos\omega t = 0 \tag{8.306}$$

For Equation 8.306 to be true at all instants of time we must have

$$\omega\underline{e} + \underline{A}\underline{c} + \underline{b} = 0$$
$$\omega\underline{f} + \underline{A}\underline{d} - \underline{c} = 0$$
$$-\omega\underline{c} + \underline{A}\underline{e} = 0 \tag{8.307}$$
$$-\omega\underline{d} + \underline{A}\underline{f} - \underline{e} = 0$$

Since each line in Equations 8.307 is two scalar equations this represents eight equations in eight unknowns, which on the surface of it is a difficult computational challenge. But it is not as formidable as it seems. Our solution strategy begins with the first and third lines, from which we derive

$$(\omega^2\underline{I} + \underline{A}^2)\underline{e} = -\omega\underline{b} \tag{8.308}$$

which we can solve for $\underline{e}$. Then from the third line we have

$$\underline{c} = \underline{A}\underline{e}/\omega \tag{8.309}$$

Now $\underline{e}$ and $\underline{c}$ are known and we are left with the second and fourth lines, which we can write as:

$$\omega\underline{f} + \underline{A}\underline{d} = \underline{c} \tag{8.310}$$

$$\underline{A}\underline{f} - \omega\underline{d} = \underline{e} \tag{8.311}$$

From Equation 8.310,

$$\omega\underline{f} = \underline{c} - \underline{A}\underline{d} \tag{8.312}$$

From Equation 8.311

$$\underline{A}(\omega\underline{f}) = \omega\underline{e} + \omega^2\underline{d} \tag{8.313}$$

Substituting Equation 8.312 into Equation 8.313 and rearranging we find

$$(\omega^2\underline{I} + \underline{A}^2)\underline{d} = \underline{A}\underline{c} - \omega\underline{e} \tag{8.314}$$

which we can solve for $\underline{d}$. Then, from Equation 8.312,

$$\underline{f} = (\underline{c} - A\underline{d})/\omega \qquad (8.315)$$

Equations 8.308, 8.309, 8.314, and 8.315 represent a solution strategy, so we are now prepared to tackle the problem with its specific, given values.

Step 2: Here $\omega = 2$, $A = \begin{pmatrix} 1 & -1 \\ 3 & -2 \end{pmatrix}$, $\underline{b} = \begin{pmatrix} 2 \\ -1 \end{pmatrix}$, so

$$A^2 = \begin{pmatrix} 1 & -1 \\ 3 & -2 \end{pmatrix}\begin{pmatrix} 1 & -1 \\ 3 & -2 \end{pmatrix} = \begin{pmatrix} 1\cdot 1 + (-1)\cdot 3 & 1\cdot(-1) + (-1)\cdot(-2) \\ 3\cdot 1 + (-2)\cdot 3 & 3\cdot(-1) + (-2)\cdot(-2) \end{pmatrix} = \begin{pmatrix} -2 & 1 \\ -3 & 1 \end{pmatrix}$$

$$\omega^2 I + A^2 = \begin{pmatrix} 4-2 & 1 \\ -3 & 4+1 \end{pmatrix} = \begin{pmatrix} 2 & 1 \\ -3 & 5 \end{pmatrix} \qquad (8.316)$$

From Equation 8.308,

$$\begin{pmatrix} 2 & 1 \\ -3 & 5 \end{pmatrix}\begin{pmatrix} e_1 \\ e_2 \end{pmatrix} = -2\begin{pmatrix} 2 \\ -1 \end{pmatrix} = \begin{pmatrix} -4 \\ 2 \end{pmatrix} \qquad (8.317)$$

which has solution

$$\begin{pmatrix} e_1 \\ e_2 \end{pmatrix} = \begin{pmatrix} -22/13 \\ -8/13 \end{pmatrix} \qquad (8.318)$$

From Equation 8.309,

$$\begin{pmatrix} c_1 \\ c_2 \end{pmatrix} = \left(\frac{1}{2}\right)\begin{pmatrix} 1 & -1 \\ 3 & -2 \end{pmatrix}\begin{pmatrix} -22/13 \\ -8/13 \end{pmatrix} = \begin{pmatrix} -7/13 \\ -25/13 \end{pmatrix} \qquad (8.319)$$

From Equations 8.314 and 8.316,

$$\begin{pmatrix} 2 & 1 \\ -3 & 5 \end{pmatrix}\begin{pmatrix} d_1 \\ d_2 \end{pmatrix} = \begin{pmatrix} 1 & -1 \\ 3 & -2 \end{pmatrix}\begin{pmatrix} -7/13 \\ -25/13 \end{pmatrix} - 2\begin{pmatrix} -22/13 \\ -8/13 \end{pmatrix} = \begin{pmatrix} 62/13 \\ 45/13 \end{pmatrix} \qquad (8.320)$$

which has solution

$$\begin{pmatrix} d_1 \\ d_2 \end{pmatrix} = \begin{pmatrix} 265/169 \\ 276/169 \end{pmatrix} \qquad (8.321)$$

From Equation 8.312

$$\begin{pmatrix} f_1 \\ f_2 \end{pmatrix} = \left(\frac{1}{2}\right)\begin{pmatrix} -7/13 \\ -25/13 \end{pmatrix} - \left(\frac{1}{2}\right)\begin{pmatrix} 1 & -1 \\ 3 & -2 \end{pmatrix}\begin{pmatrix} 265/169 \\ 276/169 \end{pmatrix} = \begin{pmatrix} -40/169 \\ -284/169 \end{pmatrix} \qquad (8.322)$$

Therefore the steady-state solution to Equation 8.302 is

$$\underline{x}(t) = (\underline{c}t + \underline{d})\sin 2t + (\underline{e}t + \underline{f})\cos 2t$$

where $\underline{c}, \underline{d}, \underline{e}, \underline{f}$ are given by Equations 8.318, 8.319, 8.321, and 8.322.

**Problems to Solve**

**P8.1** Use the Euler method to compute to four decimal places the first three time steps ($t = 0.001, 0.002, 0.003$) in the numerical integration of

$$\frac{dz}{dt} = (vz)^2 \sin(50t + \pi/4) \qquad z(0) = 3$$

$$\frac{dv}{dt} = -10 \cdot \ln\left(\frac{v}{z}\right)(1 + 20t) \qquad v(0) = 6$$

**P8.2** Use the Euler method to compute to four decimal places the first three time steps ($t = 0.10, 0.20, 0.30$) in the numerical integration of

$$\frac{dz}{dt} = \frac{e^{2v} + e^{-2v}}{e^{2z} - e^{-2z}} \qquad z(0) = 0.4$$

$$\frac{dv}{dt} = -\frac{v^2 + z^2}{e^{3t} + e^{-3t}} \qquad v(0) = 0.2$$

**P8.3** Use the Euler method to compute to four decimal places the first three time steps ($t = 0.05, 0.10, 0.15$) in the numerical integration of

$$\frac{dz}{dt} = -(v \ln z + z \ln v)\ln(2 + t) \qquad z(0) = 5$$

$$\frac{dv}{dt} = (4 + t)^2 \sin\left(\left(\frac{z}{v}\right)\left(\frac{\pi}{4}\right)\right) \qquad v(0) = 7$$

**P8.4** Solve by the eigenvalue/eigenvector method

$$\begin{pmatrix} \dot{x}_1 \\ \dot{x}_2 \end{pmatrix} = \begin{pmatrix} -5 & 1 \\ 2 & -6 \end{pmatrix}\begin{pmatrix} x_1 \\ x_2 \end{pmatrix}, \qquad \begin{pmatrix} x_1(0) \\ x_2(0) \end{pmatrix} = \begin{pmatrix} 2 \\ -1 \end{pmatrix}$$

**P8.5** Solve by the eigenvalue/eigenvector method

$$\begin{pmatrix} \dot{x}_1 \\ \dot{x}_2 \end{pmatrix} = \begin{pmatrix} -5 & -2 \\ 1 & -2 \end{pmatrix}\begin{pmatrix} x_1 \\ x_2 \end{pmatrix}, \qquad \begin{pmatrix} x_1(0) \\ x_2(0) \end{pmatrix} = \begin{pmatrix} 3 \\ 2 \end{pmatrix}$$

**P8.6** Solve by the eigenvalue/eigenvector method

$$\begin{pmatrix} \dot{x}_1 \\ \dot{x}_2 \end{pmatrix} = \begin{pmatrix} 1 & -1 \\ -5 & -3 \end{pmatrix} \begin{pmatrix} x_1 \\ x_2 \end{pmatrix}, \qquad \begin{pmatrix} x_1(0) \\ x_2(0) \end{pmatrix} = \begin{pmatrix} 4 \\ -2 \end{pmatrix}$$

**P8.7** Solve by the eigenvalue/eigenvector method

$$\begin{pmatrix} \dot{x}_1 \\ \dot{x}_2 \end{pmatrix} = \begin{pmatrix} -2 & 5 \\ -1 & -4 \end{pmatrix} \begin{pmatrix} x_1 \\ x_2 \end{pmatrix}, \qquad \begin{pmatrix} x_1(0) \\ x_2(0) \end{pmatrix} = \begin{pmatrix} -1 \\ 3 \end{pmatrix}$$

**P8.8** Solve by the eigenvalue/eigenvector method

$$\begin{pmatrix} \dot{x}_1 \\ \dot{x}_2 \end{pmatrix} = \begin{pmatrix} -6 & 4 \\ -5 & 2 \end{pmatrix} \begin{pmatrix} x_1 \\ x_2 \end{pmatrix}, \qquad \begin{pmatrix} x_1(0) \\ x_2(0) \end{pmatrix} = \begin{pmatrix} -2 \\ 1 \end{pmatrix}$$

**P8.9** Solve by the eigenvalue/eigenvector method

$$\begin{pmatrix} \dot{x}_1 \\ \dot{x}_2 \end{pmatrix} = \begin{pmatrix} -1 & -4 \\ 9 & -1 \end{pmatrix} \begin{pmatrix} x_1 \\ x_2 \end{pmatrix}, \qquad \begin{pmatrix} x_1(0) \\ x_2(0) \end{pmatrix} = \begin{pmatrix} 3 \\ 2 \end{pmatrix}$$

**P8.10** Solve by the eigenvalue/eigenvector method

$$\begin{pmatrix} \dot{x}_1 \\ \dot{x}_2 \end{pmatrix} = \begin{pmatrix} 2 & -5 \\ 5 & -8 \end{pmatrix} \begin{pmatrix} x_1 \\ x_2 \end{pmatrix}, \qquad \begin{pmatrix} x_1(0) \\ x_2(0) \end{pmatrix} = \begin{pmatrix} 2 \\ 4 \end{pmatrix}$$

**P8.11** Solve by the eigenvalue/eigenvector method

$$\begin{pmatrix} \dot{x}_1 \\ \dot{x}_2 \end{pmatrix} = \begin{pmatrix} -9 & 2 \\ -2 & -5 \end{pmatrix} \begin{pmatrix} x_1 \\ x_2 \end{pmatrix}, \qquad \begin{pmatrix} x_1(0) \\ x_2(0) \end{pmatrix} = \begin{pmatrix} 5 \\ -2 \end{pmatrix}$$

**P8.12** By the eigenvalue/eigenvector method, show that the solution to

$$\begin{pmatrix} \dot{x}_1 \\ \dot{x}_2 \\ \dot{x}_3 \end{pmatrix} = \begin{pmatrix} -\gamma & -1 & -1 \\ -1 & -\gamma & -1 \\ -1 & -1 & -\gamma \end{pmatrix} \begin{pmatrix} x_1 \\ x_2 \\ x_3 \end{pmatrix}, \qquad \begin{pmatrix} x_1(0) \\ x_2(0) \\ x_3(0) \end{pmatrix} = \begin{pmatrix} x_{10} \\ x_{20} \\ x_{30} \end{pmatrix}$$

is

$$\begin{pmatrix} x_1(t) \\ x_2(t) \\ x_3(t) \end{pmatrix} = \left( \frac{x_{10} + x_{20} + x_{30}}{3} \right) \begin{pmatrix} 1 \\ 1 \\ 1 \end{pmatrix} e^{-(\gamma+2)t} +$$

$$\left( \frac{x_{10} + x_{20} - 2x_{30}}{3} \right) \begin{pmatrix} 1 \\ 0 \\ -1 \end{pmatrix} e^{-(\gamma-1)t} + \left( \frac{x_{10} - 2x_{20} + x_{30}}{3} \right) \begin{pmatrix} 1 \\ -1 \\ 0 \end{pmatrix} e^{-(\gamma-1)t}$$

Hint: Show that

$$\det(r\underline{I} - \underline{A}) = (r + (\gamma - 1))^2 (r + (\gamma + 2)).$$

**P8.13** By the eigenvalue/eigenvector method, show that the solution to

$$
\begin{pmatrix} \dot{x}_1 \\ \dot{x}_2 \\ \dot{x}_3 \end{pmatrix} = \begin{pmatrix} -\gamma & 1 & -1 \\ 1 & -\gamma & -1 \\ -1 & -1 & -\gamma \end{pmatrix} \begin{pmatrix} x_1 \\ x_2 \\ x_3 \end{pmatrix}, \qquad \begin{pmatrix} x_1(0) \\ x_2(0) \\ x_3(0) \end{pmatrix} = \begin{pmatrix} x_{10} \\ x_{20} \\ x_{30} \end{pmatrix}
$$

is

$$
\begin{pmatrix} x_1(t) \\ x_2(t) \\ x_3(t) \end{pmatrix} = \left( \frac{x_{10} + x_{20} - x_{30}}{3} \right) \begin{pmatrix} 1 \\ 1 \\ -1 \end{pmatrix} e^{-(\gamma - 2)t}
$$

$$
+ \left( \frac{x_{10} + x_{20} + 2x_{30}}{3} \right) \begin{pmatrix} 1 \\ 0 \\ 1 \end{pmatrix} e^{-(\gamma + 1)t} + \left( \frac{x_{10} - 2x_{20} - x_{30}}{3} \right) \begin{pmatrix} 1 \\ -1 \\ 0 \end{pmatrix} e^{-(\gamma + 1)t}
$$

Hint: Show that

$$\det(r\underline{I} - \underline{A}) = (r + (\gamma + 1))^2 (r + (\gamma - 2)).$$

**P8.14** Find the state transition matrix $\underline{\Phi}(t)$ for the system

$$
\begin{pmatrix} \dot{x}_1 \\ \dot{x}_2 \end{pmatrix} = \begin{pmatrix} -5 & -2 \\ 1 & -8 \end{pmatrix} \begin{pmatrix} x_1 \\ x_2 \end{pmatrix}
$$

**P8.15** Find the state transition matrix $\underline{\Phi}(t)$ for the system

$$
\begin{pmatrix} \dot{x}_1 \\ \dot{x}_2 \end{pmatrix} = \begin{pmatrix} -5 & 5 \\ -1 & -3 \end{pmatrix} \begin{pmatrix} x_1 \\ x_2 \end{pmatrix}
$$

**P8.16** Find the state transition matrix $\underline{\Phi}(t)$ for the system

$$
\begin{pmatrix} \dot{x}_1 \\ \dot{x}_2 \end{pmatrix} = \begin{pmatrix} 3 & -4 \\ 4 & -5 \end{pmatrix} \begin{pmatrix} x_1 \\ x_2 \end{pmatrix}
$$

**P8.17** Find a transformation $\underline{z} = \underline{T}\underline{x}$ to convert the system

$$
\dot{\underline{x}} = \begin{pmatrix} -3 & 1 \\ 1 & -3 \end{pmatrix} \underline{x}
$$

into a system in the form

$$\dot{\underline{z}} = \begin{pmatrix} r_1 & 0 \\ 0 & r_2 \end{pmatrix} \underline{z}$$

and show that the system matrix is indeed diagonalized with the transformation.

**P8.18** Find a transformation $\underline{z} = \underline{T}\,\underline{x}$ to convert the system

$$\dot{\underline{x}} = \begin{pmatrix} -5 & -4 \\ 2 & -11 \end{pmatrix} \underline{x}$$

into a system in the form

$$\dot{\underline{z}} = \begin{pmatrix} r_1 & 0 \\ 0 & r_2 \end{pmatrix} \underline{z}$$

and show that the system matrix is indeed diagonalized with the transformation.

**P8.19** Find a transformation $\underline{z} = \underline{T}\,\underline{x}$ to convert the system

$$\dot{\underline{x}} = \begin{pmatrix} -6 & 4 \\ -5 & 2 \end{pmatrix} \underline{x}$$

into a system in the form

$$\dot{\underline{z}} = \begin{pmatrix} r_1 & 0 \\ 0 & r_2 \end{pmatrix} \underline{z}$$

and show that the system matrix is indeed diagonalized with the transformation.

**P8.20** Solve by the matrix kernel method

$$\begin{pmatrix} \dot{x}_1 \\ \dot{x}_2 \end{pmatrix} = \begin{pmatrix} 3 & 2 \\ 7 & -2 \end{pmatrix} \begin{pmatrix} x_1 \\ x_2 \end{pmatrix} + \begin{pmatrix} 2 \\ -1 \end{pmatrix} u_0(t), \qquad \begin{pmatrix} x_1(0) \\ x_2(0) \end{pmatrix} = \begin{pmatrix} 0 \\ 0 \end{pmatrix}$$

where $u_0(t)$ is the unit step function.

**P8.21** Solve by the matrix kernel method

$$\begin{pmatrix} \dot{x}_1 \\ \dot{x}_2 \end{pmatrix} = \begin{pmatrix} -7 & 3 \\ -1 & -3 \end{pmatrix} \begin{pmatrix} x_1 \\ x_2 \end{pmatrix} + \begin{pmatrix} 1 \\ 2 \end{pmatrix} e^{-2t}, \qquad \begin{pmatrix} x_1(0) \\ x_2(0) \end{pmatrix} = \begin{pmatrix} 0 \\ 0 \end{pmatrix}$$

**P8.22** Solve by the matrix kernel method

$$\begin{pmatrix} \dot{x}_1 \\ \dot{x}_2 \end{pmatrix} = \begin{pmatrix} -6 & -3 \\ 1 & -10 \end{pmatrix} \begin{pmatrix} x_1 \\ x_2 \end{pmatrix} + \begin{pmatrix} -1 \\ 3 \end{pmatrix} t, \qquad \begin{pmatrix} x_1(0) \\ x_2(0) \end{pmatrix} = \begin{pmatrix} 0 \\ 0 \end{pmatrix}$$

**P8.23** Solve by the matrix kernel method

$$\begin{pmatrix} \dot{x}_1 \\ \dot{x}_2 \end{pmatrix} = \begin{pmatrix} -8 & -3 \\ 3 & -2 \end{pmatrix} \begin{pmatrix} x_1 \\ x_2 \end{pmatrix} + \begin{pmatrix} 2 \\ 4 \end{pmatrix} e^{-5t}, \qquad \begin{pmatrix} x_1(0) \\ x_2(0) \end{pmatrix} = \begin{pmatrix} 0 \\ 0 \end{pmatrix}$$

**P8.24** Solve by the matrix kernel method

$$\begin{pmatrix} \dot{x}_1 \\ \dot{x}_2 \end{pmatrix} = \begin{pmatrix} -14 & -4 \\ 13 & -6 \end{pmatrix} \begin{pmatrix} x_1 \\ x_2 \end{pmatrix} + \begin{pmatrix} 2 \\ 1 \end{pmatrix} u_0(t), \qquad \begin{pmatrix} x_1(0) \\ x_2(0) \end{pmatrix} = \begin{pmatrix} 0 \\ 0 \end{pmatrix}$$

**P8.25** Solve by the matrix kernel method

$$\begin{pmatrix} \dot{x}_1 \\ \dot{x}_2 \end{pmatrix} = \begin{pmatrix} -3 & 5 \\ -2 & -10 \end{pmatrix} \begin{pmatrix} x_1 \\ x_2 \end{pmatrix} + \begin{pmatrix} 1 \\ 3 \end{pmatrix} e^{-4t} \sin 2t, \qquad \begin{pmatrix} x_1(0) \\ x_2(0) \end{pmatrix} = \begin{pmatrix} 0 \\ 0 \end{pmatrix}$$

**P8.26** Solve by the matrix Laplace transform method

$$\begin{pmatrix} \dot{x}_1 \\ \dot{x}_2 \end{pmatrix} = \begin{pmatrix} -3 & -4 \\ 2 & -9 \end{pmatrix} \begin{pmatrix} x_1 \\ x_2 \end{pmatrix}, \qquad \begin{pmatrix} x_1(0) \\ x_2(0) \end{pmatrix} = \begin{pmatrix} 2 \\ 3 \end{pmatrix}$$

**P8.27** Solve by the matrix Laplace transform method

$$\begin{pmatrix} \dot{x}_1 \\ \dot{x}_2 \end{pmatrix} = \begin{pmatrix} -8 & -3 \\ 6 & -2 \end{pmatrix} \begin{pmatrix} x_1 \\ x_2 \end{pmatrix}, \qquad \begin{pmatrix} x_1(0) \\ x_2(0) \end{pmatrix} = \begin{pmatrix} 3 \\ -1 \end{pmatrix}$$

**P8.28** Solve by the matrix Laplace transform method

$$\begin{pmatrix} \dot{x}_1 \\ \dot{x}_2 \end{pmatrix} = \begin{pmatrix} -2 & 8 \\ -18 & -2 \end{pmatrix} \begin{pmatrix} x_1 \\ x_2 \end{pmatrix} + \begin{pmatrix} 2 \\ -3 \end{pmatrix} u_0(t), \qquad \begin{pmatrix} x_1(0) \\ x_2(0) \end{pmatrix} = \begin{pmatrix} 0 \\ 0 \end{pmatrix}$$

**P8.29** Solve by the matrix Laplace transform method

$$\begin{pmatrix} \dot{x}_1 \\ \dot{x}_2 \end{pmatrix} = \begin{pmatrix} -9 & 5 \\ -9 & 3 \end{pmatrix} \begin{pmatrix} x_1 \\ x_2 \end{pmatrix} + \begin{pmatrix} 2 \\ -2 \end{pmatrix} e^{-3t}, \qquad \begin{pmatrix} x_1(0) \\ x_2(0) \end{pmatrix} = \begin{pmatrix} 0 \\ 0 \end{pmatrix}$$

**P8.30** By the matrix Laplace transform method, find the first component $x_1(t)$ in the solution to

$$\begin{pmatrix} \dot{x}_1 \\ \dot{x}_2 \end{pmatrix} = \begin{pmatrix} -6 & 3 \\ -6 & -12 \end{pmatrix} \begin{pmatrix} x_1 \\ x_2 \end{pmatrix} + \begin{pmatrix} 3 \\ 1 \end{pmatrix} e^{-6t} \sin 2t, \qquad \begin{pmatrix} x_1(0) \\ x_2(0) \end{pmatrix} = \begin{pmatrix} 0 \\ 0 \end{pmatrix}$$

Hint: Use web resources to solve simultaneous linear equations.

**P8.31** By the matrix undetermined coefficients method, find the steady-state solution to

$$\begin{pmatrix} \dot{x}_1 \\ \dot{x}_2 \end{pmatrix} = \begin{pmatrix} -6 & -5 \\ 6 & 3 \end{pmatrix} \begin{pmatrix} x_1 \\ x_2 \end{pmatrix} + \begin{pmatrix} 4 \\ 8 \end{pmatrix} t$$

**P8.32** By the matrix undetermined coefficients method, find the steady-state solution to

$$\begin{pmatrix} \dot{x}_1 \\ \dot{x}_2 \end{pmatrix} = \begin{pmatrix} -3 & -4 \\ 1 & -2 \end{pmatrix} \begin{pmatrix} x_1 \\ x_2 \end{pmatrix} + \begin{pmatrix} 2 \\ -1 \end{pmatrix} e^{6t}$$

**P8.33** By the matrix undetermined coefficients method, find the steady-state solution to

$$\begin{pmatrix} \dot{x}_1 \\ \dot{x}_2 \end{pmatrix} = \begin{pmatrix} -4 & 3 \\ -6 & 2 \end{pmatrix} \begin{pmatrix} x_1 \\ x_2 \end{pmatrix} + \begin{pmatrix} 2 \\ 3 \end{pmatrix} te^{2t}$$

**P8.34** By the matrix undetermined coefficients method, find the steady-state solution to

$$\begin{pmatrix} \dot{x}_1 \\ \dot{x}_2 \end{pmatrix} = \begin{pmatrix} -8 & 7 \\ -5 & 4 \end{pmatrix} \begin{pmatrix} x_1 \\ x_2 \end{pmatrix} + \begin{pmatrix} 3 \\ 4 \end{pmatrix} \sin 4t$$

**P8.35** By the matrix undetermined coefficients method, find the steady-state solution to

$$\begin{pmatrix} \dot{x}_1 \\ \dot{x}_2 \end{pmatrix} = \begin{pmatrix} 2 & -4 \\ 5 & -4 \end{pmatrix} \begin{pmatrix} x_1 \\ x_2 \end{pmatrix} + \begin{pmatrix} 3 \\ -2 \end{pmatrix} e^{2t} \cos 3t$$

## Word Problems

**WP8.1** Three identical spheres are connected by identical thin rods of low mass and specific heat capacity, as shown in Figure 8.30.

The spheres are initially at different temperatures $T_{10}, T_{20}, T_{30}$. The thermal energy in each sphere is give by $E_i = mcT_i$, where $m$ is the mass of each sphere and $c$ is its specific heat capacity. The rate of thermal energy loss or gain in sphere $i$ is given by

$$\frac{dE_i}{dt} = q_i$$

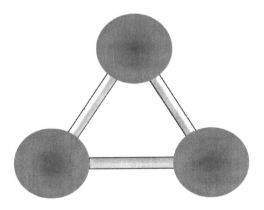

**FIGURE 8.30**   Three spheres in thermal energy exchange.

where $q_i$ is the rate of heat conducted to or from the other spheres, taken together. The rate of heat transferred from sphere $i$ to sphere $j$ is given by

$$q_{ij} = \frac{ka}{l}(T_i - T_j)$$

where:

$k$ is the rod's thermal conductivity

$a$ is the cross-sectional area

$l$ is the length

Determine the temperature as a function of time for each sphere. Show that

$$\begin{pmatrix} T_1 \\ T_2 \\ T_3 \end{pmatrix} = \frac{T_{10} + T_{20} + T_{30}}{3}\begin{pmatrix} 1 \\ 1 \\ 1 \end{pmatrix} + \frac{T_{10} - 2T_{20} + T_{30}}{3}\begin{pmatrix} 1 \\ -1 \\ 0 \end{pmatrix}e^{-3\beta t} + \frac{T_{10} + T_{20} - 2T_{30}}{3}\begin{pmatrix} 1 \\ 0 \\ -1 \end{pmatrix}e^{-3\beta t}$$

where $\beta = ka/mcl$.

**WP8.2** Consider the electrical circuit in Figure 8.31.

The input voltage is $v_{in}(t)$ and the output is the voltage over the capacitor, $v_{out}(t)$. By Kirchhoff's law the sum of the voltage drops around a closed loop is zero. The voltage drop over a resistor is $Ri$, where $i$ is the current running through it. The voltage drop over an inductor is $L(di/dt)$ and the current running through a capacitor is given by $i = C(dv/dt)$, where $v$ is the voltage drop over it. In the left-hand loop of the circuit, the voltage *drop* over the voltage source in the direction of current $i_1$ is $-v_{in}(t)$. If $R = 2$, $L = 1$, and $C = 1$, show that the state equations are

$$\frac{d}{dt}\begin{pmatrix} i_1 \\ v_{out} \end{pmatrix} = \begin{pmatrix} -2 & -1 \\ 1 & -1/2 \end{pmatrix}\begin{pmatrix} i_1 \\ v_{out} \end{pmatrix} + \begin{pmatrix} 1 \\ 0 \end{pmatrix}v_{in}$$

If $v_{in} = 15\sin 2t$, find the steady-state output $v_{out}(t)$.

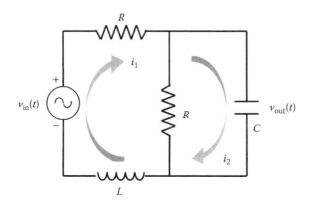

**FIGURE 8.31**  Electrical circuit with input and output voltages.

**WP8.3** Consider again Word Problem WP7.1 in Chapter 7, which analyzed a space-craft landing on an asteroid using thrust vector control. Write the system equations in state space form. Using the parameter values given in part (c) of WP7.1, find the eigenvalues and eigenvectors. Show that the mode associated with the real eigenvalue is a pure rotation of the spacecraft (i.e., that the eigenvector component associated with $v$ is zero). Show that in the oscillatory mode the angle $\theta$ is in phase with the velocity $v$ (i.e., that the ratio of the eigenvector component associated with $\theta$ to the eigenvector component associated with $v$ is a positive real number. This means, physically, that the motions of $\theta$ and $v$ are in unison.)

**WP8.4** The beautiful nighttime displays of the aurora borealis (northern lights) are due to electrically charged particles (mainly electrons and protons) from the Sun's solar wind interacting with molecules in the Earth's upper atmosphere. How do these charged particles move as they approach Earth? The principal force acting on them is the Earth's magnetic field. The particles experience a force

$$\underline{F} = q\underline{V} \times \underline{B}$$

where:
$\underline{F}$ is the force vector
$q$ is the electrical charge on the particle
$\underline{V}$ is the particle's velocity vector
$\underline{B}$ is the Earth's magnetic field vector
"$\times$" denotes vector cross product

Consider motion over a local region in which the magnetic field is constant. If we choose a coordinate system in which the component of the magnetic field is the same for each spatial coordinate

$$B_1 = B_2 = B_3$$

and ignore gravity, the equations of motion for a particle of mass $m$ are

$$\frac{d\underline{x}}{dt} = \underline{A}\underline{x}$$

where

$$\underline{x} = \begin{pmatrix} V_1 \\ V_2 \\ V_3 \end{pmatrix}$$

$$\underline{A} = \begin{pmatrix} 0 & \beta & -\beta \\ -\beta & 0 & \beta \\ \beta & -\beta & 0 \end{pmatrix}$$

$$\beta = \frac{qB_1}{m}$$

Solve the equations and show, based on your results and physical reasoning that a better choice of coordinate system would have had one spatial coordinate aligned with the magnetic field and the other two perpendicular to it, in which case the $\underline{A}$ matrix would have been

$$\underline{A} = \begin{pmatrix} 0 & 0 & 0 \\ 0 & 0 & \sqrt{3}\beta \\ 0 & -\sqrt{3}\beta & 0 \end{pmatrix}$$

Describe the motion in words.

**WP8.5** Show that a symmetric two-dimensional system of the form $\dot{x} = \underline{A}x$, where the components of $\underline{A}$ are real, can have a repeated eigenvalue $r = \lambda$ only if

$$\underline{A} = \begin{pmatrix} \lambda & 0 \\ 0 & \lambda \end{pmatrix}$$

**WP8.6** In this problem we reconsider Example 2 in Section 8.5.1.4,

$$\underline{\dot{x}} = \begin{pmatrix} 1 & 1 & 1 \\ 2 & 1 & -1 \\ -3 & 2 & 4 \end{pmatrix} \underline{x}$$

a system with one eigenvalue and one eigenvector. We showed in Section 8.5.1.4 a solution in terms of a series of matrix equations. Here we ask you to solve the equation using Laplace transforms. You may find it an easier method. We add the initial condition

$$\underline{x}(0) = \underline{x}_0 = \begin{pmatrix} x_{10} \\ x_{20} \\ x_{30} \end{pmatrix}$$

**WP8.7** Explore the dynamics of the nonlinear system

$$\underline{\dot{x}} = \underline{A}(\underline{x}) \cdot \underline{x}$$

where

$$\underline{A}(\underline{x}) = \begin{pmatrix} 1 - (x_1^2 + x_2^2) & 1 \\ -1 & 1 - (x_1^2 + x_2^2) \end{pmatrix}$$

Part (a): Is the origin a stable equilibrium point?

Part (b): Using computer-based numerical integration, examine the long-term behavior of the system. Consider initial conditions for which

$$x_1^2(0) + x_2^2(0) < 1$$

and for which

$$x_1^2(0) + x_2^2(0) > 1$$

Do the trajectories reach steady states? If so, how do the steady-state trajectories depend on initial conditions?

### Challenge Problems

**CP8.1** Two identical masses slide on a frictionless floor as shown in Figure 8.32. They have equilibrium positions $y_{10}$ and $y_{20}$. When out of these equilibrium positions they experience forces due to two identical springs and two identical dashpots.

Each spring exerts a force $F = -kz$ where $z$ is the extension of the spring and each dashpot exerts a damping force $F = -\mu\dot{z}$. The equations of motion are

$$m\ddot{y}_1 = -\mu(\dot{y}_1 - \dot{y}_2) - k((y_1 - y_{10}) - (y_2 - y_{20})) - \mu\dot{y}_1 - k(y_1 - y_{10})$$

$$m\ddot{y}_2 = -\mu(\dot{y}_2 - \dot{y}_1) - k((y_2 - y_{20}) - (y_1 - y_{10}))$$

We are given that the damping force is sufficiently small that the system is underdamped. The challenge is to find the frequencies and damping factors describing the system's motion.

*Guidance*: Put the equations of motion into state space form and find the system's eigenvalues. Let $x_1 = y_1 - y_{10}$, $x_2 = \dot{y}_1$, $x_3 = y_2 - y_{20}$, and $x_4 = \dot{y}_2$. For simplicity define a nondimensional time $t_* = t\sqrt{k/m}$ and a nondimensional parameter $v = \mu/\sqrt{km}$. Then the state space representation is

$$\frac{d\underline{x}}{dt_*} = \underline{A}\,\underline{x}$$

where

$$\underline{A} = \begin{pmatrix} 0 & 1 & 0 & 0 \\ -2 & -2v & 1 & v \\ 0 & 0 & 0 & 1 \\ 1 & v & -1 & -v \end{pmatrix}$$

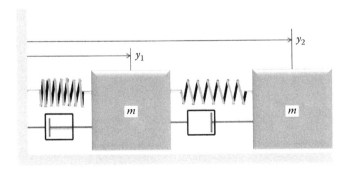

**FIGURE 8.32**    A dual mass–spring–damper system.

Show that the equation $\det(r\underline{I} - \underline{A}) = 0$ leads to

$$r^4 + 3vr^3 + (3 + v^2)r^2 + 2vr + 1 = 0$$

Since the system is underdamped the polynomial can be factored into a form that can be written

$$r^4 + 3vr^3 + (3 + v^2)r^2 + 2vr + 1 = (r^2 + 2\varsigma_1\omega_1 r + \omega_1^2)(r^2 + 2\varsigma_2\omega_2 r + \omega_2^2)$$

Multiply the two factors on the right-hand side and equate coefficients of like powers of $r$ on right- and left-hand sides. The result is four nonlinear equations in four unknowns, but guessing (correctly) that $\omega_1$ and $\omega_2$ are independent of $v$ makes the solution possible. Discuss how the guess is motivated by physical reasoning. Show that

$$\omega_1 = \left(\sqrt{5} + 1\right)\!/2 \qquad \omega_2 = \left(\sqrt{5} - 1\right)\!/2$$

$$\varsigma_1 = v\left(\sqrt{5} + 1\right)\!/4 \qquad \varsigma_2 = v\left(\sqrt{5} - 1\right)\!/4$$

With these parameters in hand, determine the system's dimensional frequencies and damping rates.

**CP8.2** The challenge here is to analyze an aircraft's lateral flight dynamics. The model we consider is simplified in that it assumes no translation to either side of the line marking the aircraft's intended direction. Figures 8.33 and 8.34 define the aircraft heading, or yaw angle, $\psi$, and roll angle $\phi$. The aircraft experiences no moment due to the roll angle itself, but it does with respect to the roll angle rate $\dot{\phi} = p$. The yaw rate $\dot{\psi} = r$ is also important in the dynamics.

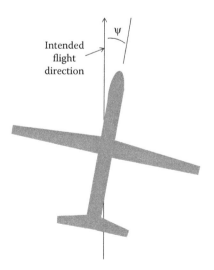

**FIGURE 8.33** Defining the yaw angle $\psi$.

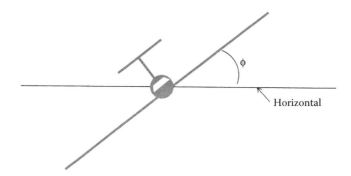

**FIGURE 8.34** Defining the roll angle $\phi$.

The equations of motion are

$$\begin{pmatrix} \dot\psi \\ \dot r \\ \dot p \end{pmatrix} = \begin{pmatrix} 0 & 1 & 0 \\ N_\psi & N_r & N_p \\ L_\psi & L_r & L_p \end{pmatrix} \begin{pmatrix} \psi \\ r \\ p \end{pmatrix}$$

where $N_\psi, L_\psi$, and so on are the lateral "aerodynamic derivatives," which depend on aircraft design and flight condition. For a Boeing 747 aircraft in final approach, [16] gives the following values for the derivatives: $N_\psi = -0.3349$, $N_r = -0.2314$, $N_p = -0.0933$, $L_\psi = 1.535$, $L_r = 0.2468$, and $L_p = -1.0994$. Determine the eigenvalues and eigenvectors for this problem. Show that a real eigenvalue pertains to a motion that involves primarily the roll rate. Show that the other eigenvalues pertain to a lightly damped oscillatory mode. By the way, the oscillatory mode is called "Dutch roll," given the similarity of the motion to that of an ice skater, swinging side-to-side.

*Guidance*: Search first for the real root, using the iterative method described in Solved Problem 6.1. Let $P(s)$ denote the polynomial whose roots you seek. When you have found a trial $s_n$ such that $P(s_n)$ is close to zero, let the next trial solution $s_{n+1}$ be given by the solution to

$$P(s_n) + \frac{dP}{ds}(s_n)(s_{n+1} - s_n) = 0$$

Once you have the real root $s_*$ find $P(s)/(s - s_*)$ by long division. Check your results with an online cubic equation calculator.

**CP8.3** Consider a particle moving under gravity $g$ in a frictionless hemispherical shell with radius $R$. Let $x$ and $y$ be the coordinates of the particle's position projected onto the horizontal plane. Defining the state vector as

$$\underline{x} = \begin{pmatrix} x \\ \dot x \\ y \\ \dot y \end{pmatrix}$$

it can be shown that the equations of motion are

$$\dot{\underline{x}} = \underline{A}(\underline{x}) \cdot \underline{x}$$

where

$$\underline{A}(\underline{x}) = \begin{pmatrix} 0 & 1 & 0 & 0 \\ -\Omega2(\underline{x}) & 0 & 0 & 0 \\ 0 & 0 & 0 & 1 \\ 0 & 0 & -\Omega2(\underline{x}) & 0 \end{pmatrix}$$

and

$$\Omega2(\underline{x}) = \left(\frac{g}{R}\right)\Gamma(\underline{x}) + \frac{1}{R^2}\frac{\Sigma(\underline{x})}{x_1^2 + x_3^2}$$

$$\Sigma(\underline{x}) = \frac{(x_1 x_2 + x_3 x_4)^2}{\Gamma^2(\underline{x})} + (x_3 x_2 - x_1 x_4)^2$$

$$\Gamma(\underline{x}) = \sqrt{1 - \left(\frac{x_1^2 + x_3^2}{R^2}\right)}$$

Explore the dynamics of this system. Specifically:

a. What are the equations of motion when $x_1^2 + x_3^2 \ll R^2$ and $x_2^2 + x_4^2 \ll gR$? What are the system dynamics in this case?

b. Using computer-based numerical integration, explore the changes in the trajectory of the particle as nonlinearity is increased. Run each simulation for the period $0 \le t \le 20\sqrt{R/g}$. Choose whatever units you desire for $g$ and any value of $R$ satisfying $g/100 < R < 100g$. Explore the effect of increasing nonlinearity in the following two studies:

    i.  Set as initial condition

$$\underline{x}(0) = \begin{pmatrix} x_0 \\ 0 \\ 0 \\ (x_0/2)\sqrt{g/R} \end{pmatrix}$$

    Plot the particle's trajectory in the $(x_1, x_3)$ plane. What specific aspect in the shape of the trajectory changes most significantly as the value of $x_0$ increases as prescribed in Table 8.6 below?

    ii. Set as initial condition

$$\underline{x}(0) = \begin{pmatrix} x_0 \\ 0 \\ 0 \\ px_0\sqrt{g/R} \end{pmatrix}$$

**TABLE 8.6**

**Initial Positions for Study of**
**Particle Motion in a Frictionless**
**Hemispherical Bowl**

| Case | $x_0/R$ |
|------|---------|
| 1 | 1/50 |
| 2 | 1/5 |
| 3 | 1/2 |
| 4 | 4/5 |
| 5 | 9/10 |

and find the value of $p$ creating a circle in the $(x_1, x_3)$ plane for each case in Table 8.6. You are given that $p$ is essentially unity for Cases 1 and 2. A caution: setting too high a value of $p$ may cause the modeled particle to shoot out of the bowl and the equation for $\Gamma(x)$ to fail.

**CP8.4** Consider a particle moving under gravity $g$ in a shallow bowl with constant radius of curvature $R$; that is, in a slice of a spherical shell. The bowl is rotating with angular frequency $\omega$.

Examine the motion of the particle from the point of view of an observer sitting on the bowl. Let $x$ and $y$ be the coordinates of the particle's position projected onto the rotating horizontal plane. Define the state vector as

$$\underline{x} = \begin{pmatrix} x \\ \dot{x} \\ y \\ \dot{y} \end{pmatrix}$$

*Part* (a): Let the bowl be frictionless. Then it can be shown that, from the point of view of the observer in the rotating frame of reference, the particle obeys the linearized equations of motion

$$\dot{\underline{x}} = \underline{A}\underline{x}$$

where

$$\underline{A} = \begin{pmatrix} 0 & 1 & 0 & 0 \\ \omega^2 - g/R & 0 & 0 & 2\omega \\ 0 & 0 & 0 & 1 \\ 0 & -2\omega & \omega^2 - g/R & 0 \end{pmatrix}$$

Show that the eigenvalues of the system are

$$r = \pm i \left( \left( \frac{g}{R} \right)^{1/2} \pm \omega \right)$$

When $\omega = \sqrt{g/R}$, what does the eigenvalue $r = 0$ imply about possible solution trajectories?

Using computer-based numerical integration, explore the trajectory of the particle in the rotating frame of reference as the rate of rotation $\omega$ is increased. Choose whatever units you desire for $g$ and any value of $R$ satisfying $g/100 < R < 100g$. For an initial condition, assume that the particle is placed with zero velocity *relative to inertial space* at $x_1(0) = R/5$ and $x_3(0) = R/5$. This means the observer on the rotating bowl would see an initial velocity $x_2(0) = \omega R/5$ and $x_4(0) = -\omega R/5$. Use the computer to numerically integrate the system equations for the period $0 \le t \le 50\sqrt{R/g}$. Plot the particle trajectory in the $x_1, x_3$ phase plane for $\omega/\sqrt{g/R} = 0.5$ and $1.5$. For comparison, look at the particle trajectory in the inertial plane (by programming the computer as just described but with $\omega = 0$).

*Part* (b): Let the bowl and particle have the coefficient of friction $\mu$. Then the equations of motion become

$$\underline{\dot{x}} = \underline{A}\underline{x} - \frac{\mu g}{v(\underline{x})} \underline{B}\underline{x}$$

where $v(\underline{x}) = \sqrt{x_2^2 + x_4^2}$ and

$$\underline{B} = \begin{pmatrix} 0 & 0 & 0 & 0 \\ 0 & 1 & 0 & 0 \\ 0 & 0 & 0 & 0 \\ 0 & 0 & 0 & 1 \end{pmatrix}$$

As in Part (a), assume that the particle is placed at $x_1(0) = R/5$ and $x_3(0) = R/5$ with initial velocity $x_2(0) = \omega R/5$ and $x_4(0) = -\omega R/5$. Use the computer to numerically integrate the system equations for the period $0 \le t \le 50\sqrt{R/g}$. Plot the particle trajectory in the $x_1, x_3$ phase plane for $\omega/\sqrt{g/R} = 0.5$ and $1.5$ and $\mu = 0.05$. For a fixed value of $\mu$, determine empirically (i.e., by computer experiment) the critical value of $\omega/\sqrt{g/R}$ (call it $\omega_C^*$) such that the particle comes to rest inside the bowl if $\omega/\sqrt{g/R} < \omega_C^*$ and is flung outside it if $\omega/\sqrt{g/R} > \omega_C^*$. Plot $\omega_C^*$ versus $\mu$ for $0.025 \le \mu \le 0.7$.

A note of caution: You will have to adjust the system equations to avoid asking the computer to divide by zero. Let

$$\underline{\dot{x}} = \underline{A}\underline{x} - \frac{\mu g}{v(\underline{x}) + \varepsilon} \underline{B}\underline{x}$$

where $\varepsilon$ is as small as possible; for example, $10^{-8}$. Show empirically that neither equilibrium positions nor $\omega_c^*$ depend on $\varepsilon$ if it is small enough.

*Part* (c): Consider the case $\omega = \sqrt{g/R}$ for the conditions addressed in Part (b). Let $\varepsilon = 0$. Show analytically that if the particle is given an initial speed $V_0$ relative to the rotating platform it comes to rest at a time $t_E = V_0/\mu g$. (Hint: examine $x_2 \dot{x}_2 + x_4 \dot{x}_4$). Verify your result via computer simulation.

**CP8.5** The challenge here is to derive the differential equations describing perturbations from straight and level flight in an aircraft's longitudinal dynamics, as given in Section 8.5.1.3.

*Guidance*: Begin with translational equations given in Computing Project B, Phase 1 (at the end of Chapter 3), augmented with the equation for rotation in the pitch plane (aircraft nose up and down):

$$\dot{V} = (T \cos\alpha - D - W \sin\gamma)/m$$

$$\dot{\gamma} = (T \sin\alpha + L - W \cos\gamma)/mV$$

$$\dot{q} = M/I_p$$

Here, $M$ is the torque and $I_p$ the moment of inertia in the pitch plane. See Section 8.5.1.3 or the Computing Project B, Phase 1, for definitions of other terms. Assume that the nominal trajectory is straight and level, constant-speed flight with negligible steady-state angle of attack.

Write $\alpha = \gamma - \theta$ and $W = mg$. Assume perturbations $u, \gamma, \theta$ and $q$ about the nominal values. Assume thrust, mass, and gravity are constant. Show that, in the steady state, $T = D_0$ and $W = L_0$. Given that

$$L = \frac{1}{2}\rho S V^2 C_L$$

$$D = \frac{1}{2}\rho S V^2 C_D$$

where $\rho$ and $S$ are constant, and given that the lift and drag coefficients are well described by

$$C_L = C_{L0} + C_{L\alpha}\alpha$$

$$C_D = C_{D0} + C_{D\alpha}\alpha$$

write *first-order* expressions for the right- and left-hand sides of the first two basic equations. That is, ignore squares or cross products of perturbation values. Given that

$$T << \frac{1}{2}\rho SV^2 C_{L\alpha}$$

show that your first two first-order equations reduce to the first two equations in Equations 8.77, where we identify

$$L_\alpha = \frac{\rho SV_0 C_{L\alpha}}{2m} \qquad D_\alpha = \frac{\rho SV_0 C_{D\alpha}}{2m}$$

Modeling the pitch moment $M$ in physical detail would take us beyond the scope of this text, but it suffices for our purposes to note that the moment can be approximated by an expression in the form

$$M = \left(\frac{\partial M}{\partial \alpha}\right)\alpha + \left(\frac{\partial M}{\partial q}\right)q$$

where the partial derivatives are constant. Show that, if we identify

$$M_\alpha = \frac{1}{I_p}\left(\frac{\partial M}{\partial \alpha}\right) \qquad M_q = \frac{1}{I_p}\left(\frac{\partial M}{\partial q}\right)$$

then the final two equations of Equations 8.77 follow.

**CP8.6** As discussed in Section 8.5.1.9, if an $n$-dimensional system LTI system has fewer than $n$ independent eigenvectors, then part of the system can be transformed into one or more series of cascaded identical first-order systems. This follows from the representation of any matrix in Jordan canonical form. For $n$-by-$n$ matrices with $n$ independent eigenvectors the Jordan canonical form is a diagonalized matrix. For $n$-by-$n$ matrices with $n - p$ independent eigenvectors the Jordan canonical form has an $n - p$ by $n - p$ segment that is diagonalized and in the rest, $p$ locations where a one appears above the diagonal. The challenge here is to analyze an $m$th order series of cascaded identical first-order systems, as shown in Figure 8.35.

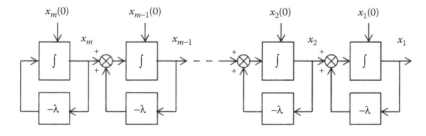

**FIGURE 8.35** Cascaded identical first-order systems.

The equations are

$$\dot{x}_m = -\lambda x_m$$

$$\dot{x}_n = -\lambda x_n + x_{n+1} \qquad 1 \le n \le m-1$$

Show that the series has:

a.  One eigenvalue
b.  One eigenvector
c.  Transition matrix $\Phi_{ij}(t) = 0$ if $j < i$ and $\Phi_{ij}(t) = e^{-\lambda t} t^{j-i}/(j-i)!$ if $j \ge i$

# 9 Partial Differential Equations

In this chapter, our march toward higher and higher dimensions reaches an end: systems modeled by partial differential equations (PDEs) can be considered infinite dimensional. You will recall from the opening chapter that PDEs have two or more independent variables, whereas ordinary differential equations (ODEs) have only one. We will see how the additional independent variables are a logical extension of an $n$-dimensional state vector.

PDEs play many key roles in engineering. We will have the opportunity here to meet only a few of the most important: the heat equation, the wave equation, Laplace's equation, and the beam equation. We will also see that PDE problems can be formulated in different ways. Our study of ODEs focused on problems with *initial conditions*, where the value of the dependent variable and its derivatives are specified at a given moment in time. Most of the PDEs we will consider portray variation in both time and space. *Boundary conditions* describe values of the dependent variable and its derivatives at given spatial locations, independent of time. Some PDE problems come with initial conditions, some with boundary conditions and some with both.

Another objective of this chapter is to introduce different forms that PDEs and their solutions can take depending on the geometry of the problem. Some important engineering problems require cylindrical or spherical coordinates. We will use separation of variables to solve these problems and in the process encounter several particularly significant ODEs with variable coefficients. We could have studied these consequential ODEs earlier but, in keeping with our focus on applications, we address them in the setting in which they naturally arise.

## 9.1 HEAT EQUATION

Figure 9.1 depicts a long rod of uniform cross section of area $A$ through which heat is flowing. Assume that the rod is insulated against heat loss to the environment along its length.

We divide the rod into segments of length $\Delta x$. Consider the interaction of segments $n-1$, $n$, and $n+1$. Energy is conserved, so in time $\Delta t$ the energy flowing into segment $n$ from segment $n-1$ equals the energy flowing out of segment $n$ into segment $n+1$ plus an addition to the energy stored in segment $n$:

$$q_{in}\Delta t = q_{out}\Delta t + \Delta e_{stored}$$

Rearranging:

$$q_{in} - q_{out} = \frac{\Delta e_{stored}}{\Delta t} \tag{9.1}$$

Fourier's law of heat conduction is

$$q = kA\frac{\Delta T}{\Delta x} \tag{9.2}$$

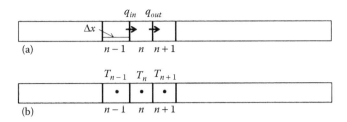

**FIGURE 9.1** Heat transfer in a long rod: (a) heat flowing in rod and (b) temperature variation around segment $n$.

where:
    $T$ is the temperature
    $k$ is the thermal conductivity, a property of the rod material

Here

$$q_{\text{in}} = kA\frac{T_{n-1} - T_n}{\Delta x} \tag{9.3}$$

$$q_{\text{out}} = kA\frac{T_n - T_{n+1}}{\Delta x} \tag{9.4}$$

The change in energy stored in segment $n$ results in a temperature change there:

$$\Delta e_{\text{stored}} = \rho c A \Delta x \Delta T_n \tag{9.5}$$

where:
    $\rho$ is the density of the rod material
    $c$ is its specific heat

Combining Equations 9.1 through 9.5 and rearranging we have

$$\frac{T_{n+1} - 2T_n + T_{n-1}}{\Delta x^2} = \left(\frac{\rho c}{k}\right)\frac{\Delta T_n}{\Delta t} \tag{9.6}$$

In the limit as $\Delta x$ and $\Delta t$ go to zero, Equation 9.6 becomes

$$\frac{\partial^2 T}{\partial x^2} = \left(\frac{\rho c}{k}\right)\frac{\partial T}{\partial t} \tag{9.7}$$

Equation 9.7 is the so-called heat equation, one of the fundamental equations of the subject of heat transfer. It is usually written as

$$\frac{\partial^2 T}{\partial x^2} = \frac{1}{\alpha}\frac{\partial T}{\partial t} \tag{9.8}$$

where $\alpha$ is called the *thermal diffusivity*.

    The heat equation is classified as a *parabolic* PDE, which for our purposes distinguishes it from hyperbolic PDE's and elliptic PDE's, which we will meet later. For now it suffices to say that solutions to parabolic, hyperbolic, and elliptic equations have very different properties. Parabolic PDEs portray motion that we call diffusion. Diffusion spreads and levels a distribution, smoothing out peaks and valleys.

## 9.1.1 HEAT EQUATION IN INITIAL VALUE PROBLEMS

As we discussed briefly in Chapter 1, an initial value problem is a differential equation combined only with initial conditions; in the case of the heat equation, a specification of the spatial distribution of the temperature at $t = 0$. This section examines the heat equation in initial value problems and shows a strong analogy with results we have seen in ODEs.

### 9.1.1.1 Heat Equation in an Example Initial Value Problem

Consider a very long rod initially at a uniform temperature $T_0$, to the center of which is applied a large amount of heat in a very short time, so that the temperature spikes in a very small region $\Delta x$ around the center and is essentially unchanged everywhere else. Assume that the rod is insulated against heat loss to the environment along its length. Figure 9.2 graphs the temperature distribution across the bar at $t = 0^+$.

For simplicity let $T_0 = 0$. Formally, we can state the problem using the delta function:

$$\frac{\partial^2 T}{\partial x^2} = \frac{1}{\alpha} \frac{\partial T}{\partial t} \qquad T(x,0) = T_s \Delta x \delta(x) \qquad (9.9)$$

Until the temperature at the ends of the rod begins to rise appreciably, this is an initial value problem. For a period of time the dynamics depend only on initial conditions, not on what is happening at the boundaries. For all intents and purposes initially, the rod could be infinite in length.

The student can easily verify by direct substitution that the solution to Equation 9.9 is

$$T(x,t) = \frac{1}{2\sqrt{\pi \alpha t}} \exp\left(-x^2/4\alpha t\right)\left(T_S \Delta x\right) \qquad (9.10)$$

Note that $T_s \Delta x$ is just the coefficient of the delta function in the initial condition. Deriving Equation 9.10 from first principles requires techniques beyond the scope of this text. We examine this problem because, as we will show, it illustrates that some initial value problems in PDEs are strongly analogous to initial value problems in ODEs. Figure 9.3 graphs $T(x,t)$ as it evolves over time.

It can further be shown that for an infinite length rod and an arbitrary initial temperature distribution $T(x,0) = T_0(x)$ the solution to the heat equation is

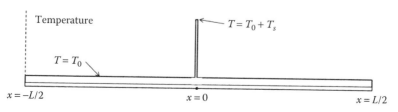

**FIGURE 9.2** Delta function initial temperature distribution across a very long rod.

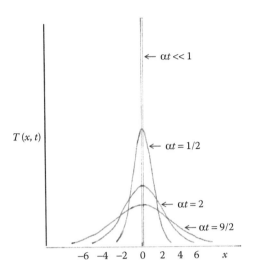

**FIGURE 9.3**   Solution to heat equation with delta function initial condition.

$$T(x,t) = \int_{-\infty}^{\infty} \Gamma(x - x',t)T_0(x')dx'$$

where:

$$\Gamma(x - x',t) = \frac{1}{2\sqrt{\pi\alpha t}} \exp\left(-(x - x')^2/4\alpha t\right) \tag{9.11}$$

Compare Equations 9.10 and 9.11. More generally yet, if there are heat sources (let $q_S(x,t)$ be the energy added per unit volume and unit time), then the heat equation is

$$\frac{\partial^2 T}{\partial x^2} + \frac{q_S(x,t)}{k} = \frac{1}{\alpha}\frac{\partial T}{\partial t}$$

and the solution is

$$T(x,t) = \int_{-\infty}^{\infty} \Gamma(x - x',t)T_0(x')dx' + \int_0^t \int_{-\infty}^{\infty} \Gamma(x - x',t - \tau)\left(\frac{q_S(x',\tau)}{k}\right)dx'd\tau \tag{9.12}$$

where $\Gamma(x,t)$ is again given by Equation 9.11.

Equation 9.12 is completely analogous to solutions of initial value problems in ODEs. The function $\Gamma(x,t)$ is equivalent to the state transition matrix. To demonstrate the analogy, we divide the rod into $N$ segments and identify

$$T(x_i,t) \rightarrow T_i(t)$$

$$\Gamma(x_i - x_j',t)\Delta x \rightarrow \Phi_{ij}(t)$$

$$\frac{q_S(x_j,\tau)}{k} \rightarrow \underline{B}\underline{u} \rightarrow u_j(\tau)$$

Making these substitutions in Equation 9.12, we arrive at

$$T_i(t) = \sum_{j=1}^{N} \Phi_{ij}(t)T_{0j} + \int_0^t \sum_{j=1}^{N} \Phi_{ij}(t-\tau)u_j(\tau)d\tau$$

which of course can be written in matrix notation as

$$\underline{T}(t) = \underline{\Phi}(t)\underline{T}_0 + \int_0^t \underline{\Phi}(t-\tau)\underline{u}(\tau)d\tau$$

Compare with Equation 8.147. Note that the additional independent variable $x$ can be thought of as a state vector of infinite dimension.

### 9.1.1.2 Existence and Uniqueness of Solutions to Parabolic Partial Differential Equations in Initial Value Problems

Consider the more general linear parabolic equation

$$\frac{\partial y}{\partial t} + \frac{\partial}{\partial x}(ay) - \frac{\partial^2}{\partial x^2}(by) + cy = f \quad 0 \le t \le t_F \quad y(x,0) = p(x) \tag{9.13}$$

where:

$$a = a(x,t)$$

$$b = b(x,t)$$

$$c = c(x,t)$$

$$f = f(x,t)$$

and $t_F$ is finite and there is no boundary on $x$. Then under certain conditions on $a,b,c,f$, and $P$, smoothness conditions and bounds akin to those imposed on ODEs in Section 8.2, it can be shown that there exists a unique solution to Equation 9.13 [17].

### 9.1.2 HEAT EQUATION IN BOUNDARY VALUE PROBLEMS

Figure 9.4 depicts a long rod with arbitrary initial temperature distribution $T_0(x)$. Let the temperature at the rod's endpoints be held at $T = 0$ for $0 < t < t_F$ and assume that the rod is

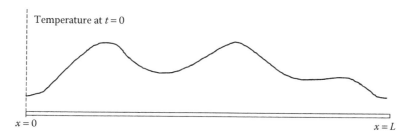

**FIGURE 9.4** Long rod with initial temperature distribution and boundary conditions $T = 0$ at $x = 0$ and $x = L$ for $t > 0$.

insulated against heat loss to the environment along its length. We seek to determine the temperature distribution across the rod as a function of time.

The equations are

$$\frac{\partial^2 T}{\partial x^2} = \frac{1}{\alpha} \frac{\partial T}{\partial t}$$

$$T(x,0) = T_0(x) \tag{9.14}$$

$$T(0,t) = T(L,t) = 0$$

We try the solution

$$T(x,t) = \Phi(x)\Theta(t) \tag{9.15}$$

You may immediately object based on your physical intuition that the solution cannot have so simple a form and you would be correct. In the end the solution will be a *combination* of terms in the form of Equation 9.15. For now we substitute Equation 9.15 into the PDE in Equation 9.14 and find

$$\frac{\partial^2}{\partial x^2}\left(T(x,t)\right) = \left(\frac{d^2}{dx^2}\Phi(x)\right)\Theta(t) = \frac{1}{\alpha}\frac{\partial}{\partial t}\left(T(x,t)\right) = \frac{1}{\alpha}\Phi(x)\left(\frac{d}{dt}\Theta(t)\right)$$

Dividing through by $T(x,t) = \Phi(x)\Theta(t)$ leads to

$$\frac{1}{\Phi(x)}\frac{d^2\Phi(x)}{dx^2} = \frac{1}{\alpha\Theta(t)}\frac{d\Theta(t)}{dt} \tag{9.16}$$

The left-hand side of Equation 9.16 is a function only of $x$ and the right side a function only of $t$, so the two sides must equal a constant. We will later determine that this constant must be a negative number:

$$\frac{1}{\Phi(x)}\frac{d^2\Phi(x)}{dx^2} = \frac{1}{\alpha\Theta(t)}\frac{d\Theta(t)}{dt} = -k \quad k > 0$$

Then we have two *ordinary* differential equations:

$$\frac{d^2\Phi(x)}{dx^2} + k\Phi(x) = 0 \tag{9.17}$$

$$\frac{d\Theta(t)}{dt} + \alpha k\Theta(t) = 0 \tag{9.18}$$

We know that the general solutions to these two equations are

$$\Phi(x) = c_1 \cos\sqrt{k}x + c_2 \sin\sqrt{k}x \tag{9.19}$$

$$\Theta(t) = c_3 e^{-\alpha k t} \tag{9.20}$$

To meet the boundary condition $T(0,t) = T(L,t) = 0$ we must have

$$\Phi(0) = c_1 \cos(0) + c_2 \sin(0) = 0 \tag{9.21}$$

$$\Phi(L) = c_1 \cos(\sqrt{k}L) + c_2 \sin(\sqrt{k}L) = 0 \tag{9.22}$$

From Equation 9.21, $c_1 = 0$. From Equation 9.22, since $c_2$ cannot be zero,

$$\sin \sqrt{k}L = 0 \tag{9.23}$$

Note that if we had chosen $k$ to be negative in Equation 9.17 then Equations 9.21 and 9.22 would have become

$$\Phi(0) = c_1 e^{\sqrt{k}(0)} + c_2 e^{-\sqrt{k}(0)} = c_1 + c_2 = 0$$

$$\Phi(L) = c_1 e^{-\sqrt{k}L} + c_2 e^{\sqrt{k}L} = 0$$

which would have required both $c_1$ and $c_2$ to be zero, which is not a viable solution. Returning to Equation 9.23, we see that we must have

$$\sqrt{k}L = n\pi$$

$$k = \left(\frac{n\pi}{L}\right)^2 \tag{9.24}$$

These values of $k$ are the *eigenvalues* of the problem. There are an infinite number of them. This is more evidence that the additional independent variable $x$ can be considered a state vector of infinite dimension.

Given Equation 9.15 and Equations 9.19 through 9.24, the *eigenfunctions* of the problem stated by Equation 9.14 are

$$\Theta_n(t)\Phi_n(x) = c_{2n}c_{3n}e^{-\alpha(n\pi/L)^2 t} \sin\left(n\pi x/L\right)$$

$$= c_n e^{-\alpha(n\pi/L)^2 t} \sin\left(n\pi x/L\right) \tag{9.25}$$

We have no reason to choose one value of $n$ over another. Therefore we try a sum over all $n$:

$$T(x,t) = \sum_{n=1}^{\infty} c_n e^{-\alpha(n\pi/L)^2 t} \sin\left(n\pi x/L\right) \tag{9.26}$$

The final condition we must meet is $T(x,0) = T_0(x)$, or

$$\sum_{n=1}^{\infty} c_n \sin\left(n\pi x/L\right) = T_0(x) \tag{9.27}$$

Can the coefficients $c_n$ be chosen to satisfy Equation 9.27? Under relatively weak conditions on the initial distribution $T_0(x)$ (a sufficient condition is that it and its derivative be piecewise continuous) the answer is yes.

To determine the coefficient $c_m$ we multiply Equation 9.27 by $\sin(m\pi x/L)$ and integrate from $x = 0$ to $x = L$:

$$\int_0^L \sum_{n=1}^{\infty} c_n \sin\left(n\pi x/L\right)\sin\left(m\pi x/L\right)dx = \int_0^L T_0(x)\sin\left(m\pi x/L\right)dx \tag{9.28}$$

The set of functions $\sin(n\pi x/L)$ is orthogonal, meaning that

$$\int_0^L \sin\left(n\pi x/L\right)\sin\left(m\pi x/L\right)dx = 0 \qquad m \neq n \tag{9.29}$$

Here

$$\int_0^L \sin\left(m\pi x/L\right)\sin\left(m\pi x/L\right)dx = L/2 \tag{9.30}$$

From Equations 9.28 to 9.30,

$$c_m = \frac{2}{L}\int_0^L T_0(x)\sin\left(m\pi x/L\right)dx \tag{9.31}$$

Equations 9.27 and 9.31 together define a *Fourier series*.

### Example:

Consider a long rod of length $L$ whose ends are controlled to temperature $T=0$ and to which is applied a point heat source at $x = L/4$, raising the temperature there to $+T_1$, and a point heat sink at $x = 3L/4$, reducing the temperature there to $-T_1$. We state without proof that the steady-state solution to the heat equation in this case is as shown in Figure 9.5.

At time $t = 0$, the heat source and sink are removed, whereas the ends of the rod continue to be controlled to a temperature of zero. The steady-state temperature profile in the instants before $t = 0$ becomes the initial condition for the dynamic behavior we want to analyze. Our task is to determine the temperature distribution across the rod as a function of time for $t \geq 0$.

From Equations 9.26 and 9.31,

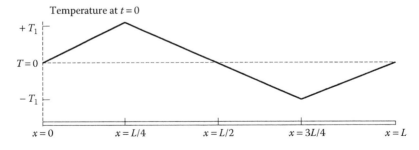

**FIGURE 9.5**   Rod with heat source and sink in steady state prior to $t = 0$.

$$T(x,t) = \sum_{n=1}^{\infty} c_n e^{-\alpha(n\pi/L)^2 t} \sin(n\pi x/L) \tag{9.32}$$

$$c_n = \frac{2}{L} \int_0^L T_0(x) \sin(n\pi x/L) dx \tag{9.33}$$

and all that remains is to compute the integral in Equation 9.33. Now $T_0(x)$, the temperature distribution at time $t = 0$, is an odd function about the midpoint of the rod. Since $\sin(\pi x/L)$, $\sin(3\pi x/L)$, $\sin(5\pi x/L)$, and so on are even functions about the midpoint, $c_n = 0$ for odd $n$. Figure 9.6 helps us see this. For instance, the product $T_0(x)\sin(\pi x/L)$ equals $T_1/\sqrt{2}$ at $x = L/4$ and $-T_1/\sqrt{2}$ at $x = 3L/4$. For every product at $x = L/2 - w$ there is another of the opposite sign at $x = L/2 + w$.

To compute $c_n$ for $n$ even,

$$c_n = \frac{2}{L} \int_0^L T_0(x) \sin(n\pi x/L) dx$$

$$= 2T_1 \int_0^{L/4} (4x/L) \sin(n\pi x/L) dx/L$$

$$+ 2T_1 \int_{L/4}^{3L/4} (2 - 4x/L) \sin(n\pi x/L) dx/L$$

$$+ 2T_1 \int_{3L/4}^{L} (4x/L - 4) \sin(n\pi x/L) dx/L$$

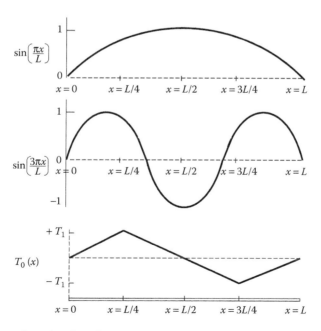

**FIGURE 9.6** Comparing sine functions even about $x = L/2$ with initial temperature distribution $T_0(x)$ odd about $x = L/2$.

The computations are long but straightforward. We find that $c_n = 0$ for $n$ any multiple of four. The only nonzero terms are

$$c_{4n+2} = (-1)^n \frac{8}{\pi^2} T_1 \frac{1}{(2n+1)^2} \quad n = 0,1,2,\ldots \tag{9.34}$$

From Equations 9.32 and 9.34,

$$T(x,t) = \frac{8}{\pi^2} T_1 \sum_{n=0}^{\infty} \frac{(-1)^n}{(2n+1)^2} e^{-\alpha((4n+2)\pi/L)^2 t} \sin\big((4n+2)\pi x/L\big) \tag{9.35}$$

As always after a long computation we should seek ways to check our results. Let us evaluate Equation 9.35 at the point $x = L/4$, $t = 0$, where we should find $T(L/4,0) = T_1$.

Now $\sin((n + 1/2)\pi) = (-1)^n$, so Equation 9.35 becomes

$$T(L/4,0) = \frac{8}{\pi^2} T_1 \sum_{n=0}^{\infty} \frac{1}{(2n+1)^2} \tag{9.36}$$

In yet another result attributed to Leonhard Euler, the sum of the inverse squares of the odd integers is

$$\sum_{n=0}^{\infty} \frac{1}{(2n+1)^2} = \frac{\pi^2}{8} \tag{9.37}$$

Hence $T(L/4,0) = T_1$, as it should.

It takes quite a few terms of the infinite series to achieve an accurate approximation when $t = 0$. However, for even a relatively small lapse of time the first term in the series dominates all others. Consider just the first three terms at $x = L/4$:

$$T\big(L/4,t\big) \cong \frac{8}{\pi^2} T_1 \left[ e^{-4\pi^2 \alpha t/L^2} + \frac{1}{9} e^{-36\pi^2 \alpha t/L^2} + \frac{1}{25} e^{-100\pi^2 \alpha t/L^2} \right] \tag{9.38}$$

If only enough time has elapsed to reduce the first exponential term to

$$e^{-4\pi^2 \alpha t/L^2} = 0.8$$

then

$$\frac{1}{9} e^{-36\pi^2 \alpha t/L^2} = \frac{1}{9}(0.8)^9 = 0.0149$$

and

$$\frac{1}{25} e^{-100\pi^2 \alpha t/L^2} = \frac{1}{25}(0.8)^{25} = 0.0002$$

The spatial distribution across the bar is then well represented by $\sin(2\pi x/L)$. Clearly, the diffusion represented by the heat equation quickly smoothes out peaks.

Two key lessons from this section are

1. In the mechanical engineering discipline of heat transfer, solving the heat equation often involves separation of variables and expansion in a Fourier series.
2. It is typically the case that, after just a brief lapse of time, engineers can employ only the first term in the infinite series as an accurate representation of the temperature distribution.

## 9.2   WAVE EQUATION AND POWER SERIES SOLUTIONS

The wave equation in three-dimensional Cartesian coordinates is

$$\frac{\partial^2 u}{\partial x^2} + \frac{\partial^2 u}{\partial y^2} + \frac{\partial^2 u}{\partial z^2} - \frac{1}{a^2}\frac{\partial^2 u}{\partial t^2} = 0 \tag{9.39}$$

It describes oscillations propagating through space, for example: electromagnetic waves, flexure of membranes, waves on water surfaces, and acoustic vibrations. The parameter $a$ in Equation 9.39 is the speed of propagation.

The wave equation is called a hyperbolic PDE. The terms *parabolic, hyperbolic,* and *elliptic* were originally assigned to classify PDEs with two independent variables written in the form

$$A\frac{\partial^2 u}{\partial x^2} + B\frac{\partial^2 u}{\partial x \partial t} + C\frac{\partial^2 u}{\partial t^2} + D\frac{\partial u}{\partial t} + E\frac{\partial u}{\partial x} + F = 0 \tag{9.40}$$

with the classification depending on the values of the coefficients $A, B,$ and $C$. The classification names were chosen by analogy with conic sections, those geometrical figures formed in the intersection of a plane and a cone: the parabola, hyperbola, and ellipse. But we need not dwell on the geometrical analogy: if the coefficients were themselves functions of $x$ and $t$ the classification could change point-to-point, and mathematicians apply the terms today to a much wider range of PDEs, including some with higher numbers of independent variables and some with higher-order derivatives. What engineers need to understand is how the behaviors of these three types of PDEs differ. We will see examples of all three types in this chapter.

### 9.2.1   WAVE EQUATION IN INITIAL VALUE PROBLEMS

This section presents examples of the wave equation in initial value problems; that is, when boundary conditions do not affect the solution. The examples are intended to give the student a clear sense of the nature and behavior of wave equation solutions at locations distant from any source or boundary. They are not intended as a lesson in how to solve PDEs; in fact our mathematical treatments of them are necessarily incomplete.

We will be more thorough in our treatment of the wave equation in boundary value problems, but solutions there are typically more difficult to visualize.

### 9.2.1.1 Traveling Plane Wave

Consider an electromagnetic wave propagating in free space (a vacuum). Maxwell's equations show that each component of the electric field vector satisfies

$$\frac{\partial^2 E_i}{\partial x^2} + \frac{\partial^2 E_i}{\partial y^2} + \frac{\partial^2 E_i}{\partial z^2} - \frac{1}{c^2}\frac{\partial^2 E_i}{\partial t^2} = 0 \tag{9.41}$$

where $x$, $y$, and $z$ are the spatial coordinates, $t$ is time, and $c$ is a constant of nature equal to $3 \times 10^8$ m/s. Now assume that the wave is far from its transmitter and any other boundary. To solve Equation 9.41 we try the form

$$E_i(x, y, z, t) = f(w)$$
$$w = k_1 x + k_2 y + k_3 z + k_4 t \tag{9.42}$$

Then

$$\frac{\partial E_i}{\partial x} = \frac{df}{dw} \cdot \frac{\partial w}{\partial x} = k_1 \frac{df}{dw}$$

$$\frac{\partial^2 E_i}{\partial x^2} = k_1^2 \frac{d^2 f}{dw^2}$$

Similarly,

$$\frac{\partial^2 E_i}{\partial y^2} = k_2^2 \frac{d^2 f}{dw^2}, \qquad \frac{\partial^2 E_i}{\partial z^2} = k_3^2 \frac{d^2 f}{dw^2}, \qquad \frac{\partial^2 E_i}{\partial t^2} = k_4^2 \frac{d^2 f}{dw^2}$$

Then Equation 9.41 becomes

$$\left( k_1^2 + k_2^2 + k_3^2 - \frac{k_4^2}{c^2} \right) \frac{d^2 f}{dw^2} = 0$$

The function $f(w)$ can be of any arbitrary form if

$$k_1^2 + k_2^2 + k_3^2 - k_4^2/c^2 = 0 \tag{9.43}$$

If we *set* $k_4 = \pm c$ then Equation 9.43 becomes

$$k_1^2 + k_2^2 + k_3^2 = 1$$

and the argument $w$ in Equation 9.42 becomes either

$$w = k_1 x + k_2 y + k_3 z - ct \tag{9.44}$$

or

$$w = k_1 x + k_2 y + k_3 z + ct \tag{9.45}$$

Equations 9.44 and 9.45 define *plane waves* traveling at speed $c$ either in the direction of the unit vector

$$\underline{k} = \begin{pmatrix} k_1 \\ k_2 \\ k_3 \end{pmatrix}$$

or opposite it. If the electric field component has the waveform

$$E_i(x, y, z, 0) = f(k_1 x + k_2 y + k_3 z)$$

at time $t = 0$ it satisfies

$$E_i(x, y, z, t) = f_1(k_1 x + k_2 y + k_3 z - ct) + f_2(k_1 x + k_2 y + k_3 z + ct) \qquad (9.46)$$

for $t > 0$, where

$$f_1(w) + f_2(w) = f(w)$$

The locus of points where the argument $w$ equals a constant is called a *wavefront*. Here, as we have said, the wavefront is a geometric plane.

Figure 9.7 depicts a case where there is only an outgoing wave; that is, $f_2 = 0$. The figure graphs the electric field as the plane wave travels along the spatial $x$-axis in time (i.e., $k_1 = 1$, $k_2 = 0$, and $k_3 = 0$), also showing the wave's presence along the spatial $y$-axis. The wave's presence along the spatial $z$-axis is not shown. The vertical axis in Figure 9.7 is not the spatial $z$-axis; it is the scale for the value of the electric field. Remember that the electric field vector does not have spatial extent; it is an attribute of a point in space and time.

Equation 9.46 is useful as a local solution, but it is only approximate. The next section takes another step toward an accurate solution by assuming a specific transmitted waveform and relating the wave to the location of the transmitter.

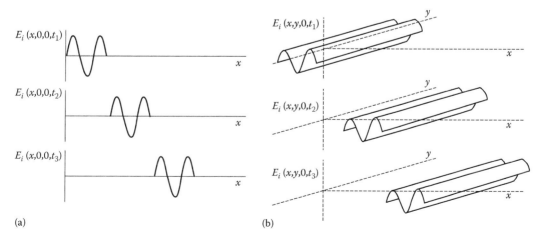

(a)                                                        (b)

**FIGURE 9.7**   A traveling plane wave at three different instants in time ($t_1 < t_2 < t_3$): (a) showing the wave along the $x$-axis only and (b) showing the wave along the $x$- and $y$- axes.

### 9.2.1.2  Traveling Wave in Spherical Coordinates

Figure 9.8 depicts a wavefront in an electromagnetic wave emanating from a distant transmitter. At a large distance the transmitter's antenna pattern is not locally observable; the wave appears to be uniform in azimuth and elevation. Hence, we can describe the value of the electric field locally by the wave equation in spherical coordinates assuming radial symmetry. Stated without proof, the wave equation is then

$$\frac{\partial^2 E_i}{\partial R^2} + \frac{2}{R}\frac{\partial E_i}{\partial R} - \frac{1}{c^2}\frac{\partial^2 E_i}{\partial t^2} = 0 \tag{9.47}$$

We assume that the transmitter is emitting a sinusoidal signal of radian frequency $\omega$, and hence that the electric field can be represented at a distance $R$ from the transmitter as

$$E_i(R,t) = \Phi_1(R)\cos\omega t + \Phi_2(R)\sin\omega t \tag{9.48}$$

Substituting Equation 9.48 into Equation 9.47 we find that

$$\left(\frac{d^2\Phi_1}{dR^2} + \frac{2}{R}\frac{d\Phi_1}{dR} + \left(\frac{\omega}{c}\right)^2\Phi_1\right)\cos\omega t + \left(\frac{d^2\Phi_2}{dR^2} + \frac{2}{R}\frac{d\Phi_2}{dR} + \left(\frac{\omega}{c}\right)^2\Phi_2\right)\sin\omega t = 0 \tag{9.49}$$

Since Equation 9.49 must be true for all $t$, it implies that

$$\frac{d^2\Phi_n}{dR^2} + \frac{2}{R}\frac{d\Phi_n}{dR} + k^2\Phi_n = 0 \tag{9.50}$$

for $n = 1, 2$, where we have substituted $k = \omega/c$. Letting

**FIGURE 9.8**    An electromagnetic wavefront at a large distance from its transmitter.

$$\Phi_n(R) = \frac{\Psi_n(R)}{R} \tag{9.51}$$

we have

$$\frac{d\Phi_n}{dR} = -\frac{1}{R^2}\Psi_n + \frac{1}{R}\frac{d\Psi_n}{dR} \tag{9.52}$$

$$\frac{d^2\Phi_n}{dR^2} = \frac{2}{R^3}\Psi_n - \frac{2}{R^2}\frac{d\Psi_n}{dR} + \frac{1}{R}\frac{d^2\Psi_n}{dR^2} \tag{9.53}$$

Substituting Equations 9.51, 9.52, and 9.53 into Equation 9.50 we find, after some cancellations, that $\Psi_n$ satisfies the familiar equation

$$\frac{d^2\Psi_n}{dR^2} + k^2\Psi_n = 0 \tag{9.54}$$

Hence

$$\Psi_1 = A_1\cos kR + B_1\sin kR$$
$$\Psi_2 = A_2\cos kR + B_2\sin kR \tag{9.55}$$

Using Equations 9.48, 9.51, and 9.55

$$E_i(R,t) = \left(\left(A_1\cos kR + B_1\sin kR\right)\cos\omega t + (A_2\cos kR + B_2\sin kR)\sin\omega t\right)\Big/R \tag{9.56}$$

Recalling the trigonometric identities

$$\cos kR\cos\omega t = \left(\cos(kR+\omega t) + \cos(kR-\omega t)\right)/2$$

$$\sin kR\cos\omega t = \left(\sin(kR+\omega t) + \sin(kR-\omega t)\right)/2$$

$$\cos kR\sin\omega t = \left(\sin(kR+\omega t) - \sin(kR-\omega t)\right)/2$$

$$\sin kR\sin\omega t = \left(-\cos(kR+\omega t) + \cos(kR-\omega t)\right)/2$$

and using, as we have often done before,

$$c_1\cos\alpha + c_2\sin\alpha = G\sin(\alpha - \theta)$$

where

$$G = \sqrt{c_1^2 + c_2^2}$$

$$\theta = \arctan(c_1, c_2)$$

Equation 9.56 becomes

$$E_i(R,t) = G_1\cos(kR-\omega t-\theta_1)/R + G_2\cos(kR+\omega t-\theta_2)/R \tag{9.57}$$

where

$$G_1 = \sqrt{(A_1 + B_2)^2 + (B_1 - A_2)^2} \Big/ 2$$

$$G_2 = \sqrt{(A_1 - B_2)^2 + (B_1 + A_2)^2} \Big/ 2$$

$$\theta_1 = \arctan(B_1 - A_2, A_1 + B_2)$$

$$\theta_2 = \arctan(B_1 + A_2, A_1 - B_2)$$

Equation 9.57 represents two traveling waves, one headed away from the transmitter and the other toward. Choosing $A_1 = B_2$ and $B_1 = -A_2$ nulls $G_2$, leaving only the outward-bound wave, which is the correct result for a signal from a distant transmitter. This can be proved by solving the nonhomogeneous problem, considering the radiation's source, which is how this problem is normally posed. For simplicity we have posed it as homogeneous; considering the physics of the antenna and its radiation properties is beyond the scope of this text.

Note that Equation 9.57 improves on the plane-wave model (Equation 9.46) by correctly showing that the magnitude of the electric field diminishes by $1/R$ as a function of the wave's distance from the transmitter. Note also that the argument of the cosine in the outgoing wave can be written as $kR - \omega t - \theta_1 = k(R - ct) - \theta_1$, showing that the wave's speed of propagation is $c$.

Again, the purpose of these traveling wave examples was to illustrate to students the nature and behavior of certain solutions to the wave equation. Our mathematical treatment has been necessarily incomplete. In the next section we will return to our primary focus on how to solve differential equations.

### 9.2.2 Wave Equation in Boundary Value Problems

The wave equation describes the dynamics of many phenomena and the derivation of the equation differs for each of them. Here we consider a thin elastic membrane under tension, fixed securely around its circumference (e.g., a drum head). (A membrane is a sheet with no resistance to bending. A steel plate, for example, is not a membrane.) The membrane is distorted by a force and then the force is removed. What are the subsequent vertical motions?

#### 9.2.2.1 Derivation of the Wave Equation for a Taut Membrane

The vertical displacement of the membrane is given by $u(x, y, t)$. We divide the membrane into infinitesimally small cells of width $\Delta x$ and length $\Delta y$. The mass of each cell is $\rho \Delta x \Delta y$. By Newton's law the displacement of the cell at location $x, y$ is given by

$$\rho \Delta x \Delta y \frac{\partial^2 u}{\partial t^2} = F(x, y) \tag{9.58}$$

The vertical force on the cell is the sum of the vertical forces on its four sides, each due to the tension $T$ in the membrane having a vertical component. See Figure 9.9.

The force due to tension on the left side of cell $i, j$, at location $x, y$, is $T \Delta y$. The vertical component of this force (assuming small angles) is

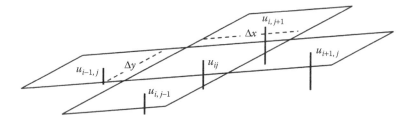

**FIGURE 9.9** Displacement of adjoining cells in membrane.

$$-T\Delta y\left(\frac{u_{ij}-u_{i-1,j}}{\Delta x}\right)$$

Continuing in this way around the four sides of cell $i, j$ we arrive at

$$F(x,y)=T\Delta y\left(\frac{u_{i+1,j}-u_{i,j}}{\Delta x}-\frac{u_{i,j}-u_{i-1,j}}{\Delta x}\right)+T\Delta x\left(\frac{u_{i,j+1}-u_{i,j}}{\Delta y}-\frac{u_{i,j}-u_{i,j-1}}{\Delta y}\right) \quad (9.59)$$

Combining Equations 9.58 and 9.59, dividing by $\Delta x\Delta y$ and letting $\Delta x$ and $\Delta y$ go to zero, we have

$$\frac{\partial^2 u}{\partial x^2}+\frac{\partial^2 u}{\partial y^2}-\left(\frac{\rho}{T}\right)\frac{\partial^2 u}{\partial t^2}=0 \quad (9.60)$$

This is the wave equation in two dimensions and Cartesian coordinates. The ratio $T/\rho = a^2$, which we will take to be constant, is the square of the speed of propagation of vertical displacements across the membrane.

### 9.2.2.2 Solving the Wave Equation in a Boundary Value Problem in Cylindrical Coordinates

The motions of a taut *circular* membrane fixed at its boundary $r_0$ are best described by the wave equation in cylindrical coordinates:

$$\frac{\partial^2 u}{\partial r^2}+\frac{1}{r}\frac{\partial u}{\partial r}+\frac{1}{r^2}\frac{\partial^2 u}{\partial \varphi^2}-\frac{1}{a^2}\frac{\partial^2 u}{\partial t^2}=0 \quad (9.61)$$

(Stated without proof.)

A drumhead is the first application that comes to mind for this model, excited by a drumstick strike. We could focus our attention on this specific application and learn a great deal, but we will learn even more, in particular about the mathematical power inherent here, if we add an initial value aspect to the problem. So, though it may seem somewhat artificial, we consider a membrane initially subject to a uniform force across it that causes the steady-state displacement

$$u(r,0)=h\left(1-r^2/r_0^2\right) \quad (9.62)$$

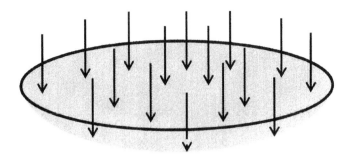

**FIGURE 9.10**   Membrane under steady-state uniform force.

See Figure 9.10.

Since the membrane is in steady state prior to $t = 0$ we have

$$\frac{\partial u}{\partial t}(r,0) = 0 \tag{9.63}$$

If the force immediately ceases, subsequent motion of the membrane is governed by Equation 9.61, the wave equation in cylindrical coordinates, with Equations 9.62 and 9.63 as initial conditions and

$$u(r_0,\varphi,t) = 0 \tag{9.64}$$

as boundary condition.

We can see that in this case there is no dependence on azimuth, so

$$\frac{\partial^2 u}{\partial \varphi^2} = 0 \tag{9.65}$$

but to demonstrate the form of the equations when there *is* an initial variation of the displacement with azimuth we will not immediately invoke Equation 9.65 here.

To solve the general problem we let

$$u(r,\varphi,t) = R(r)\Phi(\varphi)T(t) \tag{9.66}$$

and try to separate variables. Substituting Equation 9.66 into Equation 9.61 we have

$$\Phi T \frac{d^2 R}{dr^2} + \Phi T \frac{1}{r}\frac{dR}{dr} + RT \frac{1}{r^2}\frac{d^2\Phi}{d\varphi^2} = R\Phi \frac{1}{a^2}\frac{d^2 T}{dt^2}$$

Dividing through by $R\Phi T$ leaves

$$\frac{1}{R}\frac{d^2 R}{dr^2} + \frac{1}{R}\frac{1}{r}\frac{dR}{dr} + \frac{1}{\Phi}\frac{1}{r^2}\frac{d^2\Phi}{d\varphi^2} = \frac{1}{T}\frac{1}{a^2}\frac{d^2 T}{dt^2} \tag{9.67}$$

The right side of Equation 9.67 is a function of $t$ only and the left side a function of $r$ and $\varphi$ only, hence both sides must equal a constant. It will turn out that to satisfy boundary conditions this constant must be negative. Call it $-k^2$. Then the right side of Equation 9.67 becomes

$$\frac{d^2T}{dt^2} + a^2k^2T = 0 \tag{9.68}$$

We know that the solution of Equation 9.68 is

$$T(t) = c_1 \cos(akt) + c_2 \sin(akt) \tag{9.69}$$

Since the membrane is initially at rest, from Equation 9.63 there is no time derivative at $t = 0$ so we must have $c_2 = 0$. Also, since we will be solving for constants to meet initial conditions in the equations for $R(r)$ and $\Phi(\varphi)$ we can, without loss of generality, set $c_1 = 1$. Then

$$T(t) = \cos(akt) \tag{9.70}$$

Remember, $a$ is given but we do not yet know what $k$ is.

The left-hand side of Equation 9.66 must also equal $-k^2$:

$$\frac{1}{R}\frac{d^2R}{dr^2} + \frac{1}{R}\frac{1}{r}\frac{dR}{dr} + \frac{1}{\Phi}\frac{1}{r^2}\frac{d^2\Phi}{d\varphi^2} = -k^2$$

$$r^2\frac{1}{R}\frac{d^2R}{dr^2} + r\frac{1}{R}\frac{dR}{dr} + \frac{1}{\Phi}\frac{d^2\Phi}{d\varphi^2} = -k^2r^2$$

$$r^2\frac{1}{R}\frac{d^2R}{dr^2} + r\frac{1}{R}\frac{dR}{dr} + k^2r^2 = -\frac{1}{\Phi}\frac{d^2\Phi}{d\varphi^2} \tag{9.71}$$

The right-hand side of Equation 9.71 is a function only of $\varphi$ and the left-hand side is a function only of $r$ hence both sides must equal a constant. We would soon find that to satisfy boundary conditions it must be positive. Call it $b$. This leads to

$$-\frac{1}{\Phi}\frac{d^2\Phi}{d\varphi^2} = b$$

$$\frac{d^2\Phi}{d\varphi^2} + b\Phi = 0 \tag{9.72}$$

Equation 9.72 has solution

$$\Phi(\varphi) = c_3 \cos(\sqrt{b}\varphi) + c_4 \sin(\sqrt{b}\varphi) \tag{9.73}$$

But by the nature of the azimuth coordinate $\varphi$ we must have

$$\Phi(0) = \Phi(2\pi) \tag{9.74}$$

From Equations 9.73 and 9.74, we must have $\sqrt{b} = m$, an integer. Then the left-hand side of Equation 9.71 becomes

$$r^2 \frac{1}{R}\frac{d^2R}{dr^2} + r\frac{1}{R}\frac{dR}{dr} + k^2r^2 = m^2$$

$$r^2 \frac{d^2R}{dr^2} + r\frac{dR}{dr} + (k^2r^2 - m^2)R = 0$$

$$\frac{d^2R}{dr^2} + \frac{1}{r}\frac{dR}{dr} + \left(k^2 - \frac{m^2}{r^2}\right)R = 0 \tag{9.75}$$

We now introduce a new dimensionless independent variable $x = kr$. (Do not confuse $x = kr$ with the Cartesian coordinate $x$.) This change of variable transforms Equation 9.75 into

$$\frac{d^2R}{dx^2} + \frac{1}{x}\frac{dR}{dx} + \left(1 - \frac{m^2}{x^2}\right)R = 0 \tag{9.76}$$

Equation 9.76 is known as *Bessel's equation of order m*. The solution with finite value and finite derivative at the origin is called the *Bessel function of the first kind of order m* and denoted as $J_m(x)$. A later section will discuss the family of Bessel equations and functions more generally. We see that higher-order ($m > 0$) Bessel equations arise when we have cylindrical geometry but not azimuthal symmetry.

In our problem (Equations 9.61, 9.62, and 9.63) there is no dependence on azimuth, so Equation 9.76 becomes *Bessel's equation of order zero*:

$$\frac{d^2R}{dx^2} + \frac{1}{x}\frac{dR}{dx} + R = 0 \tag{9.77}$$

The solution to Equation 9.77 with finite value and derivative at the origin is the *Bessel function of the first kind of order zero*, which, as we will show momentarily, is defined by its power series:

$$R(x) = J_0(x) = \sum_{n=0}^{\infty} \frac{(-1)^n}{(n!)^2}\left(\frac{x}{2}\right)^{2n} \tag{9.78}$$

But there is also a boundary condition. We must have $R(kr_0) = 0$ that implies

$$J_0(kr_0) = 0 \tag{9.79}$$

Equation 9.79 tells us that not all values of $k$ will work. Those that do are called eigenvalues $k_n$ and the corresponding functions $J_0(k_nr)$ called *eigenfunctions*. The first three zeros of $J_0(x)$ are approximately $x_1 = 2.405$, $x_2 = 5.520$, and $x_3 = 8.654$ and for $n > 3$ the zeros are given, approximately but accurately, by $x_n = n\pi - \pi/4$.

Recall that we selected a negative number $(-k^2)$ as the separation constant in Equation 9.67. If we had selected a positive number $(+k^2)$ Equation 9.77 would have read

$$\frac{d^2R}{dx^2} + \frac{1}{x}\frac{dR}{dx} - R = 0$$

and its solutions could not have met the condition $R(kr_0) = 0$. See Problem P9.5 in the Problems to Solve section.

Recalling that $u(r,t) = R(r)T(t)$ and that $T(t) = \cos(akt)$ and also recognizing that an infinite number of eigenvalues $k_n = x_n/r_0$ exist, we have

$$u(r,t) = \sum_{n=0}^{\infty} c_n J_0\left(x_n r/r_0\right)\cos\left(x_n at/r_0\right) \tag{9.80}$$

We determine the coefficients $c_n$ from the initial condition $u(r,0) = h(1 - r^2/r_0^2)$, which implies that

$$\sum_{n=0}^{\infty} c_n J_0(x_n r/r_0) = h(1 - r^2/r_0^2) \tag{9.81}$$

Recall that in Section 9.1.2, in a problem defined along one Cartesian spatial axis with boundary conditions at each end, we sought coefficients $c_n$ to satisfy

$$\sum_{n=1}^{\infty} c_n \sin\left(\frac{n\pi x}{L}\right) = f_0(x)$$

where $f_0(x)$ was a given function. We learned that under relatively weak smoothness conditions on $f_0(x)$ we could indeed accomplish this, because the sine functions are orthogonal under a certain integral; that is,

$$\int_0^L \sin\left(\frac{n\pi x}{L}\right)\sin\left(\frac{m\pi x}{L}\right)dx = 0 \quad m \neq n$$

$$\int_0^L \sin\left(\frac{n\pi x}{L}\right)\sin\left(\frac{m\pi x}{L}\right)dx = \frac{L}{2} \quad m = n$$

That is also the case in this boundary value problem with cylindrical geometry. We can find $c_n$ to satisfy Equation 9.81 because Bessel functions also form an orthogonal set:

$$\int_0^{r_0} J_0(x_n r/r_0) J_0(x_m r/r_0)r\,dr = 0 \quad m \neq n$$

$$\int_0^{r_0} J_0(x_n r/r_0)J_0(x_m r/r_0)r\,dr = \frac{r_0^2}{2} J_1^2(x_n) \quad m = n$$

where $J_1(x)$ is the Bessel function of the first kind of order 1. (See Equations 9.1.28 and 11.4.5 of Reference [18], available on the Internet.) To find $c_n$ we multiply Equation 9.81 by $J_0(x_m r/r_0)$ and integrate:

$$\sum_{n=0}^{\infty} c_n J_0(x_n r/r_0)J_0(x_m r/r_0) = h(1 - r^2/r_0^2)J_0(x_m r/r_0)$$

$$\sum_{n=0}^{\infty} c_n \int_0^{r_0} J_0\left(x_n r/r_0\right)J_0\left(x_m r/r_0\right)r\,dr = \int_0^{r_0} h\left(1 - r^2/r_0^2\right)J_0(x_m r/r_0)r\,dr$$

$$c_m \left( \frac{r_0^2}{2} J_1^2(x_m) \right) = \int_0^{r_0} h\left(1 - r^2/r_0^2\right) J_0\left(x_m r/r_0\right) r \, dr$$

$$c_m = \frac{\int_0^{r_0} h\left(1 - r^2/r_0^2\right) J_0\left(x_m r/r_0\right) r \, dr}{\left( \dfrac{r_0^2}{2} J_1^2(x_m) \right)} \tag{9.82}$$

Challenge Problem CP9.1 guides the student through this calculation, which is not as difficult as it may seem. The result is

$$c_m = h \frac{8}{x_m^3 J_1(x_m)} \tag{9.83}$$

Using Equations 9.80 and 9.83, the membrane displacement, the solution to Equations 9.61 through 9.65, is

$$u(r,t) = h \sum_{n=0}^{\infty} \left( \frac{8}{x_n^3 J_1(x_n)} \right) J_0\left(x_n r/r_0\right) \cos\left(x_n at/r_0\right) \tag{9.84}$$

We can check this result by evaluating the solution at $r = 0$ and $t = 0$, where according to the initial condition Equation 9.62 we should have $u(0,0) = h$. That is, we expect

$$\lim_{n\to\infty} \sum_{m=1}^{n} c_m = h \tag{9.85}$$

where $c_m$ is given by Equation 9.83. Table 9.1 records the partial sums. It appears the series is indeed converging to the correct result.

At this point you may well be daunted by the apparent complexity of the wave equation in cylindrical coordinates, with its unfamiliar mathematical solutions. Do not be; one of the

**TABLE 9.1**

**Checking the Accuracy of the Wave Equation Solution**

| $n$ | $x_n$ | $J_1(x_n)$ | $c_n/h$ | $\sum_{m=1}^{n} c_m/h$ |
|---|---|---|---|---|
| 1 | 2.405 | 0.5191 | 1.1079 | 1.1079 |
| 2 | 5.520 | −0.3403 | −0.1398 | 0.9681 |
| 3 | 8.654 | 0.2714 | 0.0455 | 1.0136 |
| 4 | 11.792 | −0.2325 | −0.0210 | 0.9926 |
| 5 | 14.931 | 0.2065 | 0.0116 | 1.0042 |
| 6 | 18.071 | −0.1877 | −0.0072 | 0.9970 |
| 7 | 21.212 | 0.1733 | 0.0048 | 1.0018 |
| 8 | 24.352 | −0.1617 | −0.0034 | 0.9984 |
| 9 | 27.493 | 0.1522 | 0.0025 | 1.0009 |
| 10 | 30.635 | −0.1442 | −0.0019 | 0.9990 |

objectives of this chapter is to demonstrate that Bessel functions are useful tools for engineers and almost as easy to use as the sine and exponential functions. One reason for this is that, as we shall see, simple approximations are available that are readily computed by hand. An even greater reason is the existence of substantial resources on the Internet. Finally, for vibration problems of the sort we are considering here, engineers are often concerned only with eigenvalues. That is, for many practical engineering applications it suffices to know the natural frequencies of the system. In such problems it is unnecessary to calculate the coefficients $c_m$.

### 9.2.3 POWER SERIES SOLUTIONS: AN INTRODUCTION

Our objective here is to show how to solve problems exemplified by Bessel's equation of order zero:

$$\frac{d^2 y}{dx^2} + \frac{1}{x}\frac{dy}{dx} + y = 0 \tag{9.86}$$

Equation 9.86 is a linear ODE with variable coefficients. Earlier in the text we called such an equation "time-varying," although time is not the independent variable here.

It suffices to solve Equation 9.86 with the initial condition $y(0) = 1$. We will have more to say about the initial value of the derivative momentarily.

In general we are unable to solve equations with variable coefficients in terms of standard functions like the sine and exponential, as we were able to do with linear equations with constant coefficients. We solve Equation 9.86 in terms of an infinite series:

$$y(x) = \sum_{n=0}^{\infty} c_n x^n \tag{9.87}$$

The initial condition $y(0) = 1$ implies that

$$c_0 = 1 \tag{9.88}$$

Differentiating Equation 9.88 term-by-term,

$$\frac{1}{x}\frac{dy}{dx} = \frac{1}{x}\sum_{n=1}^{\infty} n c_n x^{n-1} = \sum_{n=1}^{\infty} n c_n x^{n-2} \tag{9.89}$$

$$\frac{d^2 y}{dx^2} = \sum_{n=2}^{\infty} n(n-1) c_n x^{n-2} \tag{9.90}$$

Substituting Equations 9.87, 9.89, and 9.90 into Equation 9.86 we have

$$\sum_{n=2}^{\infty} n(n-1) c_n x^{n-2} + \sum_{n=1}^{\infty} n c_n x^{n-2} + \sum_{n=0}^{\infty} c_n x^n = 0$$

$$\sum_{n=2}^{\infty} n^2 c_n x^{n-2} - \sum_{n=2}^{\infty} n c_n x^{n-2} + \left( \sum_{n=2}^{\infty} n c_n x^{n-2} + c_1 x^{-1} \right) + \sum_{n=0}^{\infty} c_n x^n = 0$$

$$\sum_{n=2}^{\infty} n^2 c_n x^{n-2} + \sum_{n=0}^{\infty} c_n x^n + c_1 x^{-1} = 0 \tag{9.91}$$

For Equation 9.91 to be true for all $x$, the coefficients of each power must be zero. From this fact we can immediately see that

$$c_1 = 0 \tag{9.92}$$

which implies that

$$\frac{dy}{dx}(0) = 0$$

We are left with

$$\sum_{n=2}^{\infty} n^2 c_n x^{n-2} + \sum_{n=0}^{\infty} c_n x^n = 0 \tag{9.93}$$

The next step is an important one in solving this problem and others of its type; we must manipulate the indices so that Equation 9.93 can be written as a single sum:

$$\sum_{m=0}^{\infty} k_m x^m = 0$$

To do this in this case we must replace $n-2$ with $m$ in the first sum in Equation 9.93 and $n$ with $m$ in the second. The result is

$$\sum_{m=0}^{\infty} \left( (m+2)^2 c_{m+2} + c_m \right) x^m = 0 \tag{9.94}$$

Again, for this equation to be true for all $x$ we must have the coefficient of each power of $x$ equal to zero. This implies the *recursion relation*

$$c_{m+2} = -\frac{1}{(m+2)^2} c_m \tag{9.95}$$

Using Equations 9.88, 9.92, and 9.95, we find that $c_m = 0$ for odd $m$ and

$$c_2 = -\frac{1}{(0+2)^2} c_0 = -\frac{1}{2^2}$$

$$c_4 = -\frac{1}{(2+2)^2} c_2 = \frac{1}{2^4 2^2} \tag{9.96}$$

$$c_6 = -\frac{1}{(4+2)^2} c_4 = -\frac{1}{2^6 (3 \cdot 2)^2}$$

In general, the solution to Equation 9.86 with $y(0) = 1$ can be written as

$$y(x) = \sum_{n=0}^{\infty} \frac{(-1)^n}{(n!)^2} \left(\frac{x}{2}\right)^{2n} \tag{9.97}$$

or as

$$y(x) = J_0(x) \tag{9.98}$$

where $J_0(x)$, defined by the power series in Equation 9.97, is, as we have seen, named the Bessel function of the first kind of order zero. As its name suggests, $J_0(x)$ is a member of a family of functions, about which we will learn more in the next section.

### 9.2.4 BESSEL EQUATIONS AND BESSEL FUNCTIONS

*Bessel equations* are ODEs that typically derive, as in the example in Section 9.2.2.2, from solving PDEs in cylindrical coordinates via separation of variables. They are also sometimes found in other venues. Bessel's equation of order $m$ is

$$\frac{d^2 y}{dx^2} + \frac{1}{x}\frac{dy}{dx} + \left(1 - \frac{m^2}{x^2}\right) y = 0 \tag{9.99}$$

It is a second-order linear ODE and has two fundamental solutions, called *Bessel functions of the first and second kinds*. They are in a category called *special functions* that are not as commonly used as, say, the logarithm, exponential or sine, but are nevertheless frequently encountered in physics and engineering.

#### 9.2.4.1 Bessel Functions of the First Kind

The solution of Equation 9.99 having finite value at the origin is called the Bessel function of the first kind. It is described by

$$J_m(x) = \sum_{n=0}^{\infty} \frac{(-1)^n}{(n+m)!n!} \left(\frac{x}{2}\right)^{2n+m} \tag{9.100}$$

Figure 9.11 presents a graph of $J_0(x)$ through $J_5(x)$.

Bessel functions of the first kind are easily computed. For small $x$ the first few terms of the infinite series in Equation 9.100 suffice. For example,

$$J_0(x) = 1 - \left(\frac{x}{2}\right)^2 + \frac{1}{4}\left(\frac{x}{2}\right)^4 - \frac{1}{36}\left(\frac{x}{2}\right)^6 + \dots$$

$$J_1(x) = \left(\frac{x}{2}\right) - \frac{1}{2}\left(\frac{x}{2}\right)^3 + \frac{1}{12}\left(\frac{x}{2}\right)^5 - \frac{1}{144}\left(\frac{x}{2}\right)^7 + \dots$$

For large $x$ the approximation

$$J_m(x) \cong \sqrt{\frac{2}{\pi x}} \cos\left(x - m\frac{\pi}{2} - \frac{\pi}{4}\right) \tag{9.101}$$

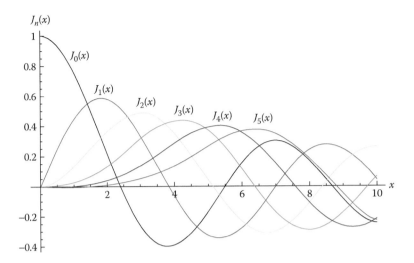

**FIGURE 9.11** Bessel functions of the first kind. (From Weisstein, E. W. "Bessel Function of the First Kind." From MathWorld–A Wolfram Web Resource. http://mathworld.wolfram.com/BesselFunctionoftheFirstKind.html [accessed June 09, 2017]).

is quite accurate. Furthermore, on the Internet one can find a number of handy "calculators" providing quick and accurate values for Bessel functions and their relatives for any argument. One can also find, on the web, compendia of relationships among Bessel functions and their relatives that are useful to engineers. See, for example, [18].

In Section 9.2.2.2 we saw that the zeros of $J_0(x)$ are of interest; that is, those values $x_n$ for which $J_0(x_n) = 0$. So, in a similar way, when azimuthal symmetry is absent, are the zeros of $J_m(x)$. The web's Bessel function "calculators" can provide these.

### 9.2.4.2 Bessel Functions of the Second Kind

From Sections 5.2 and 5.4 we learned that the equation

$$\frac{d^2 y}{dt^2} + p(t)\frac{dy}{dt} + q(t)y = 0 \tag{9.102}$$

with specified initial conditions

$$y(t_0) = y_0 \atop \dot{y}(t_0) = \dot{y}_0 \tag{9.103}$$

has a solution of the form

$$y(t) = c_1 y_1(t) + c_2 y_2(t) \tag{9.104}$$

with $c_1$ and $c_2$ given by the solution to

$$c_1 y_1(t_0) + c_2 y_2(t_0) = y_0 \atop c_1 \dot{y}_1(t_0) + c_2 \dot{y}_2(t_0) = \dot{y}_0$$

*provided* that $p(t)$ and $q(t)$ are continuous in some open interval $I$ containing $t = t_0$.

The Bessel equation does not meet this proviso when $t_0 = 0$. Writing the independent variable as $t$ instead of $x$ in Equation 9.99 and comparing to Equation 9.102, $p(t) = 1/t$ is discontinuous at $t = 0$ for all orders and $q(t) = 1 - m^2/t^2$ is discontinuous at $t = 0$ for all except order zero. In fact it is not possible to solve the Bessel equation with arbitrary initial conditions as in Equation 9.103 when $t_0 = 0$. We already saw in Section 9.2.3 that a solution $y(t)$ of the Bessel equation of order zero that is valid at $t = 0$ must satisfy $dy/dt = 0$ at $t = 0$. A solution to the Bessel equation with arbitrary initial conditions *is* possible if those conditions are specified at a point $t_0 > 0$ and we are interested in a solution only for $t \geq t_0$.

As mentioned earlier, the second fundamental solution to Bessel's equation is called a Bessel function of the second kind. Returning to the independent variable $x$, the Bessel equation

$$\frac{d^2y}{dx^2} + \frac{1}{x}\frac{dy}{dx} + \left(1 - \frac{m^2}{x^2}\right)y = 0$$

has a second solution valid for $x > 0$ that can be written in the form

$$y_m(x) = J_m(x)\ln x + \sum_{n=1}^{\infty} c_n x^n \qquad (9.105)$$

Note that the sum begins with $n = 1$.

Rather than use Equation 9.105 as the second solution, it is traditional to employ instead a certain linear combination of it and $J_m(x)$. That traditional form for order zero is

$$Y_0(x) = \frac{2}{\pi}\left[\left(\gamma + \ln\left(\frac{x}{2}\right)\right)J_0(x) + \sum_{n=1}^{\infty}\frac{(-1)^{n+1}}{(n!)^2}\left(\sum_{m=1}^{n}\frac{1}{m}\right)\left(\frac{x}{2}\right)^{2n}\right]$$

where

$$\gamma = \lim_{n\to\infty}\left(\sum_{m=1}^{n}\frac{1}{m} - \ln n\right) \cong 0.5772$$

The higher-order functions $Y_m(x)$ are defined similarly and as a result the asymptotic relationships are complementary. That is, for large $x$,

$$J_m(x) \cong \sqrt{\frac{2}{\pi x}}\cos\left(x - m\frac{\pi}{2} - \frac{\pi}{4}\right)$$

$$Y_m(x) \cong \sqrt{\frac{2}{\pi x}}\sin\left(x - m\frac{\pi}{2} - \frac{\pi}{4}\right)$$

## 9.2.5 POWER SERIES SOLUTIONS: ANOTHER EXAMPLE

This section considers Bessel's equation of order $m$, where $m$ is an integer greater than zero:

$$\frac{d^2y}{dx^2} + \frac{1}{x}\frac{dy}{dx} + \left(1 - \frac{m^2}{x^2}\right)y = 0 \qquad (9.106)$$

We have seen a power series solution already. The solution with finite value and derivative at $x = 0$ is

$$J_m(x) = \sum_{n=0}^{\infty} \frac{(-1)^n}{(n+m)!n!}\left(\frac{x}{2}\right)^{2n+m} \qquad (9.107)$$

We seek to derive Equation 9.107 from first principles.

If we were to try the form

$$y(x) = \sum_{n=0}^{\infty} c_n x^n$$

that worked for order zero we would find that it fails for order $m$. We try, instead,

$$y(x) = x^\alpha \sum_{n=0}^{\infty} c_n x^n \qquad (9.108)$$

We must solve for the exponent $\alpha$ as well as the coefficients $c_n$.

Substituting Equation 9.108 into Equation 9.106 we find

$$\sum_{n=0}^{\infty}(n+\alpha)(n+\alpha-1)c_n x^{n+\alpha-2} + \sum_{n=0}^{\infty}(n+\alpha)c_n x^{n+\alpha-2} + \sum_{n=0}^{\infty}c_n x^{n+\alpha} - \sum_{n=0}^{\infty}m^2 c_n x^{n+\alpha-2} = 0 \quad (9.109)$$

The coefficient of $x^{\alpha-2}$ in Equation 9.109 is $(\alpha(\alpha-1)+\alpha-m^2)c_0 = (\alpha^2-m^2)c_0$, which must equal zero, as must all coefficients. Either $\alpha^2-m^2 = 0$ or $c_0 = 0$. We will see shortly that $c_0$ cannot be zero. Hence

$$\alpha^2 - m^2 = 0 \qquad (9.110)$$

Equation 9.110 is the *indicial equation* for the ODE Equation 9.106. We will discuss the indicial equation in more generality shortly.

For $y(x)$ to be finite at the origin we must have $\alpha = +m$.

The coefficient of $x^{\alpha-1}$ in Equation 9.109 is

$$((1+\alpha)(1+\alpha-1)+(1+\alpha)-m^2)c_1$$

$$= ((1+m)(1+m-1)+(1+m)-m^2)c_1$$

$$= (2m+1)c_1$$

Again the coefficient must be zero, which implies that $c_1 = 0$.

To evaluate the higher-order coefficients we rearrange the remaining terms of Equation 9.109 into a more convenient form:

$$\sum_{n=2}^{\infty}(n+\alpha)(n+\alpha-1)c_n x^{n+\alpha-2} + \sum_{n=2}^{\infty}(n+\alpha)c_n x^{n+\alpha-2} + \sum_{n=0}^{\infty}c_n x^{n+\alpha} - \sum_{n=2}^{\infty}m^2 c_n x^{n+\alpha-2} = 0$$

$$\sum_{n=2}^{\infty}((n+m)^2 - m^2)c_n x^{n+m-2} + \sum_{n=0}^{\infty}c_n x^{n+m} = 0$$

$$\sum_{n=0}^{\infty}(n^2 + 2nm)c_n x^{n+m-2} + \sum_{n=0}^{\infty}c_n x^{n+m} = 0 \qquad (9.111)$$

We need to represent Equation 9.111 in the form

$$\sum_{k=0}^{\infty}a_k x^{k+m} = 0$$

To do this we replace $n$ in the first sum with $k + 2$ and replace $n$ in the second sum with $k$. This yields

$$\sum_{k=0}^{\infty}(((k+2)^2 + 2(k+2)m)c_{k+2} + c_k)x^{k+m} = 0$$

and we immediately see the recursion relation

$$c_{k+2} = -\frac{c_k}{(k+2)^2 + 2(k+2)m} \qquad (9.112)$$

Letting $k = 1$ in Equation 9.112 and recalling that $c_1 = 0$ shows us that $c_k = 0$ for odd $k$. Letting $k = 0$ in Equation 9.112 shows us that to have a nonzero solution to the ODE we must have $c_0 \neq 0$. We have

$$c_2 = -\frac{1}{2^2 + 2\cdot 2\cdot m}c_0 = (-1)\left(\frac{1}{2}\right)^2 \frac{1}{m+1}c_0$$

$$c_4 = -\frac{1}{4^2 + 2\cdot 4\cdot m}c_2 = (-1)^2\left(\frac{1}{2}\right)^4\left(\frac{1}{(m+2)(m+1)}\right)\left(\frac{1}{2}\right)c_0$$

$$c_6 = -\frac{1}{6^2 + 2\cdot 6\cdot m}c_4 = (-1)^3\left(\frac{1}{2}\right)^6\left(\frac{1}{(m+3)(m+2)(m+1)}\right)\left(\frac{1}{3\cdot 2}\right)c_0$$

It is convenient and useful to let

$$c_0 = \frac{1}{2^m m!}$$

Then the general coefficient can be written as

$$c_{2k} = (-1)^k \left(\frac{1}{2}\right)^{2k+m} \frac{1}{(k+m)!k!}$$

and the solution is

$$J_m(x) = \sum_{n=0}^{\infty} \frac{(-1)^n}{(n+m)!n!} \left(\frac{1}{2}\right)^{2n+m} x^{2n+m}$$

$$J_m(x) = \sum_{n=0}^{\infty} \frac{(-1)^n}{(n+m)!n!} \left(\frac{x}{2}\right)^{2n+m}$$

as was to be shown.

### 9.2.6 POWER SERIES SOLUTIONS: SOME GENERAL RULES

We have now seen three different forms of power series that satisfy different second-order linear ODEs:

$$y_1(x) = \sum_{n=0}^{\infty} a_n x^n$$

$$y_2(x) = y_1 \ln x + \sum_{n=1}^{\infty} b_n x^n$$

$$y_3(x) = x^{\alpha} \sum_{n=0}^{\infty} c_n x^n$$

It is time to state some general rules: Given a specific form of ODE, what form of solution should we try?

#### 9.2.6.1 A Few Definitions

A function $f(x)$ is said to be *analytic* at $x = x_0$ if it can be expanded in a power series

$$f(x) = \sum_{n=0}^{\infty} a_n (x - x_0)^n$$

that converges in some interval around $x_0$.
Consider the ODE

$$\frac{d^2 y}{dx^2} + p(x)\frac{dy}{dx} + q(x)y = 0 \tag{9.113}$$

The point $x = x_0$ is said to be an *ordinary point* of the ODE if $p(x)$ and $q(x)$ are analytic at $x_0$. If the point $x = x_0$ is not an ordinary point it is called a *singular point*.

A singular point is called a *regular singular point* of the ODE (Equation 9.113) if

$$\lim_{x \to x_0} (x - x_0) p(x) = p_0 \tag{9.114}$$

$$\lim_{x \to x_0} (x - x_0)^2 q(x) = q_0 \tag{9.115}$$

and $p_0$ and $q_0$ are finite. The *indicial equation* is

$$\alpha(\alpha - 1) + p_0 \alpha + q_0 = 0 \tag{9.116}$$

With these definitions, some general rules follow, which we state without proof.

### 9.2.6.2  Some General Rules

Before applying the rules, first divide by the coefficient of the highest derivative in the ODE to make that coefficient equal to one. Then determine if the coefficients of $dy/dx$ and $y$ are analytic.

If $x = x_0$ is an ordinary point of the ODE (Equation 9.113) then *each* of the two fundamental solutions can be written in the form

$$y_i = \sum_{n=0}^{\infty} c_n (x - x_0)^n \tag{9.117}$$

The series in Equation 9.117 and in equations written below converge in some region around $x_0$.

If $x = x_0$ is a regular singular point and $\alpha_1$ and $\alpha_2$ are real solutions to the indicial equation with $\alpha_1 \geq \alpha_2$, then one solution of the ODE can be written in the form

$$y_1 = (x - x_0)^{\alpha_1} \sum_{n=0}^{\infty} c_n (x - x_0)^n$$

If $\alpha_1 = \alpha_2$ then the second fundamental solution of the ODE can be written in the form

$$y_2 = y_1 \ln(x - x_0) + (x - x_0)^{\alpha_1} \sum_{n=1}^{\infty} b_n (x - x_0)^n$$

If $\alpha_1 \neq \alpha_2$ then:

If $\alpha_1 - \alpha_2$ is not an integer then the second ODE solution can be written in the form

$$y_2 = (x - x_0)^{\alpha_2} \sum_{n=0}^{\infty} b_n (x - x_0)^n$$

If $\alpha_1 - \alpha_2$ is an integer then the second solution can be written in the form

$$y_2 = a y_1 \ln(x - x_0) + (x - x_0)^{\alpha_2} \sum_{n=0}^{\infty} b_n (x - x_0)^n$$

where the coefficient $a$ may or may not be zero. The easiest way to determine whether $a$ must be nonzero is to try for a solution without it.

Note that the starting point in summations in this section is sometimes $n = 0$ and sometimes $n = 1$.

ODEs with singular points that are not regular are beyond the scope of this text.

### 9.2.6.3 Special Cases

There are several special cases that deserve attention because, while they conform to the general rules given in the last Section 9.2.6.2, they can be confusing.

The Cauchy–Euler equation has the form

$$x^2 \frac{d^2 y}{dx^2} + bx \frac{dy}{dx} + cy = 0$$

One finds that neither of the two fundamental solutions to this equation is expressed in an infinite series. If the two solutions $\alpha_1$ and $\alpha_2$ to the indicial equation are real and $\alpha_1 \geq \alpha_2$ then one of the solutions is simply $y_1(x) = x^{\alpha_1}$. The second fundamental solution conforms to the rules of the previous Section 9.2.6.2, but again there is no infinite series. Problem 9.15 provides an example.

The second special case is when the indicial equation has complex roots. What does $x^{\alpha}$ mean when $\alpha = \lambda + i\omega$? Given that the constants in the ODE are real, the roots, if complex, will be complex conjugates. Using the Cauchy–Euler equation as an example, Problem to Solve P9.16 motivates that $x^{\lambda + i\omega}$ and $x^{\lambda - i\omega}$ are replaced in the real solutions by $x^{\lambda} \cos(\omega \ln x)$ and $x^{\lambda} \sin(\omega \ln x)$. It should be noted that these functions oscillate wildly for small $x$, decidedly unrealistic behavior.

## 9.3 LAPLACE'S EQUATION

This section addresses Laplace's equation. It represents the steady state of the heat equation and the wave equation and is sometimes called the potential equation because it governs situations, for example, in electrostatics, gravitation, or fluid flow, where derivatives of the scalar solution determine a vector field. In three-dimensional Cartesian coordinates (Figure 9.12a) it is

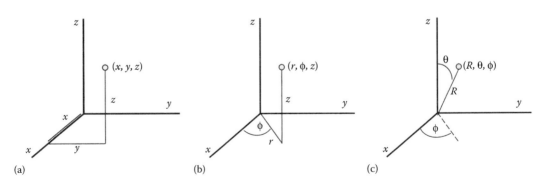

**FIGURE 9.12**   Three coordinate systems: (a) cartesian, (b) cylindrical, and (c) spherical.

$$\frac{\partial^2 V}{\partial x^2} + \frac{\partial^2 V}{\partial y^2} + \frac{\partial^2 V}{\partial z^2} = 0$$

In three-dimensional cylindrical coordinates (Figure 9.12b) it is

$$\frac{\partial^2 V}{\partial r^2} + \frac{1}{r}\frac{\partial V}{\partial r} + \frac{1}{r^2}\frac{\partial^2 V}{\partial \varphi^2} + \frac{\partial^2 V}{\partial z^2} = 0$$

In spherical coordinates (Figure 9.12c) it is

$$\frac{\partial^2 V}{\partial R^2} + \frac{2}{R}\frac{\partial V}{\partial R} + \frac{1}{R^2}\frac{\partial^2 V}{\partial \theta^2} + \frac{\cot\theta}{R^2}\frac{\partial V}{\partial \theta} + \frac{1}{R^2\sin^2\theta}\frac{\partial^2 V}{\partial \varphi^2} = 0 \qquad (9.118)$$

Laplace's Equation is called an elliptic PDE. A change in the value of $V$ at any point on the boundary changes the value of $V$ everywhere inside the boundary and $V$ takes on its maximum and minimum values on the boundary.

### 9.3.1  LAPLACE'S EQUATION EXAMPLE IN SPHERICAL COORDINATES

Figure 9.13 depicts a solid sphere of radius $R_0$ with uniform heat conduction properties whose surface is maintained at a temperature that depends on the angle $\theta$ (latitude). What is the temperature $T(R,\theta)$ inside the sphere?

Given the azimuthal (longitudinal) symmetry, Equation 9.118 becomes

$$\frac{\partial^2 T}{\partial R^2} + \frac{2}{R}\frac{\partial T}{\partial R} + \frac{1}{R^2}\frac{\partial^2 T}{\partial \theta^2} + \frac{\cot\theta}{R^2}\frac{\partial T}{\partial \theta} = 0 \qquad (9.119)$$

Challenge Problem CP9.2 guides the student through the solution of 9.119 with the boundary condition $T(R_0,\theta) = T_S(\theta)$. As in this chapter's previous boundary value problems,

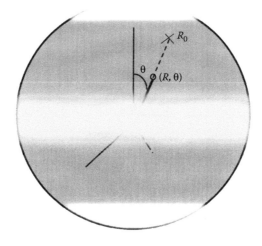

**FIGURE 9.13**   Temperature distribution on surface of sphere.

the solution approach is to separate variables and develop an infinite series of eigenfunctions. Letting

$$T(R,\theta) = G(R)H(\theta)$$

the resulting ODEs are

$$R^2 \frac{d^2G}{dR^2} + 2R\frac{dG}{dR} - n(n+1)G = 0, \qquad \frac{d^2H}{d\theta^2} + \cot\theta\frac{dH}{d\theta} + n(n+1)H = 0$$

With the substitution $x = \cos\theta$ the second equation becomes

$$(1-x^2)\frac{d^2H}{dx^2} - 2x\frac{dH}{dx} + n(n+1)H = 0$$

which is *Legendre's equation of order n.*

Following the general rules of Section 9.2.6, the result is

$$T(R,\theta) = \sum_{n=1}^{\infty} c_n \left(\frac{R}{R_0}\right)^n P_n(\cos\theta)$$

where the functions $P_n(x)$ are Legendre polynomials, the first five of which are

$$P_0(x) = 1$$

$$P_1(x) = x$$

$$P_2(x) = (3x^2 - 1)/2$$

$$P_3(x) = (5x^3 - 3x)/3$$

$$P_4(x) = (35x^4 - 30x^2 + 3)/8$$

and

$$c_n = \frac{2n+1}{2} \int_0^\pi T_S(\theta)P_n(\cos\theta)\sin\theta d\theta$$

Similar to the sine and Bessel functions, Legendre polynomials are orthogonal. Here,

$$\int_0^\pi P_n(\cos\theta)P_m(\cos\theta)\sin\theta d\theta = 0 \quad m \neq n$$

$$\int_0^\pi P_n(\cos\theta)P_m(\cos\theta)\sin\theta d\theta = \frac{2}{2n+1} \quad m = n$$

## 9.4 BEAM EQUATION

This section addresses the last of the example PDEs we shall consider, describing the motion of a cantilevered structural beam. There are, of course, many more; we have chosen these as representative and among the most important in engineering. The beam equation differs from the other examples in that it is a fourth-order equation and its boundary conditions require a different solution approach. We will again find that the eigenfunctions of the beam equation form an orthogonal set and in proving this fact discover that the kinds of eigenfunctions that can be used in infinite series expansions go well beyond the sine, Bessel functions, and the Legendre polynomials we have thus far encountered.

### 9.4.1 DERIVING THE BEAM EQUATION

A cantilever beam is a long, thin structural member fixed at one end and otherwise unsupported along its length.

Consider a cantilever beam of length $h$ and uniform cross section initially depressed with a point load at the end and then released to vibrate freely (Figure 9.14).

This model is somewhat similar to an aircraft wing in turbulence and a skyscraper after the shock of an earthquake.

Scientific analysis of bending motions in a continuous beam is found in *solid mechanics*. In that discipline, the forces and moments in the *interior* of the beam and its resultant deformations are rigorously examined. The formulation presented here is due to Leonhard Euler and Daniel Bernoulli. There are more advanced formulations.

Once released, the force along the beam is given by

$$\frac{\partial F}{\partial x} = \rho \frac{\partial^2 y}{\partial t^2} \tag{9.120}$$

where:

$y$ is the beam deflection (displacement of the beam from vertical)
$\rho = m/h$
$m$ is the mass of the beam. See Figure 9.15.

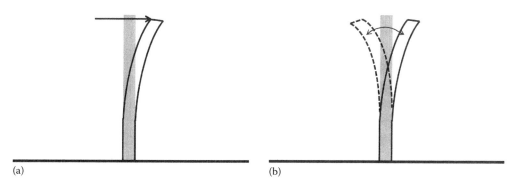

(a)　　　　　　　　　　　　　　(b)

**FIGURE 9.14** Motion of cantilever beam: (a) beam shape prior to release and (b) beam shape after release.

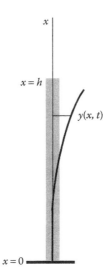

**FIGURE 9.15**  Definition of variables in analysis of motions of a cantilever beam.

The increase in torque along the beam is given by

$$\frac{\partial \tau}{\partial x} = -F \tag{9.121}$$

After considerable analysis of the interior of the beam, one arrives at

$$\tau = EI \frac{\partial^2 y}{\partial x^2} \tag{9.122}$$

where:
   $E$ is Young's modulus for the beam material
   $I$ is the beam's cross-sectional moment of inertia

Combining Equations 9.120, 9.121, and 9.122, we arrive at

$$\frac{\partial^4 y}{\partial x^4} = -\left(\frac{\rho}{EI}\right)\frac{\partial^2 y}{\partial t^2} \tag{9.123}$$

Equation 9.123 is the Euler–Bernoulli beam equation for a uniform cantilever beam in free vibration.

### 9.4.2  BEAM EQUATION IN A BOUNDARY VALUE PROBLEM

Suppose, as shown in Figure 9.14, the initial external force on the beam is a point load at its end. Then the initial deflection along the beam is

$$y(x,0) = \frac{P}{6EI}(3hx^2 - x^3) \tag{9.124}$$

(Stated without proof.)

We also assume that the beam is in steady state prior to its release, so that

$$\frac{\partial y}{\partial t}(x,0) = 0 \tag{9.125}$$

What are the motions of the beam after its release?

The equations to be solved are Equation 9.123 with initial conditions Equations 9.124 and 9.125 together with boundary conditions (for $t > 0$) whose statement requires a knowledge of solid mechanics:

$$\text{The base is held stationary: } y(0,t) = 0 \tag{9.126}$$

$$\text{The beam remains vertical at its base: } \frac{\partial y}{\partial x}(0,t) = 0 \tag{9.127}$$

$$\text{There is no torque at the tip: } \frac{\partial^2 y}{\partial x^2}(h,t) = 0 \tag{9.128}$$

$$\text{There is no shear stress at the tip: } \frac{\partial^3 y}{\partial x^3}(h,t) = 0 \tag{9.129}$$

To solve Equations 9.123 through 9.129 we try separation of variables:

$$y(x,t) = w(x)u(t) \tag{9.130}$$

and then Equation 9.123 becomes

$$u\frac{d^4 w}{dx^4} = -w\left(\frac{\rho}{EI}\right)\frac{d^2 u}{dt^2}$$

Dividing through by $w(x)u(t)$ we have

$$\frac{1}{w}\frac{d^4 w}{dx^4} = -\frac{1}{u}\left(\frac{\rho}{EI}\right)\frac{d^2 u}{dt^2} \tag{9.131}$$

The left side of Equation 9.131 is a function only of $x$ and the right side a function only of $t$, hence both sides must equal a constant $k$. It will turn out that to meet boundary conditions this constant must be positive. Then we have two ODEs

$$\frac{d^4 w}{dx^4} - kw = 0 \tag{9.132}$$

$$\frac{d^2 u}{dt^2} + \left(\frac{kEI}{\rho}\right)u = 0 \tag{9.133}$$

We immediately recognize that the solution to Equation 9.133 is

$$u(t) = A\cos\omega t + B\sin\omega t \tag{9.134}$$

where

$$\omega = \sqrt{\frac{kEI}{\rho}} \tag{9.135}$$

From Equation 9.125 we see that $B = 0$. Also, without loss of generality we can set $A = 1$, since we will deal with the initial condition as we solve Equation 9.132.

Trying $w(x) = e^{rx}$ in Equation 9.132 we arrive at the characteristic equation $r^4 - k = 0$, which has solutions $r_1 = k^{1/4}$, $r_2 = -k^{1/4}$, $r_3 = ik^{1/4}$, and $r_4 = -ik^{1/4}$. Therefore the fundamental solutions to Equation 9.132 are

$$\varphi_1(x) = e^{px}$$

$$\varphi_2(x) = e^{-px}$$

$$\varphi_3(x) = \cos px \tag{9.136}$$

$$\varphi_4(x) = \sin px$$

where

$$p = k^{1/4} \tag{9.137}$$

The general solution to Equation 9.132 can be written as

$$w(x) = \sum_{i=1}^{4} a_i\varphi_i(x) \tag{9.138}$$

The coefficients $a_i$ must be chosen to satisfy the boundary conditions, which, from Equations 9.126 through 9.129, are

$$w(0) = 0$$

$$\frac{dw}{dx}(0) = 0 \tag{9.139}$$

$$\frac{d^2w}{dx^2}(h) = 0$$

$$\frac{d^3w}{dx^3}(h) = 0$$

Substituting Equations 9.136 and 9.138 into Equation 9.139, we arrive at the matrix equation

$$\begin{pmatrix} 1 & 1 & 1 & 0 \\ 1 & -1 & 0 & 1 \\ e^{ph} & e^{-ph} & -\cos ph & -\sin ph \\ e^{ph} & -e^{-ph} & \sin ph & -\cos ph \end{pmatrix} \begin{pmatrix} a_1 \\ a_2 \\ a_3 \\ a_4 \end{pmatrix} = \begin{pmatrix} 0 \\ 0 \\ 0 \\ 0 \end{pmatrix} \qquad (9.140)$$

For Equation 9.140 to be true for nonzero $a_i$ we must have

$$\det \begin{pmatrix} 1 & 1 & 1 & 0 \\ 1 & -1 & 0 & 1 \\ e^{ph} & e^{-ph} & -\cos ph & -\sin ph \\ e^{ph} & -e^{-ph} & \sin ph & -\cos ph \end{pmatrix} = 0 \qquad (9.141)$$

After some algebra Equation 9.141 results in

$$-2(1+\cos ph \cosh ph) = 0$$

or

$$\cos ph \cosh ph = -1 \qquad (9.142)$$

where

$$\cosh ph = \frac{e^{ph} + e^{-ph}}{2} \qquad (9.143)$$

is called the *hyperbolic cosine*.

Equation 9.142 defines the eigenvalues for this problem, the particular values of $ph$ satisfying the boundary conditions (Equation 9.139). We can use nonlinear problem solvers available on the web to solve Equation 9.142. We find that there are an infinite number of eigenvalues, the first four given by $p_1h = 1.87510$, $p_2h = 4.69409$, $p_3h = 7.85476$, and $p_4h = 10.99554$. Note that larger values of $p_nh$ are given, approximately but accurately, by the zeros of $\cos ph$, since $\cosh ph$ becomes very large.

Recall that we selected a positive number ($k$) as the separation constant in Equation 9.131. If we had selected a negative number then Equation 9.132 would have become

$$\frac{d^4w}{dx^4} + kw = 0$$

and its solutions could not have met the boundary conditions in 9.139.

Substituting solutions of Equation 9.142 into Equation 9.140 and applying the results in Equation 9.138 provides an infinite number of eigenfunctions, which can be most compactly written as

$$w_n(x) = \cosh p_n x - \cos p_n x - \sigma_n(\sinh p_n x - \sin p_n x) \qquad (9.144)$$

where

$$\sigma_n = \frac{\cosh p_n h + \cos p_n h}{\sinh p_n h + \sin p_n h} \qquad (9.145)$$

and

$$\sinh z = \frac{e^z - e^{-z}}{2} \tag{9.146}$$

is the *hyperbolic sine.*

Note that, via Equations 9.135, 9.137, and 9.142, the allowed frequencies of vibration in time are also distinct and infinite in number:

$$\omega_n = \left(\frac{EI}{\rho h^4}\right)^{1/2} (p_n h)^2 \tag{9.147}$$

To summarize results to this point, we tried the solution $y(x,t) = w(x)u(t)$ and found an infinite number of solutions. We now set

$$y(x,t) = \sum_{n=1}^{\infty} c_n w_n(x) u_n(t) \tag{9.148}$$

and determine $c_n$ through the initial condition. We must have

$$y(x,0) = \sum_{n=1}^{\infty} c_n w_n(x) = \frac{P}{6EI} (3hx^2 - x^3) \tag{9.149}$$

We will prove in the next section that the eigenfunctions $w_n(x)$ are orthogonal to one another, so we can determine the coefficient $c_n$ via

$$c_n = \frac{\int_0^h w_n(x) y(x,0) dx}{\int_0^h w_n^2(x) dx} \tag{9.150}$$

where $w_n(x)$ is given by Equations 9.144 and 9.145 and $y(x,0)$ by Equation 9.149. Calculation of $c_n$ via Equation 9.150 is straightforward but tedious. As a practical matter, for engineering applications related to vibrating beams, as we discussed for those related to vibrating membranes, it often suffices to determine the natural frequencies of the system, and calculation of the coefficients $c_n$ is unnecessary. For completeness we give the result here. After some simplification that Challenge Problem CP9.4 addresses, we ultimately find that

$$c_n = \left(\frac{2Ph^3}{EI}\right) \frac{(-1)^{n+1}}{(p_n h)^4} \tag{9.151}$$

As always after a long calculation, we must find a way to check our results. In this case, we can evaluate $y(x,t)$ as given by Equations 9.142 through 9.150 at the points $x = h$ and $t = 0$ and compare to the initial condition. At the point $x = h$ the initial condition is

$$y(h,0) = \frac{P}{6EI} (3h(h)^2 - h^3) = \frac{Ph^3}{3EI} \tag{9.152}$$

Using

$$y(h,0) = \sum_{n=1}^{\infty} c_n w_n(h) u_n(0) \tag{9.153}$$

we have

$$u_n(0) = 1 \tag{9.154}$$

and again from the results of Challenge Problem CP9.4

$$w_n(h) = 2(-1)^{n+1} \tag{9.155}$$

Then

$$y(h,0) = \sum_{n=1}^{\infty} c_n w_n(h) u_n(0) = \left(\frac{4Ph^3}{EI}\right) \sum_{n=1}^{\infty} \frac{1}{(p_n h)^4} \tag{9.156}$$

Equating 9.152 and 9.156 we see that we must have

$$\sum_{n=1}^{\infty} \frac{1}{(p_n h)^4} = \frac{1}{12} \tag{9.157}$$

Table 9.2 records the partial sums and shows that the series does indeed appear to be converging to the correct value.

**TABLE 9.2**
**Checking the Accuracy of the Beam Equation Solution**

| $n$ | $p_n h$ | $1/(p_n h)^4$ | $\sum_{m=1}^{n} 1/(p_m h)^4$ | $\left(\sum_{m=1}^{n} 1/(p_m h)^4\right)^{-1}$ |
|---|---|---|---|---|
| 1 | 1.87510 | 0.0808914 | 0.0808914 | 12.3623 |
| 2 | 4.69409 | 0.0020596 | 0.0829510 | 12.0553 |
| 3 | 7.85476 | 0.0002627 | 0.0832137 | 12.0173 |
| 4 | 10.99554 | 0.0000684 | 0.0832821 | 12.0074 |
| 5 | 14.13717 | 0.0000250 | 0.0833071 | 12.0038 |
| 6 | 17.27876 | 0.0000112 | 0.0833183 | 12.0022 |
| 7 | 20.42035 | 0.0000058 | 0.0833241 | 12.0013 |
| 8 | 23.56194 | 0.0000032 | 0.0833273 | 12.0009 |
| 9 | 26.70354 | 0.0000020 | 0.0833293 | 12.0006 |
| 10 | 29.84513 | 0.0000013 | 0.0833306 | 12.0004 |

### 9.4.3   ORTHOGONALITY OF THE BEAM EQUATION EIGENFUNCTIONS

This section proves that the eigenfunction solutions to the beam equation with the given boundary conditions are orthogonal. It can also be shown that under only mild limitations the infinite series of eigenfunctions converges to the initial condition. Earlier in this chapter we saw these orthogonality and convergence properties in sine functions (Section 9.1.2), Bessel functions (Section 9.2.2.2), and Legendre polynomials (Section 9.3) that were solutions to differential equations with specific boundary conditions. We will see in this section that the class of functions with these properties is broad, and that they depend for these properties not only on the differential equation but also on the boundary conditions applied. The means of demonstrating the generality of the orthogonality property is Lagrange's identity, which we will meet in the process of our proof.

**Theorem:**

The eigenfunctions $w_n(x)$ in the solution $y(x,t) = \sum_{n=1}^{\infty} c_n w_n(x) u_n(t)$ to the beam equation

$$\frac{\partial^4 y}{\partial x^4} = -\left(\frac{\rho}{EI}\right)\frac{\partial^2 y}{\partial t^2}$$

with boundary conditions

$$y(0,t) = 0, \qquad \frac{\partial^2 y}{\partial x^2}(h,t) = 0$$

$$\frac{\partial y}{\partial x}(0,t) = 0, \qquad \frac{\partial^3 y}{\partial x^3}(h,t) = 0$$

are orthogonal.

**Proof:**

Separating variables as we did in Equations 9.130 and 9.131 led to Equation 9.132:

$$\frac{d^4 w}{dx^4} - kw = 0$$

For two different eigenfunction solutions we can write

$$\frac{d^4 w_n}{dx^4} = k_n w_n, \qquad \frac{d^4 w_m}{dx^4} = k_m w_m$$

Multiplying the first equation by $w_m$ and the second by $w_n$, subtracting the second equation from the first and integrating over the length of the beam we have

$$\int_0^h \left(w_m \frac{d^4 w_n}{dx^4} - w_n \frac{d^4 w_m}{dx^4}\right) dx = (k_n - k_m)\int_0^h w_n(x)w_m(x)dx \qquad (9.158)$$

Integrating by parts on the left-hand side, letting $w_i = u$ and $(d^4 w_j/dx^4)dx = dv$:

$$\int_0^h \left( w_m \frac{d^4 w_n}{dx^4} - w_n \frac{d^4 w_m}{dx^4} \right) dx = \left( w_m \frac{d^3 w_n}{dx^3} - w_n \frac{d^3 w_m}{dx^3} \right) \Big|_0^h$$

$$- \int_0^h \left( \frac{dw_m}{dx} \frac{d^3 w_n}{dx^3} - \frac{dw_n}{dx} \frac{d^3 w_m}{dx^3} \right) dx$$

Integrating the last term by parts, letting $dw_i/dx = u$ and $(d^3 w_j/dx^3)dx = dv$:

$$\int_0^h \left( w_m \frac{d^4 w_n}{dx^4} - w_n \frac{d^4 w_m}{dx^4} \right) dx = \left( w_m \frac{d^3 w_n}{dx^3} - w_n \frac{d^3 w_m}{dx^3} \right) \Big|_0^h$$

$$- \left( \frac{dw_m}{dx} \frac{d^2 w_n}{dx^2} - \frac{dw_n}{dx} \frac{d^2 w_m}{dx^2} \right) \Big|_0^h + \int_0^h \left( \frac{d^2 w_m}{dx^2} \frac{d^2 w_n}{dx^2} - \frac{d^2 w_n}{dx^2} \frac{d^2 w_m}{dx^2} \right) dx$$

(9.159)

Now

$$\frac{d^2 w_m}{dx^2} \frac{d^2 w_n}{dx^2} - \frac{d^2 w_n}{dx^2} \frac{d^2 w_m}{dx^2} = 0$$

(9.160)

and the boundary conditions on $y(x,t)$ imply

$$w(0) = 0, \qquad \frac{d^2 w}{dx^2}(h) = 0$$

$$\frac{dw}{dx}(0) = 0, \qquad \frac{d^3 w}{dx^3}(h) = 0$$

hence

$$\left( w_m \frac{d^3 w_n}{dx^3} - w_n \frac{d^3 w_m}{dx^3} \right) \Big|_0^h - \left( \frac{dw_m}{dx} \frac{d^2 w_n}{dx^2} - \frac{dw_n}{dx} \frac{d^2 w_m}{dx^2} \right) \Big|_0^h = 0$$

(9.161)

From Equations 9.159 through 9.161,

$$\int_0^h \left( w_m \frac{d^4 w_n}{dx^4} - w_n \frac{d^4 w_m}{dx^4} \right) dx = 0$$

(9.162)

Hence from Equation 9.158,

$$(k_n - k_m) \int_0^h w_n(x) w_m(x) dx = 0$$

(9.163)

Since the eigenvalues $k_n$ and $k_m$ are distinct,

$$\int_0^h w_n(x) w_m(x) dx = 0$$

as was to be shown.

Equation 9.162 is Lagrange's identity in the form it takes for the beam problem. For each boundary value problem we have considered in this chapter, a similar proof could be constructed to demonstrate that each of those sets of eigenfunctions is orthogonal. As we have seen, the proof depends on the form of both the differential equation and the boundary conditions. For a broad class of boundary value problems Lagrange's identity can be constructed and its left-hand side shown to be zero, as we have done here, leading to Equation 9.163. As long as eigenvalues are distinct, Lagrange's identity then shows that the eigenfunctions are orthogonal.

In a similar way, the infinite series eigenfunction solutions for a broad class of boundary value problems can be shown to converge to the initial condition, with only mild limitations on the initial condition.

As an example where these properties do not hold, consider the PDE in Problem to Solve P9.17. A Lagrange identity argument paralleling Equations 9.158 through 9.163 cannot be constructed for that PDE because it is of third order in the spatial coordinate. Word Problem 9.4 guides the student to use the third-order PDE's Lagrange identity to demonstrate that its eigenfunctions are not orthogonal over the integral $\int_0^h (...)dx$, where the boundary conditions are stated at $x = 0$ and $x = h$.

### 9.4.4 NONHOMOGENOUS BOUNDARY VALUE PROBLEMS

Challenge Problems CP9.5 and CP9.6 guide the student through solution of example nonhomogeneous PDEs in boundary value problems, considering cantilevered beams driven to oscillate. The results demonstrate that if forces are applied in the rhythm of one of the beam's natural frequencies, a resonance occurs. We see that a beam can be regarded as a system of infinite dimensions, responding to inputs in ways similar to that exhibited by finite dimensional systems.

### 9.4.5 OTHER GEOMETRIES

Just as the heat, wave and Laplace's equations take different forms in different geometries so, too, does the beam equation, although it is not known by that name in the other geometries. For example, vibrations of a circular flat metal plate clamped at its circumference are governed by a PDE with variable coefficients that is fourth order in the radial spatial dimension and second order in time. Its solutions involve Bessel functions. We will not be able to explore here vibrations of solid objects in other geometries; we are reaching the limit of our allocated space and time.

## 9.5 SUMMARY

This chapter discussed four of the most important PDEs in engineering. In two-dimensional Cartesian coordinates the heat equation is

$$\frac{\partial^2 u}{\partial x^2} + \frac{\partial^2 u}{\partial y^2} - \frac{1}{\alpha}\frac{\partial u}{\partial t} = 0$$

Note that it is first order in time. The heat equation describes diffusion and its solutions smooth initial peaks and valleys. In the same coordinates the wave equation is

$$\frac{\partial^2 u}{\partial x^2} + \frac{\partial^2 u}{\partial y^2} - \frac{1}{a^2}\frac{\partial^2 u}{\partial t^2} = 0$$

Note that it is second order in time. The wave equation portrays oscillations propagating in space at a constant speed $a$. Again in the same coordinates Laplace's equation is

$$\frac{\partial^2 u}{\partial x^2} + \frac{\partial^2 u}{\partial y^2} = 0$$

Laplace's equation governs steady-state solutions of the heat and wave equation and situations involving a potential. The beam equation

$$\frac{\partial^4 u}{\partial x^4} + \left(\frac{\rho}{EI}\right)\frac{\partial^2 u}{\partial t^2} = 0$$

describes the lateral vibrations of a cantilever structural beam. Note that it is fourth order in the spatial coordinate. These equations take different forms depending on geometry. We can solve boundary value problems using separation of variables, which lead us to important ODEs with variable coefficients: Bessel equations in cylindrical coordinates

$$\frac{d^2 v}{dr^2} + \frac{1}{r}\frac{dv}{dr} + \left(1 - \frac{m^2}{r^2}\right)v = 0$$

Legendre equations in spherical coordinates (with $x = \cos\theta$)

$$(1 - x^2)\frac{d^2 v}{dx^2} - 2x\frac{dv}{dx} + n(n+1)v = 0$$

We solve these ODEs using power series. This chapter stated general rules for computing power series solutions to second-order linear equations with variable coefficients. We learned the difference between an ordinary point and a regular singular point. When series expansions are about a regular singular point, solutions to the indicial equation determine the form of the power series. Solution of boundary value problems takes us into infinite series of eigenfunctions. There is a broad class of PDEs that, with the right boundary conditions, have eigenfunctions that are orthogonal and capable of representing well behaved functions (as initial conditions) in an infinite series. Bessel functions, in particular, are useful tools for engineers because simple approximations enable computation by hand and substantial resources exist on the Internet. Finally, this chapter presented several examples to show that the additional independent variables in PDEs can be regarded as an infinite number of states, and that PDEs can be thought of as infinite dimensional systems.

## Solved Problems

**SP9.1** *Problem*: Solve

$$\frac{\partial^2 T}{\partial x^2} = \frac{1}{\alpha}\frac{\partial T}{\partial t} \tag{9.164}$$

$$T(x,0) = T_0 \sin^5\left(\pi x/L\right) \tag{9.165}$$

$$T(0,t) = 0$$
$$T(L,t) = 0 \tag{9.166}$$

Check your results by comparing your calculated value for $T(L/2,0)$ with the given initial condition, which is $T_0$.

*Solution*: We try the solution

$$T(x,t) = \Phi(x)\Theta(t) \tag{9.167}$$

Subsequent steps follow exactly the steps taken from Equations 9.15 to 9.31 in Section 9.1.2, leading to the solution

$$T(x,t) = \sum_{n=1}^{\infty} c_n e^{-\alpha(n\pi/L)^2 t} \sin\left(n\pi x/L\right) \tag{9.168}$$

where

$$c_n = \frac{2}{L}\int_0^L T_0(x)\sin\left(n\pi x/L\right)dx = \frac{2}{L}\int_0^L T_0\sin^5\left(\pi x/L\right)\sin\left(n\pi x/L\right)dx \tag{9.169}$$

Rather than computing the integral directly, however, we can simplify matters by looking for a geometric identity of the form

$$\sin^5\theta = \sum_{n=1}^{N} a_n \sin n\theta \tag{9.170}$$

We can derive the appropriate expression by hand using the geometric identities in Section 1.1.3 and check our result by searching the internet, or vice versa. In any case, without difficulty we find that

$$\sin^5\theta = (10\sin\theta - 5\sin 3\theta + \sin 5\theta)/16 \tag{9.171}$$

From Equations 9.168, 9.169, and 9.171, and recalling that

$$\int_0^L \sin\left(n\pi x/L\right)\sin\left(m\pi x/L\right)dx = 0$$

$$\int_0^L \sin^2\left(n\pi x/L\right)dx = L/2$$

we determine that

$$c_1 = \left(\frac{10}{16}\right)T_0 \quad c_3 = \left(\frac{-5}{16}\right)T_0 \quad c_5 = \left(\frac{1}{16}\right)T_0 \tag{9.172}$$

and that all other $c_n$ are zero.

From 9.168 and 9.172, the solution to Equations 9.164 through 9.166 is

$$T(x,t) = (T_0/16)\left(10e^{-\alpha(\pi/L)^2 t}\sin(\pi x/L)\right.$$

$$-5e^{-\alpha(3\pi/L)^2 t}\sin(3\pi x/L) \tag{9.173}$$

$$\left.+e^{-\alpha(5\pi/L)^2 t}\sin(5\pi x/L)\right)$$

Checking our results, we compare Equation 9.173 with the initial condition at $x = L/2$ and find

$$T(L/2,0) = (T_0/16)\left[10\sin(\pi/2) - 5\sin(3\pi/2) + \sin(5\pi/2)\right]$$

$$T(L/2,0) = (T_0/16)(10\cdot(1) - 5\cdot(-1) + (1)) = T_0$$

as it should.

**SP9.2** *Problem*: Section 9.1.1 introduced the kernel (also called the impulse response or Green's function) for the one-dimensional heat equation. Search the mathematical literature and, for the infinite space (i.e., unbounded) case, find the kernels for the two-dimensional and three-dimensional wave equations, which are

$$\frac{\partial^2 u}{\partial x^2} + \frac{\partial^2 u}{\partial y^2} - \frac{1}{a^2}\frac{\partial^2 u}{\partial t^2} = 0 \qquad \frac{\partial^2 u}{\partial x^2} + \frac{\partial^2 u}{\partial y^2} + \frac{\partial^2 u}{\partial z^2} - \frac{1}{a^2}\frac{\partial^2 u}{\partial t^2} = 0$$

Show that wave propagation in two dimensions (e.g., ripples on the surface when a pebble is dropped into a pond) is qualitatively different from that in three (e.g., an electromagnetic pulse from an omnidirectional antenna).

*Solution*: We conduct our search of the mathematical literature on the Internet and discover that the kernel for the wave equation in two dimensions is

$$K_2(x,y,t) = \frac{1}{2\pi((at)^2 - (x^2 + y^2))^{1/2}}, \qquad (x^2 + y^2)^{1/2} < at$$

$$K_2(x,y,t) = 0, \qquad (x^2 + y^2)^{1/2} > at$$

and the kernel for the wave equation in three dimensions is

$$K_3(x,y,z,t) = \frac{\delta(at - (x^2 + y^2 + z^2)^{1/2})}{4\pi(x^2 + y^2 + z^2)^{1/2}}$$

where $\delta(w)$ is the Dirac delta function. (Recall Section 7.6.3.1). It is convenient here to consider the impulse response interpretation of these functions. Consider an impulse at the spatial origin and at time equal to zero. That is, consider the equations

$$\frac{\partial^2 u}{\partial x^2} + \frac{\partial^2 u}{\partial y^2} - \frac{1}{a^2} \frac{\partial^2 u}{\partial t^2} = \delta(x)\delta(y)\delta(t) \qquad \frac{\partial^2 u}{\partial x^2} + \frac{\partial^2 u}{\partial y^2} + \frac{\partial^2 u}{\partial z^2} - \frac{1}{a^2} \frac{\partial^2 u}{\partial t^2} = \delta(x)\delta(y)\delta(z)\delta(t)$$

Then, by definition of the impulse response, the solutions are, for the two-dimensional case,

$$u(x, y, t) = K_2(x, y, t) = \frac{1}{2\pi((at)^2 - (x^2 + y^2))^{1/2}}, \qquad (x^2 + y^2)^{1/2} < at$$

$$u(x, y, t) = K_2(x, y, t) = 0, \qquad (x^2 + y^2)^{1/2} > at$$

and for the three-dimensional case

$$u(x, y, z, t) = K_3(x, y, z, t) = \frac{\delta(at - (x^2 + y^2 + z^2)^{1/2})}{4\pi(x^2 + y^2 + z^2)^{1/2}}$$

In three dimensions there is a response to the impulse only at the wavefront; that is, at the range

$$(x^2 + y^2 + z^2)^{1/2} = at$$

In two dimensions there is no sharp trailing edge to the impulse response; the response at locations behind the wavefront lingers on. For any forcing function at any location, if the input cuts off sharply: In three dimensions the response at a distant location will eventually cease, and it will cease in an instant. In two dimensions the response will die away slowly.

**SP9.3** *Problem*: Using the general rules in Section 9.2.6.2, find a power series solution to the following equation about the point $x_0 = 0$. Find the recursive relation and calculate the first five terms in the power series.

$$\frac{d^2 y}{dx^2} + (x + 3)\frac{dy}{dx} + (x - 2)y = 0$$

$$y(0) = y_0$$

$$\frac{dy}{dx}(0) = \dot{y}_0 \qquad\qquad (9.174)$$

*Solution*: According to the general rules, the first step is to determine whether the coefficients of $dy/dx$ and $y$ are analytic at the point $x = x_0$ when the coefficient of $d^2 y/dx^2$ is unity. The coefficients are $x + 3$ and $x - 2$, respectively. Each can be expanded in a power series of about $x = 0$, so the point $x_0 = 0$ is an ordinary point and there exist two solutions of the form

$$y(x) = \sum_{n=0}^{\infty} c_n x^n \qquad (9.175)$$

The key to finding power series solutions is to work the ODE from the form

$$\frac{d^2}{dx^2}\left(\sum_{n=0}^{\infty} c_n x^n\right) + p(x)\frac{d}{dx}\left(\sum_{n=0}^{\infty} c_n x^n\right) + q(x)\left(\sum_{n=0}^{\infty} c_n x^n\right) = 0$$

into the form $\sum k_m x^m = 0$ and then find coefficients $c_n$ such that $k_m = 0$ for each $m$. From Equation 9.175,

$$\frac{dy}{dx} = \sum_{n=1}^{\infty} n c_n x^{n-1} \qquad \frac{d^2 y}{dx^2} = \sum_{n=2}^{\infty} n(n-1)c_n x^{n-2} \qquad (9.176)$$

Then we have

$$\sum_{n=2}^{\infty} n(n-1)c_n x^{n-2} + (x+3)\sum_{n=1}^{\infty} n c_n x^{n-1} + (x-2)\sum_{n=0}^{\infty} c_n x^n = 0$$

$$\sum_{n=2}^{\infty} n(n-1)c_n x^{n-2} + \sum_{n=1}^{\infty} n c_n x^n + 3\sum_{n=1}^{\infty} n c_n x^{n-1} + \sum_{n=0}^{\infty} c_n x^{n+1} - \sum_{n=0}^{\infty} 2 c_n x^n = 0 \quad (9.177)$$

In the sum $\sum_{n=2}^{\infty} n(n-1)c_n x^{n-2}$ let $n-2=m$. In the sum $3\sum_{n=1}^{\infty} n c_n x^{n-1}$ let $n-1=m$. In the sum $\sum_{n=0}^{\infty} c_n x^{n+1}$ let $n+1=m$. In the sums $\sum_{n=1}^{\infty} n c_n x^n$ and $\sum_{n=0}^{\infty} 2 c_n x^n$ let $n=m$.

Then Equation 9.177 becomes

$$\sum_{m=1}^{\infty} ((m+2)(m+1)c_{m+2} + m c_m + 3(m+1)c_{m+1} + c_{m-1} - 2 c_m)x^m \qquad (9.178)$$

$$+ (2(1)c_2 + 3(1)c_1 - 2 c_0)x^{(0)} = 0$$

where the $x^{(0)}$ (constant) term stands alone because it was not included in two of the sums and isolating it allows us to write the rest of the left-hand side of Equation 9.194 in the desired form $\sum k_m x^m$. We must have the coefficient of $x^m$ for each $m$ equal to zero. Then, from Equation 9.178,

$$c_2 = c_0 - \left(\frac{3}{2}\right)c_1 \qquad (9.179)$$

$$c_{m+2} = -\frac{1}{(m+1)(m+2)}(3(m+1)c_{m+1} + (m-2)c_m + c_{m-1}) \quad m \geq 1 \qquad (9.180)$$

From the initial conditions and Equations 9.175 and 9.176 we determine that $c_0 = y_0$ and $c_1 = \dot{y}_0$. From Equations 9.179 and 9.180 we find, after some simplification,

$$c_2 = y_0 - \left(\frac{3}{2}\right)\dot{y}_0$$

$$c_3 = -\left(\frac{7}{6}\right)y_0 + \left(\frac{5}{3}\right)\dot{y}_0$$

$$c_4 = \left(\frac{7}{8}\right)y_0 - \left(\frac{4}{3}\right)\dot{y}_0 \tag{9.181}$$

$$c_5 = -\left(\frac{31}{60}\right)y_0 + \left(\frac{19}{24}\right)\dot{y}_0$$

**SP9.4** *Problem*: Using the general rules in Section 9.2.6.2, find a power series solution to the following equation about the point $x_0 = 0$. Find the recursive relation and calculate the first five terms in the power series.

$$x^2 \frac{d^2 y}{dx^2} + 5x\frac{dy}{dx} + (x+3)y = 0 \tag{9.182}$$

*Solution*: Following the general rules, we divide Equation 9.182 through by $x^2$ to make the coefficient of the highest derivative equal unity. The resulting coefficients $p(x)$ and $q(x)$ are not analytic, so we must examine the indicial equation. We have

$$p_0 = \lim_{x\to0} xp(x) = \lim_{x\to0} x\left(\frac{5x}{x^2}\right) = 5$$

$$\tag{9.183}$$

$$q_0 = \lim_{x\to0} x^2 q(x) = \lim_{x\to0} x^2\left(\frac{x+3}{x^2}\right) = 3$$

Then the indicial equation is

$$\alpha(\alpha-1) + p_0\alpha + q_0 = \alpha(\alpha-1) + 5\alpha + 3 = \alpha^2 + 4\alpha + 3 = (\alpha+1)(\alpha+3) = 0 \tag{9.184}$$

The two roots of the indicial equation are $\alpha_1 = -1$, $\alpha_2 = -3$. The largest of these is $\alpha_1$. From the general rules, we are assured that there exists a solution of the form

$$y(x) = x^{\alpha_1} \sum_{n=0}^{\infty} c_n x^n = x^{-1} \sum_{n=0}^{\infty} c_n x^n \tag{9.185}$$

Because the difference between the two roots is an integer, the second solution may have one of two forms. We were only asked to find one solution, so our easiest option is to seek the solution in the form of Equation 9.185. Before we begin substituting this form into Equation 9.182 it is convenient to rewrite the form as

$$y(x) = x^{-1} \sum_{n=0}^{\infty} c_n x^n = \sum_{n=0}^{\infty} c_n x^{n-1} \tag{9.186}$$

Our objective is to work the ODE from the form

$$a(x)\frac{d^2}{dx^2}\left(\sum_{n=0}^{\infty}c_nx^{n+\alpha}\right)+a(x)p(x)\frac{d}{dx}\left(\sum_{n=0}^{\infty}c_nx^{n+\alpha}\right)+a(x)q(x)\left(\sum_{n=0}^{\infty}c_nx^{n+\alpha}\right)=0 \quad (9.187)$$

into the form $\sum k_m x^m = 0$ and then find coefficients $c_n$ such that $k_m = 0$ for each $m$. From Equation 9.186,

$$\frac{dy}{dx}=\sum_{n=0}^{\infty}c_n(n-1)x^{n-2}$$

$$\frac{d^2y}{dx^2}=\sum_{n=0}^{\infty}c_n(n-1)(n-2)x^{n-3}$$

$$(9.188)$$

Substituting Equations 9.186 and 9.188 into Equation 9.182 we have

$$x^2\left(\sum_{n=0}^{\infty}c_n(n-1)(n-2)x^{n-3}\right)+5x\left(\sum_{n=0}^{\infty}c_n(n-1)x^{n-2}\right)+(x+3)\left(\sum_{n=0}^{\infty}c_nx^{n-1}\right)=0$$

$$\sum_{n=0}^{\infty}((n-1)(n-2)+5(n-1)+3)c_nx^{n-1}+\sum_{n=0}^{\infty}c_nx^n=0 \quad (9.189)$$

The first sum in Equation 9.189 has an $x^{-1}$ term and has different exponents than the second (i.e., each element in the sum is written in terms of $x^{n-1}$ as compared to $x^n$). We isolate the $x^{-1}$ term from the first sum and define $m = n - 1$ in the remainder of it, and then Equation 9.189 can be written as

$$((0-1)(0-2))+5(0-1)+3)c_0x^{-1}$$

$$+\sum_{n=0}^{\infty}((n-1)(n-2)+5(n-1)+3)c_nx^{n-1}+\sum_{n=0}^{\infty}c_nx^n=0$$

$$(9.190)$$

$$(0)c_0x^{-1}+\sum_{m=0}^{\infty}(m(m-1)+5m+3)c_{m+1}x^m+\sum_{n=0}^{\infty}c_nx^n=0 \quad (9.191)$$

The coefficient of $c_0$ is zero by virtue of the indicial equation, which allows us a nonzero $c_0$. Then we can change the index in the second sum from $n$ to $m$ and Equation 9.191 becomes

$$\sum_{m=0}^{\infty}((m(m-1)+5m+3)c_{m+1}+c_m)x^m=0 \quad (9.192)$$

Equation 9.192 is in the form $\sum k_m x^m = 0$, and so we must now find coefficients $c_n$ such that $k_m = 0$ for each $m$. This gives us the recurrence relation

$$(m(m-1)+5m+3)c_{m+1}+c_m = 0 \tag{9.193}$$

which can be simplified to

$$c_{m+1} = -\frac{c_m}{(m+1)(m+3)} \tag{9.194}$$

Then

$$c_1 = -\frac{c_0}{3}$$

$$c_2 = -\frac{c_1}{8} = \frac{c_0}{24}$$

$$c_3 = -\frac{c_2}{15} = -\frac{c_0}{360} \tag{9.195}$$

$$c_4 = -\frac{c_3}{24} = \frac{c_0}{8640}$$

and

$$y(x) = \frac{c_0}{x}\left(1 - \frac{x}{3} + \frac{x^2}{24} - \frac{x^3}{360} + \frac{x^4}{8640} - \cdots\right) \tag{9.196}$$

**Problems to Solve**

**P9.1** Consider the heated rod example in Section 9.1.1.1. If the rod loses heat due to convection along its entire length, the equation for temperature is modified to

$$\frac{\partial^2 T}{\partial x^2} - \lambda T = \frac{1}{\alpha}\frac{\partial T}{\partial t} \quad T(x,0) = 0 \quad x \neq 0$$

$$T(0,0) = T_s \Delta x \delta(0)$$

Solve these equations. Hint: Try $T(x,t) = T_*(x,t)h(t)$ where $T_*(x,t)$ satisfies

$$\frac{\partial^2 T_*}{\partial x^2} = \frac{1}{\alpha}\frac{\partial T_*}{\partial t} \quad \begin{array}{l} T_*(x,0) = 0 \\ T_*(0,0) = T_s \Delta x \delta(0) \end{array}$$

**P9.2** Consider the heated rod example analyzed in Section 9.1.2. A rod of length $L$ subject to uniform heating across it and to both ends being held at zero temperature will have the steady-state temperature distribution $4T_0 x(L-x)/L^2$. See Figure 9.16. Use a Fourier series expansion to find the temperature as a function of space and time if the heating is removed. The equations to be solved are

$$\frac{\partial^2 T}{\partial x^2} = \frac{1}{\alpha}\frac{\partial T}{\partial t} \quad T(x,0) = 4T_0 x\frac{(L-x)}{L^2} \quad \begin{array}{l} T(0,t) = 0 \\ T(L,t) = 0 \end{array}$$

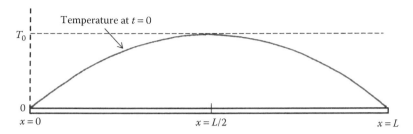

**FIGURE 9.16** Temperature distribution across a uniformly heated rod.

Check your results by evaluating your series solution for $T(L/2,0)$ and comparing it to the given initial value $T_0$. If you need to evaluate the sum of an infinite series use web resources.

**P9.3** Consider the traveling wave in spherical coordinates analyzed in Section 9.2.1.2. In the Earth's slightly conductive atmosphere the equation for the electric field is modified to

$$\frac{\partial^2 E_i}{\partial R^2} + \frac{2}{R}\frac{\partial E_i}{\partial R} - \mu\sigma\frac{\partial E_i}{\partial t} - \frac{1}{c^2}\frac{\partial^2 E_i}{\partial t^2} = 0$$

where:
 $\mu$ is the magnetic permeability
 $\sigma$ is the conductivity of the atmosphere
 Show that an approximate solution to this equation is

$$E_i(R,t) = E_0 \exp\left[-\left(\mu\sigma c/2\right)R\right]\cos\left[\left(\omega/c\right)(R-ct)\right]/R \tag{9.197}$$

Hints: Noting that the outward propagating solution in Section 9.2.1.2 was of the form

$$E_i(R,t) = E_0 \cos(kR - \omega t)/R$$

try here the complex solution

$$E_i(R,t) = E_0 \exp(i(kR - \omega t))/R$$

Show that $k$ must satisfy the equation

$$k^2 = \left(\omega/c\right)^2 + i\mu\sigma\omega$$

Use the approximation $\sqrt{1-\varepsilon} \cong 1-\varepsilon/2$ for $\varepsilon \ll 1$. Show that Equation 9.197 implies that the magnitude of the wave attenuates with distance due to the atmosphere's slight conductivity.

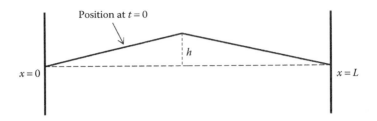

**FIGURE 9.17**   Displacement of a taut string pulled at its middle.

**P9.4** A taut string is pulled at its middle to a stationary position and then released. See Figure 9.17. Using a Fourier series expansion, determine the subsequent motion. The equations to be solved are

$$\frac{\partial^2 u}{\partial x^2} = \frac{1}{a^2}\frac{\partial^2 u}{\partial t^2} \qquad \begin{array}{l} u(0,t)=0 \\[4pt] u(L,0)=0 \end{array} \qquad \begin{array}{ll} u(x,0)=\dfrac{2hx}{L} & x<\dfrac{L}{2} \\[8pt] u(x,0)=2h\left(1-x/L\right) & x>\dfrac{L}{2} \end{array} \qquad \frac{\partial u}{\partial t}(x,0)=0$$

Check your results by evaluating your series solution for $u(L/2,0)$ and comparing it to the given initial value $h$. If you need to evaluate the sum of an infinite series use web resources.

**P9.5** Using a power series expansion, solve

$$\frac{d^2 y}{dx^2} + \frac{1}{x}\frac{dy}{dx} - y = 0, \qquad y(0)=1, \qquad \frac{dy}{dx}(0)=0$$

This is the Modified Bessel's Equation of order zero.

**P9.6** Using a power series expansion, solve

$$\frac{d^2 y}{dx^2} + (2+x)\frac{dy}{dx} + (3+x^2)y = 0, \qquad y(0)=4, \qquad \frac{dy}{dx}(0)=-5$$

Find the recursive relation and calculate the first five terms.

**P9.7** Using a power series expansion, solve

$$\frac{d^2 y}{dx^2} + (4+x^2)\frac{dy}{dx} + (5+x)y = 0, \qquad y(0)=-2, \qquad \frac{dy}{dx}(0)=3$$

Find the recursive relation and calculate the first five terms.

**P9.8** Show that substituting $y = u/\sqrt{x}$ transforms Bessel's equation of order $m$

$$\frac{d^2 y}{dx^2} + \frac{1}{x}\frac{dy}{dx} + \left(1-\frac{m^2}{x^2}\right)y = 0$$

into

$$\frac{d^2u}{dx^2} + c(x)u = 0$$

where

$$\lim_{x \to \infty} c(x) = 1$$

Show how this lends plausibility to (but does not prove) the assertion in Section 9.2.4 that the two solutions $y_i$ to Bessel's equation can be approximated for large $x$ as

$$y_i = \frac{1}{\sqrt{x}}(a_i \cos x + b_i \sin x)$$

**P9.9** Between any two adjacent zeros of $J_m(x)$ there is a zero of $J_{m+1}(x)$ and a zero of $Y_m(x)$. To demonstrate an example of this, use the computational resources available on the web to compute adjacent zeros of any Bessel function of the first kind, for example, the third and fourth zeros of $J_{10}(x)$. Use the web resources to show that there is a zero of $J_{11}(x)$ and a zero of $Y_{10}(x)$ between them.

**P9.10** Show by term-by-term addition in the power series that

$$J_{n-1}(x) + J_{n+1}(x) = \frac{2n}{x}J_n(x)$$

**P9.11** Show by term-by-term integration that

$$\int_0^x z^n J_{n-1}(z)dz = x^n J_n(x)$$

In each of Problems 9.12, 9.13, and 9.14 below, using the general rules in Section 9.2.6.2, find the solution corresponding to the largest indicial root. Find the recursive relation and calculate the first five terms in the power series.

**P9.12** $x\dfrac{d^2y}{dx^2} + (x-2)\dfrac{dy}{dx} + (x+1)y = 0$

**P9.13** $x^2\dfrac{d^2y}{dx^2} + (x+2x^2)\dfrac{dy}{dx} + (4x-4)y = 0$

**P9.14** $x^2\dfrac{d^2y}{dx^2} + (4x-3x^2)\dfrac{dy}{dx} + (3x^2 - x + 2)y = 0$

**P9.15** Using the general rules in Section 9.2.6.2, find the solutions to

$$x^2\frac{d^2y}{dx^2} + 7x\frac{dy}{dx} + 8y = 0$$

**P9.16** Using the general rules in Section 9.2.6.2, find the solutions to

$$x^2 \frac{d^2 y}{dx^2} + 7x \frac{dy}{dx} + 34y = 0$$

Then solve the equation again using the substitution $t = \ln x$ and apply the results of Chapter 5 concerning how to convert complex solutions to real ones. Show how this motivates replacing $x^{\lambda \pm i\omega}$ with $x^\lambda \cos(\omega \ln x)$ and $x^\lambda \sin(\omega \ln x)$ in the rules of 9.2.6.2.

**P9.17** Consider a very long cantilevered beam of very short, identical, and perfectly rigid segments joined with strong coiled torsional springs. See Figure 9.18. The beam is initially deflected from vertical by a distribution of forces, held, and then released in a manner similar to that described for the continuous cantilevered beam in Section 9.4. Challenge Problem CP9.3 shows that subsequent motions of the beam can be approximately modeled by the equations

$$\frac{\partial^3 y}{\partial x^3} - \left(\frac{\rho}{2k}\right)\frac{\partial^2 y}{\partial t^2} = 0 \tag{9.198}$$

$$
\begin{aligned}
y(x,0) &= y_0(x) \\
\frac{\partial y}{\partial t}(x,0) &= 0
\end{aligned}
\qquad
\begin{aligned}
y(0,t) &= 0 \\
\frac{\partial y}{\partial x}(0,t) &= 0 \\
\frac{\partial^2 y}{\partial x^2}(h,t) &= 0
\end{aligned}
\tag{9.199}
$$

where:
   $\rho = m/h$, $m$ is the mass of the beam
   $k$ is the constant of each spring

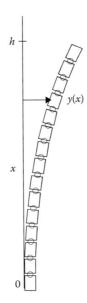

**FIGURE 9.18**   Deflection of a beam consisting of many rigid segments joined by torsional springs.

Show that the solution to this boundary value problem is of the form

$$y(x,t) = \sum_{n=1}^{\infty} c_n w_n(x) u_n(t) \tag{9.200}$$

where:

$$u_n(t) = \cos \omega_n t \tag{9.201}$$

$$\omega_n = \left( \frac{2k}{\rho h^3} \right)^{1/2} (q_n h)^{3/2} \tag{9.202}$$

$$w_n(x) = e^{-q_n x} - e^{q_n x/2} \cos\left( \sqrt{3} q_n x/2 \right) + \sqrt{3} e^{qx/2} \sin\left( \sqrt{3} q_n x/2 \right) \tag{9.203}$$

and $q_n h$ is the $n$th solution to

$$1 + 2e^{3qh/2} \cos\left( \sqrt{3} q h/2 \right) = 0 \tag{9.204}$$

Word Problem WP9.4 invites you to show that the eigenfunctions of this boundary value problem are not orthogonal. See also Challenge Problem CP9.5.

## Word Problems

**WP9.1** Consider a uniform rod of length $L$ and diffusivity $\alpha$ with an initial temperature distribution given by $T_1(1-\exp(-m\pi x/L))(1-\exp(-m\pi(L-x)/L))$, where $x$ is the distance along the rod from one end. The ends of the rod are held at zero temperature. Show that the subsequent temperature distribution as a function of time is given by

$$T(x,t) = \sum_{n=0}^{\infty} c_{2n+1} \exp\left( -\left( (2n+1)\pi/L \right)^2 \alpha t \right) \sin\left( (2n+1)\pi x/L \right)$$

where

$$c_n = \frac{4T_1}{n\pi} \left( (1+e^{-m\pi}) \left( \frac{m^2}{m^2+n^2} \right) \right)$$

**WP9.2** Show that when a guitar string is plucked, its subsequent motion can be represented as either a series of waves traveling back and forth along the string or a series of standing waves with fixed nodes (locations of zero amplitude). Recall P9.4 from the Problems to Solve section.

**WP9.3** As depicted in Figure 9.19a, a metal duct with uniform cross section in the form of a square of side $L$ has three sides held at zero voltage and the other with a voltage distribution $V_B(x)$. Given that this is a situation governed by Laplace's Equation,

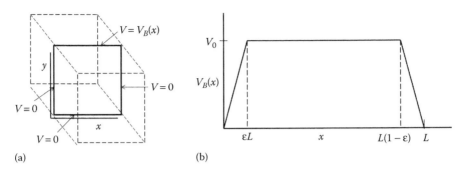

**FIGURE 9.19** Electrified duct and its voltage distribution: (a) electrified duct and (b) voltage distribution.

if $V_B(x)$ is as illustrated in Figure 9.19b, show that the voltage distribution inside the duct is given by

$$V(x,y) = \sum_{n=0}^{\infty} c_{2n+1} \sinh\big((2n+1)\pi y/L\big)\sin\big((2n+1)\pi x/L\big)$$

where

$$c_n = \frac{4V_0}{(n\pi)^2 \varepsilon} \frac{\sin(n\pi\varepsilon)}{\sinh(n\pi)}$$

**WP9.4** We consider again the boundary value problem introduced in P9.17 in the Problems to Solve section.

*Part (a):*
Show that for any sufficiently smooth functions $u(x)$ and $v(x)$ the following form of Lagrange's identity holds:

$$\frac{d}{dx}\left( u\frac{d^2v}{dx^2} - \frac{du}{dx}\frac{dv}{dx} + v\frac{d^2u}{dx^2} \right) = u\frac{d^3v}{dx^3} + v\frac{d^3u}{dx^3}$$

*Part (b):*
Use the form of Lagrange's identity in part(a) to show that if $u(x) = w_i(x)$ and $v(x) = w_j(x)$ are eigenfunctions of the boundary value problem in P9.17, they are not orthogonal under the integral $\int_0^h (...)dx$. Hint: Integrate the identity and recall that

$$\frac{d^3 w_n}{dx^3} = -\lambda_n w_n \qquad w_n(0) = 0 \qquad \frac{dw_n}{dx}(0) = 0 \qquad \frac{d^2 w_n}{dx^2}(h) = 0$$

## Challenge Problems

**CP9.1** Following on to the vibrating membrane problem in Section 9.2.2.2, evaluate

$$c_m = \frac{\displaystyle\int_0^{r_0} h\big(1 - r^2/r_0^2\big) J_0\big(x_m r/r_0\big)}{\left(\dfrac{r_0^2}{2} J_1^2(x_m)\right)} \tag{9.205}$$

*Guidance:*
To compute $c_m$, we need to evaluate

$$I = I_1 - I_2 = \int_0^{r_0} J_0\left(x_m r/r_0\right) r\, dr - \left(1/r_0^2\right) \int_0^{r_0} J_0\left(x_m r/r_0\right) r^3\, dr \qquad (9.206)$$

In both integrals let $z = x_m r/r_0$. Then

$$I_1 = \int_0^{r_0} J_0\left(x_m r/r_0\right) r\, dr = \left(r_0/x_m\right)^2 \int_0^{x_m} J_0(z) z\, dz \qquad (9.207)$$

Reference [18] provides in its Equation 11.3.20 the general formula

$$\int_0^x z^n J_{n-1}(z)\, dz = x^n J_n(x) \qquad (9.208)$$

Use Equation 9.208 to show that

$$I_1 = \left(r_0^2/x_m\right) J_1(x_m) \qquad (9.209)$$

With $z = x_m r/r_0$,

$$I_2 = \left(r_0^2/x_m^4\right) \int_0^{x_m} z^3 J_0(z)\, dz \qquad (9.210)$$

Integrating by parts, let $u = z^2$ and $dv = z J_0(z)$. Then use Equation 9.208 to show that

$$I_2 = \left(r_0^2/x_m\right) J_1(x_m) - 2\left(r_0^2/x_m^4\right) \int_0^{x_m} z^2 J_1(z)\, dz \qquad (9.211)$$

Referring a third time to Equation 9.208, show that

$$\int_0^{x_m} z^2 J_1(z)\, dz = x_m^2 J_2(x_m) \qquad (9.212)$$

From [18], Equation 9.1.27,

$$J_{n-1}(x) + J_{n+1}(x) = \frac{2n}{x} J_n(x) \qquad (9.213)$$

Use Equation 9.213 and the fact that $J_0(x_m) = 0$ to derive

$$J_2(x_m) = \frac{2}{x_m} J_1(x_m) \qquad (9.214)$$

Use Equations 9.206, 9.209 and 9.211, 9.212 and 9.214 to arrive at

$$I = 4\left(r_0^2 / x_m^3\right) J_1(x_m) \tag{9.215}$$

Show that Equations 9.205 and 9.215 result in Equation 9.83.

**CP9.2** We consider here the problem posed in Section 9.3.1: a temperature profile $T_S(\theta)$ is imposed on the surface of a sphere. See Figure 9.13. We are challenged to find the temperature distribution $T(R,\theta)$ inside the sphere. In particular, what is the temperature at the center of the sphere? The governing equation is

$$\frac{\partial^2 T}{\partial R^2} + \frac{2}{R}\frac{\partial T}{\partial R} + \frac{1}{R^2}\frac{\partial^2 T}{\partial \theta^2} + \frac{\cot \theta}{R^2}\frac{\partial T}{\partial \theta} = 0 \tag{9.216}$$

*Guidance*: Begin by separating variables. Let

$$T(R,\theta) = G(R)H(\theta)$$

For reasons that will become clear later, let the separation constant be $n(n+1)$ where $n$ is an integer. Show that the resulting ODEs are

$$R^2 \frac{d^2 G}{dR^2} + 2R\frac{dG}{dR} - n(n+1)G = 0 \tag{9.217}$$

$$\frac{d^2 H}{d\theta^2} + \cot \theta \frac{dH}{d\theta} + n(n+1)H = 0 \tag{9.218}$$

Using the general rules in Section 9.2.6, show that Equation 9.217 has a regular singular point at $R = 0$ and that the solution that is finite at the origin has the form

$$G = R^n \sum_{m=0}^{\infty} c_m R^m \tag{9.219}$$

Next substitute Equation 9.219 into Equation 9.217 and solve for the coefficients $c_m$. Show that the result is simply

$$G = c_0 R^n \tag{9.220}$$

Without loss of generality you can set $c_0 = 1$ since you will address the boundary condition as you solve Equation 9.218. To solve Equation 9.218 make the change of variable $x = \cos \theta$. Show that Equation 9.218 becomes

$$(1 - x^2)\frac{d^2 H}{dx^2} - 2x\frac{dH}{dx} + n(n+1)H = 0 \tag{9.221}$$

Equation 9.221 is called *Legendre's equation*. It is defined for $-1 \le x \le 1$.

Using the general rules in Section 9.2.6, show that the point $x = 0$ is an ordinary point of Equation 9.221, hence its two solutions can both be expressed by power series of the form

$$H = \sum_{m=0}^{\infty} a_m x^m \tag{9.222}$$

Substitute Equation 9.222 into Equation 9.221 and arrive at the recursion relation

$$a_{m+2} = \frac{m(m+1) - n(n+1)}{(m+1)(m+2)} a_m \tag{9.223}$$

You can form two independent series solutions from Equation 9.223, one with $a_0 = 1$ and $a_1 = 0$, forming an even series, and the other with $a_0 = 0$ and $a_1 = 1$, forming an odd series. Show that one of these two series terminates when $m = n$, a result of having chosen the separation constant to be $n(n+1)$. Using Equation 9.223, the terminating series solutions to Legendre's equation (Equation 9.221) are called Legendre polynomials. Scaling the polynomials so that $P_n(1) = 1$, show that first five are

$$P_0(x) = 1$$
$$P_1(x) = x$$
$$P_2(x) = (3x^2 - 1)/2$$
$$P_3(x) = (5x^3 - 3x)/2$$
$$P_4(x) = (35x^4 - 30x^2 + 3)/8$$

Show that the terminating series is finite at the end points ($x = \pm 1$) and hence a viable solution.

You tried a solution $T(R, \theta) = G(R)H(\theta)$ and found an infinite number of possibilities

$$G_n(R)H_n(\theta) = R^n P_n(\cos \theta)$$

(Recall that $x = \cos \theta$.) Now seek a solution in the form of a sum of these:

$$T(R, \theta) = \sum_{n=0}^{\infty} c_n R^n P_n(\cos \theta) \tag{9.224}$$

For a given boundary condition $T(R_0, \theta) = T_S(\theta)$, determine the coefficients $c_n$ via

$$\sum_{n=0}^{\infty} c_n R_0^n P_n(\cos \theta) = T_S(\theta) \tag{9.225}$$

*Given* that Legendre polynomials satisfy

$$\int_{-1}^{1} P_n(x)P_m(x)dx = 0 \quad m \neq n$$

$$\int_{-1}^{1} P_n(x)P_m(x)dx = \frac{2}{(2n+1)} \quad m = n$$

or equivalently,

$$\int_{0}^{\pi} P_n(\cos\theta)P_m(\cos\theta)\sin\theta d\theta = 0 \quad m \neq n$$

$$\int_{0}^{\pi} P_n(\cos\theta)P_m(\cos\theta)\sin\theta d\theta = \frac{2}{(2n+1)} \quad m = n$$

show that the coefficient $c_n$ satisfying Equation 9.225 is given by

$$c_n = \frac{(2n+1)}{2R_0^n}\int_{0}^{\pi} P_n(\cos\theta)T_S(\theta)\sin\theta d\theta \tag{9.226}$$

which completes the solution.

Show that, from Equations 9.224 and 9.226, the temperature at the center of the sphere is

$$\frac{1}{2}\int_{0}^{\pi} T_S(\theta)\sin\theta d\theta$$

**CP9.3** The challenge here is to derive the equations of motion for the segmented beam described in Problem 9.17 and pictured in Figure 9.18.

*Guidance*: Determine the equations of motion for an individual segment and then transition from a discrete model to a continuous one (i.e., from a finite-dimensional space to an infinite-dimensional one) by taking the limit as the number of segments goes to infinity.

Let the number of segments be $N$. Consider the $i$th segment. Newton's laws of translational and rotational motion are

$$m_i\ddot{y}_i = \sum \text{forces}$$

$$I_i\ddot{\theta}_i = \sum \text{moments} \tag{9.227}$$

where $m_i$ is the mass of segment $i$ and $I_i$ its moment of inertia about its center of gravity. See Figure 9.20.

Now let the force exerted on segment $i$ by the connection to the segment below be $F_i$. Then from Newton's law of action and reaction the force on segment $i$ due to its connection with the segment above it is $-F_{i+1}$. Similarly, let the moment (torque) on segment $i$ due to the torsional spring below be $\tau_i$. Then the moment on segment $i$ due to the torsional spring above it is $-\tau_{i+1}$. By definition

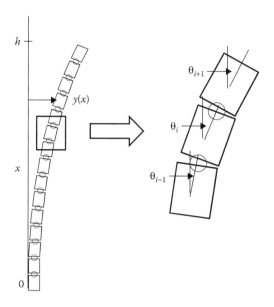

**FIGURE 9.20** Variables for analysis of segmented beam.

$$\tau_i = -k(\theta_i - \theta_{i-1})$$

$$\tau_{i+1} = -k(\theta_{i+1} - \theta_i)$$

Let $h_i = h/N$. Then Equations 9.227 become

$$m_i \ddot{y}_i = F_i - F_{i+1} \qquad (9.228)$$

$$I_i \ddot{\theta}_i = k(\theta_{i+1} - 2\theta_i + \theta_{i-1}) - (F_i - F_{i+1})(h_i/2) \qquad (9.229)$$

Using Equation 9.228 to replace the $F_i - F_{i+1}$ term in Equation 9.229 and rearranging, we have

$$k(\theta_{i+1} - 2\theta_i + \theta_{i-1}) = I_i \ddot{\theta}_i + m_i \ddot{y}_i (h_i/2) \qquad (9.230)$$

Now let $h_i = h/N = \Delta x$ and then show that

$$m_i = m/N = (m/h)(h/N) = \rho \Delta x$$

$$I_i = m_i h_i^2/12 = \rho(\Delta x)^3/12 \qquad (9.231)$$

$$y_i = \left( \sum_{j=1}^{i-1} \theta_j + \theta_i/2 \right) \Delta x$$

Show that substituting Equation 9.231 into Equation 9.230 results in

$$k(\theta_{i+1} - 2\theta_i + \theta_{i-1}) = \left(\frac{\rho}{2}\right)(\Delta x)^3 \left(\frac{2}{3}\ddot{\theta}_i + \sum_{j=1}^{i-1}\ddot{\theta}_j\right) \tag{9.232}$$

Now divide Equation 9.232 by $(\Delta x)^2$ and apply

$$\lim_{\Delta x \to 0} \frac{\theta_{i+1} - 2\theta_i + \theta_{i-1}}{(\Delta x)^2} = \frac{\partial^2 \theta(x,t)}{\partial^2 x}$$

$$\lim_{\Delta x \to 0} \frac{2}{3}\ddot{\theta}_i \Delta x + \sum_{j=1}^{i-1}\ddot{\theta}_j \Delta x = \int_0^x \frac{\partial^2 \theta(u,t)}{\partial^2 t}\,du = \frac{\partial^2 y(x,t)}{\partial^2 t}$$

$$\frac{\partial y(x,t)}{\partial x} = \theta(x,t)$$

Show that the resulting PDE is

$$\frac{\partial^3 y}{\partial x^3} - \left(\frac{\rho}{2k}\right)\frac{\partial^2 y}{\partial t^2} = 0$$

Also show that writing the equation of motion for the last segment in the beam results in

$$-k\frac{(\theta_N - \theta_{N-1})}{\Delta x} = \left(\frac{\rho}{2}\right)(\Delta x)^2 \left(\frac{2}{3}\ddot{\theta}_N + \sum_{j=1}^{N-1}\ddot{\theta}_j\right)$$

which in the limit as $\Delta x \to 0$ leads to the boundary condition

$$\frac{\partial^2 y}{\partial x^2}(h,t) = 0$$

**CP9.4** Following on to the beam vibration problem in Section 9.4.2, *Part (a):* Evaluate

$$c_n = \frac{\displaystyle\int_0^h w_n(x)y(x,0)\,dx}{\displaystyle\int_0^h w_n^2(x)\,dx}$$

where

$$w_n(x) = \cosh p_n x - \cos p_n x - \sigma_n(\sinh p_n x - \sin p_n x) \tag{9.233}$$

$$\sigma_n = \frac{\cosh p_n h + \cos p_n h}{\sinh p_n h + \sin p_n h} \tag{9.234}$$

$$y(x,0) = \frac{P}{6EI}(3hx^2 - x^3) \tag{9.235}$$

You are given that straightforward but tedious calculations lead to

$$\int_0^h w_n(x)\, y(x,0)\, dx =$$

$$\left(\frac{2Ph^4}{EI}\right)\frac{\cosh p_n h \sin p_n h - \sinh p_n h \cos p_n h}{(p_n h)^4(\sinh p_n h + \sin p_n h)} \tag{9.236}$$

and

$$\int_0^h w_n^2(x)\, dx = h \tag{9.237}$$

hence

$$c_n = \left(\frac{2Ph^3}{EI}\right)\frac{\cosh p_n h \sin p_n h - \sinh p_n h \cos p_n h}{(p_n h)^4(\sinh p_n h + \sin p_n h)} \tag{9.238}$$

The challenge is to show that this can be simplified to

$$c_n = \left(\frac{2Ph^3}{EI}\right)\frac{(-1)^{n-1}}{(p_n h)^4} \tag{9.239}$$

*Part (b):*
Use the same arguments that lead to Equation 9.239 to show that

$$w_n(h) = 2(-1)^{n-1} \tag{9.240}$$

and so

$$y(h,0) = \sum_{n=1}^{\infty} c_n w_n(h) u_n(0) = \left(\frac{4Ph^3}{EI}\right)\sum_{n=1}^{\infty}\frac{1}{(p_n h)^4} \tag{9.241}$$

*Guidance:*
Use the characteristic equation

$$\cos ph \cosh ph = -1 \tag{9.242}$$

to show that

$$\cosh p_n h \sin p_n h = (-1)^{n-1}\sinh p_n h \tag{9.243}$$

and then use Equations 9.242 and 9.243 to show that

$$-\sinh p_n h \cos p_n h = (-1)^{n-1} \sin p_n h \qquad (9.244)$$

Show that Equations 9.243 and 9.244 lead directly to Equations 9.239 through 9.241.

**CP9.5** Consider again the segmented beam analyzed in Problem 9.17 and in Challenge Problem CP9.3. Suppose its base is subject to a sideways oscillation $z(t) = z_0 \sin \omega t$. Assuming that dissipative effects not modeled in the segmented beam PDE but occurring in real life would cause solution terms due to initial conditions to die away, what is the resulting motion of the beam in the steady state? What frequencies $\omega$ cause a resonance in the beam's response? What is the shape of the beam at low input frequencies?

*Guidance*: The equations of motion and boundary conditions are as given in Problem 9.17, except that, here, $y(0,t) = z_0 \sin \omega t$. Try the steady-state solution $y(x,t) = w(x)\sin \omega t$. Substitute this trial solution into the PDE. Show that

$$w(x) = c_1 e^{-qx} + c_2 e^{qx/2} \cos\left(\sqrt{3}qx/2\right) + c_3 e^{qx/2} \sin\left(\sqrt{3}qx/2\right) \qquad (9.245)$$

where

$$q = \left(\frac{\omega^2 \rho}{2k}\right)^{1/3}$$

Given the boundary conditions, show that

$$c_1 = z_0 e^{3qh/2}\left(\frac{\cos\left(\sqrt{3}qh/2\right) + \left(1/\sqrt{3}\right)\sin\left(\sqrt{3}qh/2\right)}{1 + 2e^{3qh/2}\cos\left(\sqrt{3}qh/2\right)}\right)$$

$$c_2 = z_0 \left(\frac{1 + e^{3qh/2}\left(\cos\left(\sqrt{3}qh/2\right) - \left(1/\sqrt{3}\right)\sin\left(\sqrt{3}qh/2\right)\right)}{1 + 2e^{3qh/2}\cos\left(\sqrt{3}qh/2\right)}\right) \qquad (9.246)$$

$$c_3 = \frac{z_0}{\sqrt{3}}\left(\frac{-1 + e^{3qh/2}\left(\cos\left(\sqrt{3}qh/2\right) + \sqrt{3}\sin\left(\sqrt{3}qh/2\right)\right)}{1 + 2e^{3qh/2}\cos\left(\sqrt{3}qh/2\right)}\right)$$

Using Equation 9.246, show that input frequencies causing resonance in the beam's response are the beam's natural frequencies, as given in Problem 9.17.

Expand the solution in a Taylor series for small $q$ and show that, for low input frequencies,

$$w(x) \cong z_0 \left[ 1 + \left( \frac{\omega^2 \rho}{2k} \right) \left( \frac{hx^2}{2} - \frac{x^3}{6} \right) \right]$$

(Hint: Begin the expansion using the generic form $(c_1, c_2, c_3)$ for the coefficients rather than Equation 9.246. Look for simplifications with the coefficients. Then expand the coefficients as necessary.)

Compare with the results of Challenge Problem CP9.6, below.

**CP9.6** Consider again the continuous cantilevered beam analyzed in Section 9.4.2. Suppose its base is subject to a sideways oscillation $z(t) = z_0 \sin \omega t$. Assuming that dissipative effects not modeled in the beam equation but occurring in real life would cause solution terms due to initial conditions to die away, what is the resulting motion of the beam in the steady state? What frequencies $\omega$ cause a resonance in the beam's response? What is the shape of the beam at low input frequencies?

*Guidance*: The equations of motion and boundary conditions are as given in Section 9.4.2, except that, here, $y(0,t) = z_0 \sin \omega t$. Try the steady-state solution $y(x,t) = w(x) \sin \omega t$. Substitute this trial solution into the PDE. Show that

$$w(x) = c_1 e^{px} + c_2 e^{-px} + c_3 \cos px + c_4 \sin px \tag{9.247}$$

where

$$p = \left( \frac{\omega^2 \rho}{EI} \right)^{1/4}$$

Given the boundary conditions, show that

$$c_1 = \frac{z_0}{4} \left( \frac{1 + e^{-ph}(\cos ph - \sin ph)}{1 + \cosh ph \cdot \cos ph} \right)$$

$$c_2 = \frac{z_0}{4} \left( \frac{1 + e^{ph}(\cos ph + \sin ph)}{1 + \cosh ph \cdot \cos ph} \right)$$

$$c_3 = \frac{z_0}{2} \left( \frac{1 + \cosh ph \cdot \cos ph - \sinh ph \cdot \sin ph}{1 + \cosh ph \cdot \cos ph} \right) \tag{9.248}$$

$$c_4 = \frac{z_0}{2} \left( \frac{\cosh ph \cdot \sin ph + \sinh ph \cdot \cos ph}{1 + \cosh ph \cdot \cos ph} \right)$$

Using Equation 9.248, show that input frequencies causing resonance in the beam's response are the beam's natural frequencies, as given in Section 9.4.2. Expand the solution in a Taylor's series for small $p$ and show that, for low input frequencies,

$$w(x) \cong z_0 \left[ 1 + \left( \frac{\omega^2 \rho}{EI} \right) \left( \frac{h^2 x^2}{4} - \frac{hx^3}{6} + \frac{x^4}{24} \right) \right]$$

(Hint: Begin the expansion using the generic form $(c_1, c_2, c_3, c_4)$ for the coefficients rather than Equation 9.248. Look for simplifications with the coefficients. Then expand the coefficients as necessary.)

Compare with results of Challenge Problem CP9.5.

# Appendix

## ANSWERS TO SELECTED PROBLEMS

### CHAPTER 1

**P1.1** 2/5

**P1.4** $-3, -7$

**P1.7** $-4 + 6i, -4 - 6i$

**P1.10** $x_1 = -2, x_2 = 5$

**P1.13** $z = (t - \tau)^2 \left( \exp(-3(t - \tau)) - \exp(-5(t - \tau)) \right)$

**P1.19** $e^{-4t} \left( (1 - 4t) \sin 8t + 8t \cos 8t \right)$

**P1.22** $I(t) = -(t/2)e^{-2t} + (1/4)(1 - e^{-2t})$

**WP1.4** $R_p = 10^3$ ohms

**WP1.5** $V = \pi \left( (R_0 + r_0)^2 + R_0^2 \right) L - 2\pi (R_0 + r_0) R_0^2 \arcsin \left( L/(2R_0) \right)$

$$- \pi (R_0 + r_0) R_0 L \sqrt{1 - \left( L/(2R_0) \right)^2} - \pi L^3 / 12$$

### CHAPTER 2

**P2.2** $y(t) = 5 \exp \left( 2(1 - \sqrt{1 - t}) \right)$

**P2.5** $y(t) = 4\sqrt{\cos 2t + t^2}$

**P2.8** $y(t) = (7/2)e^{-2t} - (1/2)e^{-4t}$

**P2.11** $y(t) = \left( 4\pi^2 + 2\pi + \sin 2\pi t \right)/(4\pi^2 t) - \cos 2\pi t/(2\pi)$

**P2.14** $y(t) = 8 \exp(-6x(t)) + (4/3)x(t) - (2/9)\left( 1 - \exp(-6x(t)) \right)$

where $x(t) = \dfrac{t/2}{(1 + t^2/4)^{1/2}}$

**WP2.3** Gain $= 0.447$, phase lag $= 63.4°$

**WP2.5** (a) $T(t) = T_S + (T_0 - T_S) \exp \left( -\left( \dfrac{4\varepsilon \sigma A T_S^3}{mc} \right) t \right)$

(b) $t_C = 1.34$ h

## CHAPTER 3

**P3.2** $y(t) = \left(1 - \left(1 - (1 + 4t)e^{-4t}\right)/8\right)^{1/2} - 1$

**P3.5** $y(t) = 3(e^{4t} - 1)/(1 - (3/7)e^{4t})$

**P3.7** $y(t) = t^2/4 - (\varepsilon/112)(t^5 - 1/t^2)$

**P3.10** $y_3(t) = (1 - \varepsilon \arcsin t + (\varepsilon \arcsin t)^2)/\sqrt{1 - t^2}$

Guess $y_\infty(t) = 1/\left((1 + \varepsilon \arcsin t)\sqrt{1 - t^2}\right)$

**P3.13** $I(t) = I_0 e^{-\lambda t}/(1 + \varepsilon I_0^2(1 - e^{-2\lambda t}))^{1/2}$

**WP3.3** $h = 1.300 \cdot 10^5$ ft

**WP3.4** $t_C = 856$ s

## CHAPTER 4

$y_0(t) = -2$

$y_1(t) = -2 - 2t + t^2$

**P4.2**

$y_2(t) = -2 - 2t + t^2 + (2/3)t^3$

$y_3(t) = -2 - 2t + t^2 + (1/3)t^3 + (1/6)t^4$

**P4.6** Equilibrium points at $y = -3$ (unstable) and $y = 6$ (stable).

**P4.10** $y(t) = 1/\sqrt{1 - 8t}$, $y(t) \to \infty$ as $t \to 1/8$

**P4.14** $t_E = 1/32$

## CHAPTER 5

**P5.1** $y(t) = (13/2)e^{-3t} - (7/2)e^{-5t}$

**P5.5** $y(t) = 2(2 + 7t)e^{-3t}$

**P5.9** $y(t) = e^{4t}\left(6\cos 5t - (26/5)\sin 5t\right)$

**P5.13** $W(t) = 11e^{-7t}$

$K(t - \tau) = \left(e^{2(t-\tau)} - e^{-9(t-\tau)}\right)/11$

**P5.17** $W(t) = e^{4t}$

$K(t - \tau) = (t - \tau)e^{2(t-\tau)}$

**P5.20** $W(t) = 3e^{-10t}$

$K(t - \tau) = e^{-5(t-\tau)}\sin 3(t - \tau)/3$

**P5.21** $W(t) = 7$

$K(t - \tau) = \sin 7(t - \tau)/7$

**P5.25** $y(t) = (3/2)\left(t/4 - 3/16 + e^{-2t}/4 - e^{-4t}/16\right)$

**P5.30** $y(t) = (1/5)(t - 11/30)e^{4t}$

**P5.35** $y(t) = e^{-4t}\sin 2t - 2te^{-2t}$

**WP5.1** $\theta(t) = \theta_0 e^{-2t}\left(\cos 5t + (2/5)\sin 5t\right)$

**WP5.4** $v(t) = V_0 e^{-t_*/2}\left(\cos(\sqrt{3}t_*/2) - (1/\sqrt{3})\sin(\sqrt{3}t_*/2)\right)$

  where $t_* = 10^5 t$

## Chapter 6

**P6.1** $y(t) = -(12/5)e^{-3t} + e^{-4t}\left((2/5)\cos 2t - (13/10)\sin 2t\right)$

**P6.3** $K(t - \tau) = \left(e^{-2(t-\tau)} - 2e^{-5(t-\tau)} + e^{-8(t-\tau)}\right)/18$

**P6.5** $y(t) = (4/17)(t - (14/17))e^{2t}$

**WP6.1** $\omega_1 = \omega_0\sqrt{2 - \sqrt{2}}$
  $\omega_2 = \omega_0\sqrt{2 + \sqrt{2}}$

## Chapter 7

**P7.1** $y(t) = 4e^{-4t} - e^{-6t}$

**P7.4** $y(t) = -2e^{-4t}\cos 8t - (7/8)e^{-4t}\sin 8t$

**P7.8** $y(t) = e^{-11t}(4 + 50t)$

**P7.11** $y(t) = 3.72514 \cdot 10^{-2}\cos 2t + 1.23145 \cdot 10^{-2}\sin 2t$

  $- 3.72514 \cdot 10^{-2}e^{-10t}\cos 5t - 7.94286 \cdot 10^{-2}e^{-10t}\sin 5t$

**P7.13** $y(t) = (31e^{-t} - 8e^{-4t} + e^{-9t})/8$

**P7.14** $\dot{y}(0^+) = e + \dot{y}_0$

**P7.16** $\lim_{t \to \infty} y(t) = e/c$

**P7.19** $\dddot{u} + 13\ddot{u} + 62\dot{u} + 45u = (4/3)\delta(t) + 18$

**P7.21** $K(t) = \left(e^{-4t} - 5e^{-8t} + 4e^{-9t}\right)/20$

## Chapter 8

**P8.1**
$$
\begin{array}{lll}
t_1 = 0.001 & z_1 = 3.2291 & v_1 = 5.9931 \\
t_2 = 0.002 & z_2 = 3.5068 & v_2 = 5.9868 \\
t_3 = 0.003 & z_3 = 3.8480 & v_3 = 5.9812
\end{array}
$$

**P8.5** $\underline{x}(t) = -7\begin{pmatrix} 1 \\ -1 \end{pmatrix}e^{-3t} + 10\begin{pmatrix} 1 \\ -1/2 \end{pmatrix}e^{-4t}$

**P8.7** $\underline{x}(t) = e^{-3t}\left(\begin{pmatrix} -1 \\ 3 \end{pmatrix}\cos 2t + \begin{pmatrix} 7 \\ -1 \end{pmatrix}\sin 2t\right)$

**P8.10** $\underline{x}(t) = 2\begin{pmatrix} 1 \\ 2 \end{pmatrix}e^{-3t} - 10\begin{pmatrix} 1 \\ 1 \end{pmatrix}te^{-3t}$

**P8.16** $\underline{\Phi}(t) = \begin{pmatrix} (1+4t)e^{-t} & -4te^{-t} \\ 4te^{-t} & (1-4t)e^{-t} \end{pmatrix}$

**P8.18** $\underline{T} = \begin{pmatrix} 2 & -2 \\ -1 & 2 \end{pmatrix}$

**P8.20** $\underline{x}(t) = \begin{pmatrix} (4/15)(e^{5t}-1)+(1/6)(1-e^{-4t}) \\ (4/15)(e^{5t}-1)-(7/12)(1-e^{-4t}) \end{pmatrix}$

**P8.23** $\underline{x}(t) = \begin{pmatrix} 2t-9t^2 \\ 4t+9t^2 \end{pmatrix}e^{-5t}$

**P8.26** $\underline{x}(t) = \begin{pmatrix} -2e^{-5t}+4e^{-7t} \\ -e^{-5t}+4e^{-7t} \end{pmatrix}$

**P8.30** $x_1(t) = 0.4941e^{-9t}\cos 3t - 0.2235e^{-9t}\sin 3t - 0.4941e^{-6t}\cos 2t + 1.0764e^{-6t}\sin 2t$

**P8.31** $\underline{x}(t) = \begin{pmatrix} -13/3 \\ 6 \end{pmatrix}t + \begin{pmatrix} 17/12 \\ -5/6 \end{pmatrix}$

**P8.35** $\underline{x}(t) = e^{2t}\left[\begin{pmatrix} 1.0067 \\ 0.1281 \end{pmatrix}\cos 3t + \begin{pmatrix} 0.8292 \\ 0.7551 \end{pmatrix}\sin 3t\right]$

**WP8.2** $v_{out} = -1.0344\sin 2t - 2.5862\cos 2t$

## CHAPTER 9

**9.4** $u(x,t) = \dfrac{8h}{\pi^2}\sum_{m=0}^{\infty}\dfrac{(-1)^m}{(2m+1)^2}\sin\big((2m+1)\pi x/L\big)\cdot\cos\big((2m+1)\pi at/L\big)$

**9.6** $c_{m+2} = -\dfrac{2}{m+2}c_{m+1} - \dfrac{m+3}{(m+2)(m+1)}c_m - \dfrac{1}{(m+2)(m+1)}c_{m-2}$

$y(x) = 4 - 5x - x^2 + 4x^3 - 23x^4/12 + \ldots$

**9.12** $c_m = -\dfrac{1}{m} c_{m-1} - \dfrac{1}{m(m+3)} c_{m-2}$

$$y(x) = c_0(x^3 - x^4 + 2x^5/5 - 7x^6/90 + 13x^7/2520 + ...)$$

**9.15** $y_1(x) = x^{-2}$ $\quad y_2(x) = x^{-4}$

# References

1. Anderson, J.D., *Introduction to Flight*, McGraw-Hill, New York, 1989.
2. Whittaker, E.T., *A Treatise on the Analytical Dynamics of Particles and Rigid Bodies*, Cambridge University Press, 1917. Available from California Digital Library, https://archive.org/details/treatisanalytdyn00whitrich.
3. Coddington, E.A., *An Introduction to Ordinary Differential Equations*, Dover Publications, Mineola, NY, 1989.
4. Domenech, A., Domenech, T., and Cebrian, J., Introduction to the study of rolling friction, *American Journal of Physics*, 55(3), 235–321, 1987. Available at http://billiards.colostate.edu/physics.
5. Vallado, D., *Fundamentals of Astrodynamics and Applications*, Microcosm Press, Portland, OR, 2001.
6. The Australian Space Weather Agency, *Satellite Orbital Decay Calculations*, prepared by John Kennewell and updated by Rakesh Panwar, 1999, IPS Radio & Space Services, Sydney, Australia, Available at http://www.sws.bom.gov.au (Educational/Space Weather/Space Weather Effects).
7. Gaposchkin, E.M. and Coster, A.J., Analysis of satellite drag, *Lincoln Laboratory Journal*, 1(2), 1988. Available at www.ll.mit.edu/publications/journal.
8. Erdelyi, A., editor, *Tables of Integral Transforms*, California Institute of Technology Bateman Manuscript Project, McGraw-Hill Book Company, 1954. Available at authors.library.caltech.edu/43489/1/Volume%201.pdf.
9. Boyce, W.E. and DiPrima, R.C., *Elementary Differential Equations and Boundary Value Problems*, John Wiley & Sons, Hoboken, NJ, 2005.
10. Polking, J., Boggess, A., and Arnold, D., *Differential Equations with Boundary Value Problems*, Prentice Hall, Upper Saddle River, NJ, 2002.
11. Nagle, R.K., Saff, E.B., and Snider, A.D., *Fundamentals of Differential Equations and Boundary Value Problems*, Pearson/Addison Wesley, Boston, MA, 2008.
12. Boyd, S., *Lecture Notes for EE263, Introduction to Linear Systems*, Stanford University, 2007. Available at http://web.stanford.edu/class/archive/ee/ee263/ee263.1082/notes/ee263course reader.pdf.
13. Cornish, N.J., The Lagrange Points, created for WMAP Education and Outreach, 1998. Available at map.gsfc.nasa.gov/ContentMedia/lagrange.pdf.
14. Lorenz, E.N., Deterministic nonperiodic flow, *Journal of Atmospheric Sciences*, 20, 1963. Available at eaps4.mit.edu/research/Lorenz/Deterministic_63.pdf.
15. Franklin, G.F., Powell, J.D., and Emami-Naeini, A., *Feedback Control of Dynamic Systems*, Pearson/Prentice Hall, Upper Saddle River, NJ, 2006.
16. Caughey, D.A., *Lecture Notes for M&AE 5070, Introduction to Aircraft Stability and Control*, 2011. Available at https://courses.cit.cornell.edu/mae5070/Caughey_2011_04.pdf.
17. Friedman, A., *Partial Differential Equations of Parabolic Type*, Prentice Hall, Upper Saddle River, NJ, 1964.
18. Abramowitz, M. and Stegun, I., *Handbook of Mathematical Functions*, 1972. Available at http://people.math.sfu.ca/~cbm/aands/frameindex.htm.
19. Weisstein, E.W., *Bessel Function of the First Kind*. From MathWorld–A Wolfram Web Resource. http://mathworld.wolfram.com/BesselFunctionoftheFirstKind.html (accessed June 09, 2017).

# BIBLIOGRAPHY

## ON DIFFERENTIAL EQUATIONS

Boyce, W.E. and DiPrima, R.C., *Elementary Differential Equations and Boundary Value Problems*, John Wiley and Sons, New York, 2005.

Coddington, E.A., *An Introduction to Ordinary Differential Equations*, Dover Publications, Mineola, NY, 1989.

Edwards, H.C. and Penney, D.E., *Elementary Differential Equations with Boundary Value Problems*, Pearson/Prentice Hall, Upper Saddle River, NJ, 2004.

Farlow, S.J., *Partial Differential Equations for Scientists and Engineers*, Dover Publications, Mineola, NY, 1993.

Friedman, A., *Partial Differential Equations of Parabolic Type*, Prentice Hall, Upper Saddle River, NJ, 1964.

Kreyszig, E., *Advanced Engineering Mathematics*, John Wiley & Sons, Hoboken, NJ, 1993.

Nagle, R.K., Saff, E.B., and Snider, A.D., *Fundamentals of Differential Equations and Boundary Value Problems*, Pearson/Addison Wesley, Boston, MA, 2008.

Polking, J., Boggess, A., and Arnold, D., *Differential Equations with Boundary Value Problems*, Prentice Hall, Upper Saddle River, NJ, 2002.

Zill, D.G. and Cullen, M.R., *Differential Equations with Boundary Value Problems*, Brooks/Cole Cengage Learning, Belmont, CA, 2009.


## ON ENGINEERING

Anderson, J.D., *Introduction to Flight*, McGraw-Hill, New York, 1989.

Dorf, R.C. and Svoboda, J.A., *Introduction to Electric Circuits*, John Wiley & Sons, Hoboken, NJ, 2010.

Franklin, G.F., Powell, J.D., and Emami-Naeini, A., *Feedback Control of Dynamic Systems*, Pearson/Prentice Hall, Upper Saddle River, NJ, 2006.

Incropera, F.P., Dewitt, D.P., Bergman, T.L., and Lavine, A.S., *Fundamentals of Heat and Mass Transfer*, John Wiley & Sons, Hoboken, NJ, 2007.

Rao, S.S., *Mechanical Vibrations*, Prentice Hall, Upper Saddle River, NJ, 2011.

Stengel, R.A., *Lecture notes for MAE 331, Aircraft Flight Dynamics*, 2006. Available on the web at https://www.princeton.edu/~stengel/MAE331.html.

Ulaby, F.T., *Electromagnetics for Engineers*, Pearson/Prentice Hall, Upper Saddle River, NJ, 2005.

# Index

**Note:** Page numbers followed by f and t refer to figures and tables, respectively.